CSE SCIENCE SERIES

MATTER AND ENERGY

CSE General Science Book I

N. E. Savage and R. S. Wood

Illustrated by David and Maureen Embry

London and Boston
Routledge & Kegan Paul

*First published 1972
by Routledge & Kegan Paul Ltd
Broadway House, 68–74 Carter Lane, London EC4V 5EL
and 9 Park Street, Boston, Mass. 02108, U.S.A.
Printed in Great Britain by
Butler & Tanner Ltd, Frome and London*

© *N. E. Savage and R. S. Wood 1972*

*No part of this book may be reproduced in
any form without permission from the
publisher, except for the quotation of
brief passages in criticism*

ISBN 0 7100 7076 4

CSE SCIENCE SERIES

SERIES EDITORS

L. J. Campbell, B.Sc.
R. J. Carlton, B.Sc.

CONTRIBUTORS TO THE SERIES

L. J. Campbell, B.Sc.
  Assistant Education Officer, Isle of Wight.
  Formerly Head of Science Department, Hinchley Wood C.S. School.
  Formerly Chairman, Science Panel, South East Regional Examinations Board for the CSE.

R. J. Carlton, B.Sc.
  Head of Science Department, Ashford North Modern Boys' School.
  Correspondent, Science Panel and Physics sub-Panel, South East Regional Examinations Board for the CSE 1962 to 1966.
  Chief Moderator in Physics, South East Regional Examinations Board for the CSE.
  Member of Kent Teachers Panel for Metrication.

E. J. Ewington, B.Sc.
  Headmaster, Howard of Effingham School.
  Formerly Senior Master, Hinchley Wood C.S. School.

D. F. Moore, B.Sc., M.I.Biol.
  Senior Lecturer, Hockerill College of Education.
  Formerly member of Biology sub-Panel, South East Regional Examinations Board for the CSE.

N. E. Savage
  Senior Physics Master, Technical High School for Girls, Canterbury.
  Member, Science Panel and Physics and Integrated Science sub-Panels, South East Regional Examinations Board for the CSE.

R. H. Stone, M.Sc., A.R.I.C.
  Senior Chemistry Master, The Judd School, Tonbridge.

D. W. H. Tripp, B.Sc., A.R.C.S., Teacher's Diploma (Lond.)
  Senior Chemistry Master, Brighton, Hove and Sussex Grammar School.
  Member of the Radiochemistry Committee, Association for Science Education.

R. S. Wood
  Science Master, Faversham C.S. School.
  Chairman of the Science Panel, South East Regional Examinations Board for the CSE.
  Chairman of the Physics sub-Panel, South East Regional Examinations Board for the CSE.

Contents

Introduction	ix
1 Fundamental Ideas	1
2 Matter	15
3 Chemistry	26
4 The Measurement of Temperature	41
5 Heat Measurement	56
6 Producing Heat	64
7 Maintaining Heat	88
8 Mechanics	111
9 Engines	137
10 The Basic Foodstuffs	155
11 Food Sources	166
12 Digestion	187
13 Cooking	196
14 Atomic Structure	209
15 Electricity	225

16 Effects of Electricity 235

17 Using Electricity 252

    Appendix 1. Units of Measurement 272

    Appendix 2. The Elements 273

    Appendix 3. Solutions 274

    Appendix 4. Common Chemical Substances 275

    Analytical Contents List 276

    Analytical Contents List for Book II 279

    Index 281

Introduction

The authors believe that a sound training in basic science is essential for any educated person living in the modern world. This course has been written for pupils preparing for examinations in General Science or Integrated Science at CSE level and, for this reason, the authors have borne in mind the requirements of the various Examination Boards. The text will also prove to be of considerable value to pupils preparing for the General Certificate of Education at Ordinary level in these subjects.

This course in science is man-based in concept. It will give to the school-leaver an appreciation and understanding of the rapid developments in science and technology, and of their consequences. With this in mind, the course is intended to develop the critical faculty of the pupil, so that he can use his own judgment to enable him to evaluate a given set of circumstances. It is for this reason that an investigational approach has been used wherever it has been considered of educational advantage to do so.

The authors hope that their approach to the subject will encourage many pupils to continue their scientific studies after they have worked through the complete course which is contained in Books I and II of General Science: *Matter and Energy* and *Man and His Environment*.

The books are intended to be used by the pupils and are couched in language which they will easily understand. The SI system of units has been used throughout and a brief summary of the principal units is given in Appendix 1. Appendices 2, 3 and 4 give tables of the elements, of solutions and of common chemical substances respectively. In Book II there is, additionally, a fifth Appendix listing some famous scientists.

We wish to acknowledge the work done by our fellow authors in this series, who have found the time to read and criticize the original manuscript; and the artists, for their willing co-operation in producing excellent drawings from our rough sketches.

N. E. S.
R. S. W.

Chapter 1

Fundamental Ideas

Man is a curious animal. Since the dawn of civilization, he has looked at the world around him in wonder. Because man is an intelligent animal, he is not content to marvel, but strives to understand the *how* and *why* of what he sees.

Man is aware of his surroundings because of his senses (sight, hearing, smell, taste and touch). The critical use of one or more of these senses is **observation**. When we use our senses to compare

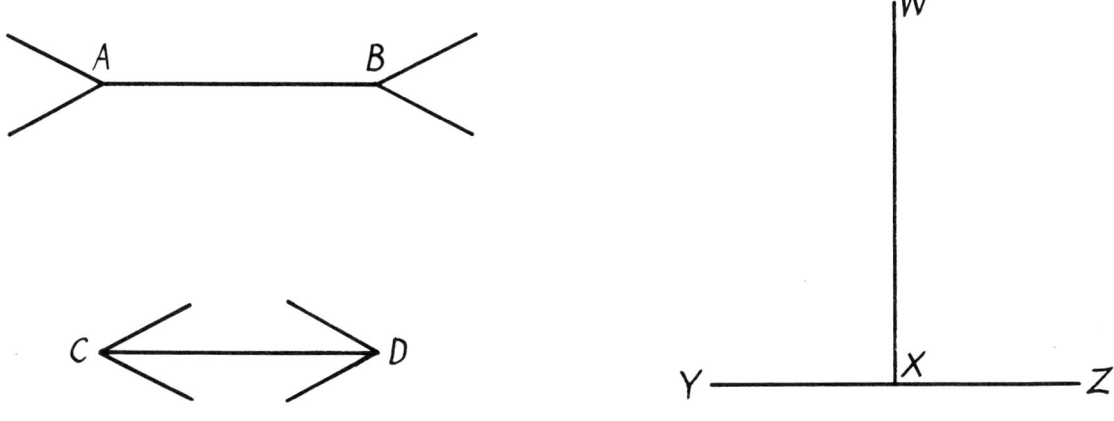

Figure 1.1 Optical illusions

similar things, we soon learn that our senses *alone* are not entirely reliable. If you step from a cold floor into a bath of warm water, the water feels much hotter than it really is.

From Figure 1.1 you will find that your sense of sight can be an unreliable guide. To find the answers to the questions, you probably used a ruler. Of course, you did not really need a ruler merely to compare the lengths. Can you suggest any other methods which you might have used? No matter which method you used, you were comparing the unknown with the known.

If you had been asked to find the lengths of the lines in Figure 1.1, you would have been compelled to use a measuring instrument, such as a ruler. When we measure any quantity, we are comparing an unknown quantity with a known standard.

MEASUREMENT

1.1. Standards of measurement

The standard of measurement is called a **unit**.

a. **The unit of length** is the **metre (m)**. For measuring large distances, we use the **kilometre (km)**, which is one thousand metres. For measuring very small distances, we use the **millimetre (mm)**, which is one-thousandth of a metre. The **centimetre (cm)** is ten millimetres.

b. **The unit of area** is the **square metre (m^2)**. From the derived units of length, we obtain the derived units of area: the **square kilometre (km^2)**, the **square millimetre (mm^2)** and the **square centimetre (cm^2)**.

c. **The unit of volume** is the **cubic metre (m^3)**. The **cubic millimetre (mm^3)** and the **cubic centimetre (cm^3)** are also used where they are more convenient than the cubic metre. The unit often used for the capacity of containers is the **litre (l)**. One litre of fluid has a volume of one thousand cubic centimetres. When dealing with small quantities of liquids, the **millilitre (ml)** may also be used. One millilitre of liquid has a volume of one cubic centimetre.

d. **The unit of mass** (see Chapter 8) is the **kilogramme (kg)**. Small masses are often measured in **grammes (g)**. One gramme is one-thousandth of a kilogramme.

e. **The unit of force** is the **newton (N)**. This is defined in Chapter 8.

f. **Energy** is the capacity for doing work. All forms of energy are measured in **joules (J)**. A **kilojoule (kJ)** is one thousand joules.

g. **Time** is usually measured, in scientific investigations, in **seconds (s)**.

These are some of the units which you will be using in your work in science. A fuller list is given in Appendix 1, which you may find useful for reference.

1.2. Methods of measurement

Investigation 1a. Measurement of length

Measure the lengths of the lines in Figure 1.2. Then measure the lengths of the lines in Figure 1.3.

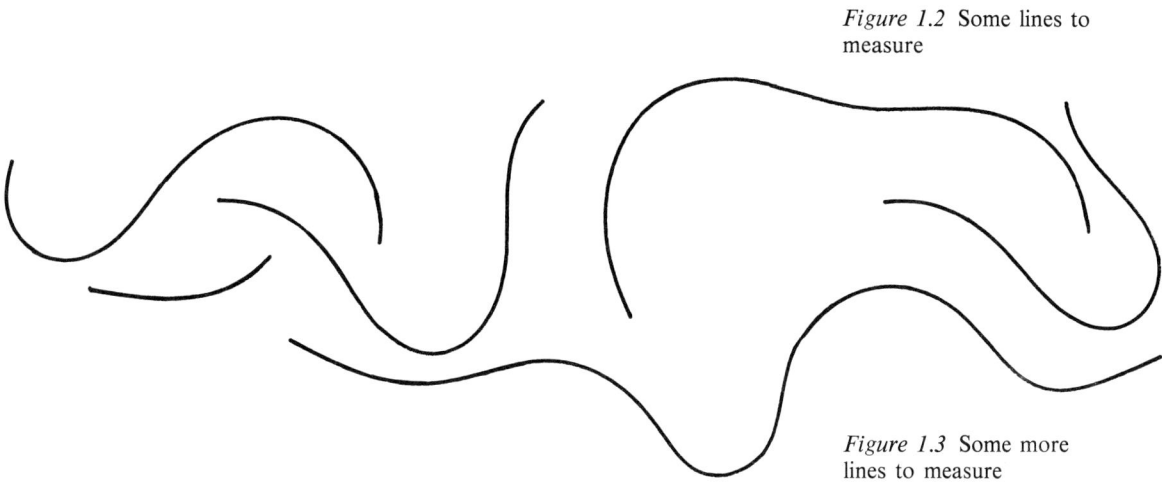

Figure 1.2 Some lines to measure

Figure 1.3 Some more lines to measure

You could not measure the lengths of these curved lines with a ruler. What did you use to help you?

Measure the diameters of the circles in Figure 1.4 as accurately as you can. You may find that a ruler and two set squares will help.

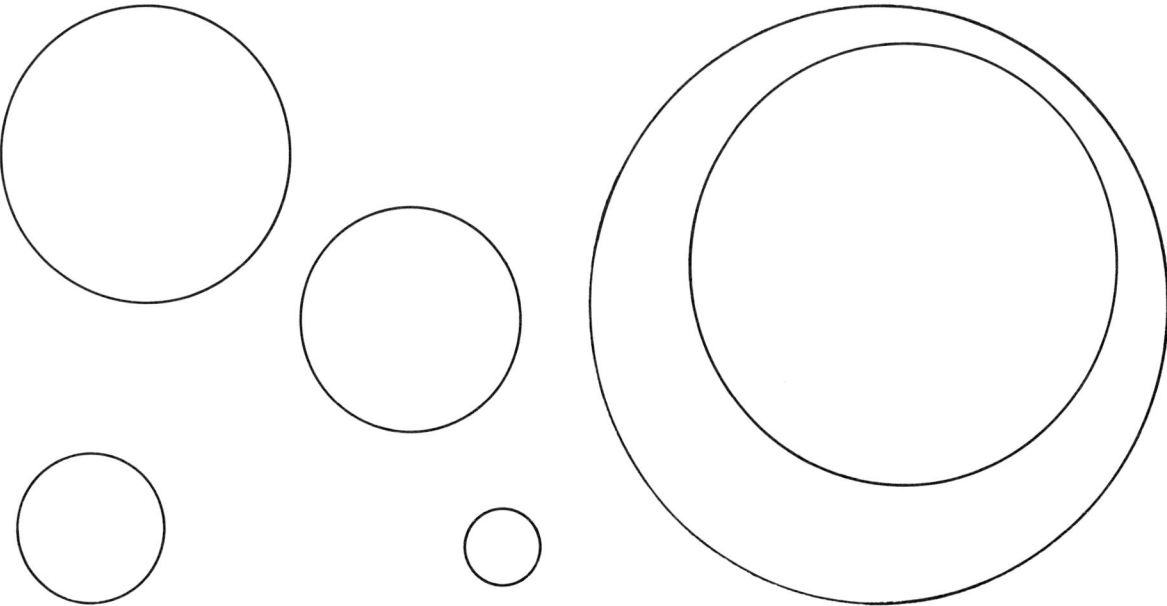

Figure 1.4 Some diameters to measure

Then measure the diameter of a test-tube and the diameter of a beaker.

Investigation 1b. Measurement of area

How many square centimetres are there in the shapes shown in Figure 1.5?

The rectangle shown in Figure 1.5(a) has been divided into centimetre squares. Because there are two columns, each three squares high, the area is calculated by multiplying the number of squares along the base by the number of squares in each column (base × height): $2 \times 3 = 6$ cm^2.

How many square millimetres are there in the rectangles shown in Figure 1.5?

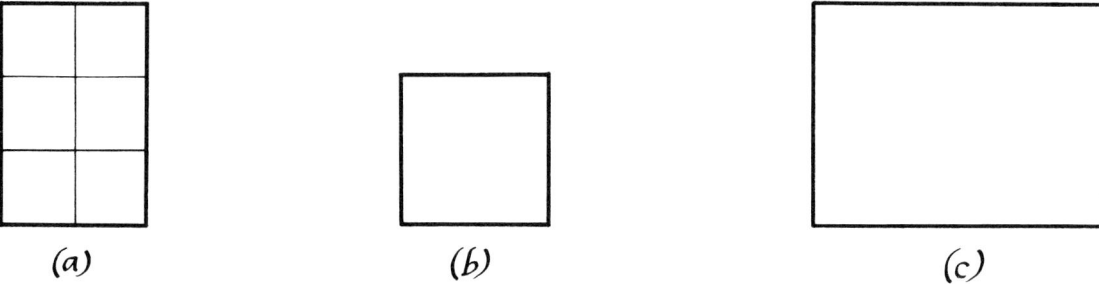

Figure 1.5 Some areas to measure

Investigation 1c. Measurement of volume

If we had a rectangular box 10 cm long, 10 cm wide and 5 cm high and fitted a single layer of centimetre cubes inside it to cover the base, the layer would consist of a hundred cubes (ten rows of ten cubes). In order to fill the box, we would need five such layers. Consequently, the volume of the inside of the box would be 10 cm × 10 cm × 5 cm = 500 cm^3, that is, the three dimensions of the box multiplied together.

Obtain a rectangular box (a chalk box or a matchbox is suitable), measure the lengths of the three sides and calculate its volume by multiplying the three dimensions together.

When we want to measure the volume of a quantity of liquid, we pour it into a container which has volumes marked at different levels up the side. Such containers include graduated beakers and measuring cylinders, as shown in Figure 1.6.

Pour some water into both a graduated beaker and a measuring cylinder. You will notice that the surface of the water is not perfectly flat, the centre being lower than the edges. This surface shape is called the **meniscus**. When reading the volume of a liquid in one of these containers, the eye must be level with the bottom of the

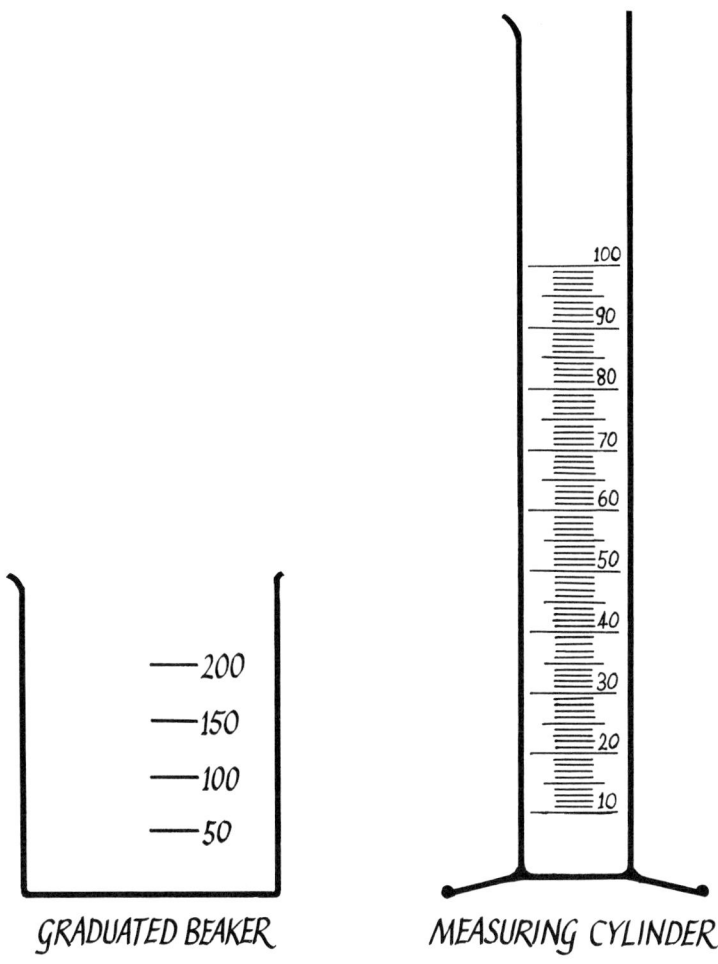

Figure 1.6 Containers for measuring volume

meniscus. The volume is read from the mark level with the bottom of the meniscus, as shown in Figure 1.7.

Pour water into a graduated beaker until the bottom of the meniscus is level with the 100 cm^3 mark. Then pour this water into a measuring cylinder. Is the bottom of the meniscus exactly level with the 100 cm^3 mark? If you had to measure out a quantity of water as accurately as possible, which of these containers would you use? Explain the reasons for your choice.

How could you find the volume of a stone?

With all of the measurements which you have made so far, you will realize that the degree of accuracy that you can achieve depends partly on practice and partly on the precision of your measuring instruments.

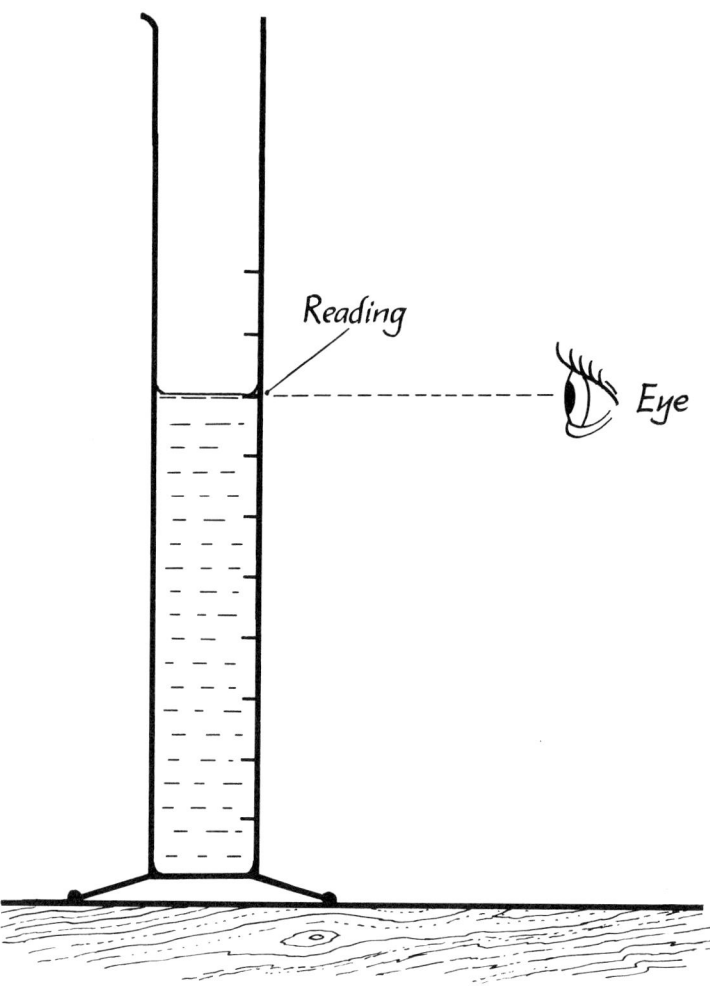

Figures 1.7 Reading the volume of a liquid in a measuring cylinder

Investigation 1d. The pendulum

Tie a large metal nut to the end of a piece of thread about 1·2 m long. Cut a cork into two halves, place the thread between the two halves and grip the cork between the jaws of a clamp fitted to a retort stand, as shown in Figure 1.8.

Move the nut a little to one side and let it go. You will notice that it swings over to the other side and then back again. For this investigation, we shall call this movement, from one side to the other and back again, one swing. Although a large swing from side to side will not affect your results, it has the disadvantage that, after a short time, your pendulum will tend to swing out of line and may collide with the retort stand or the bench.

Make a copy of Table 1.1.

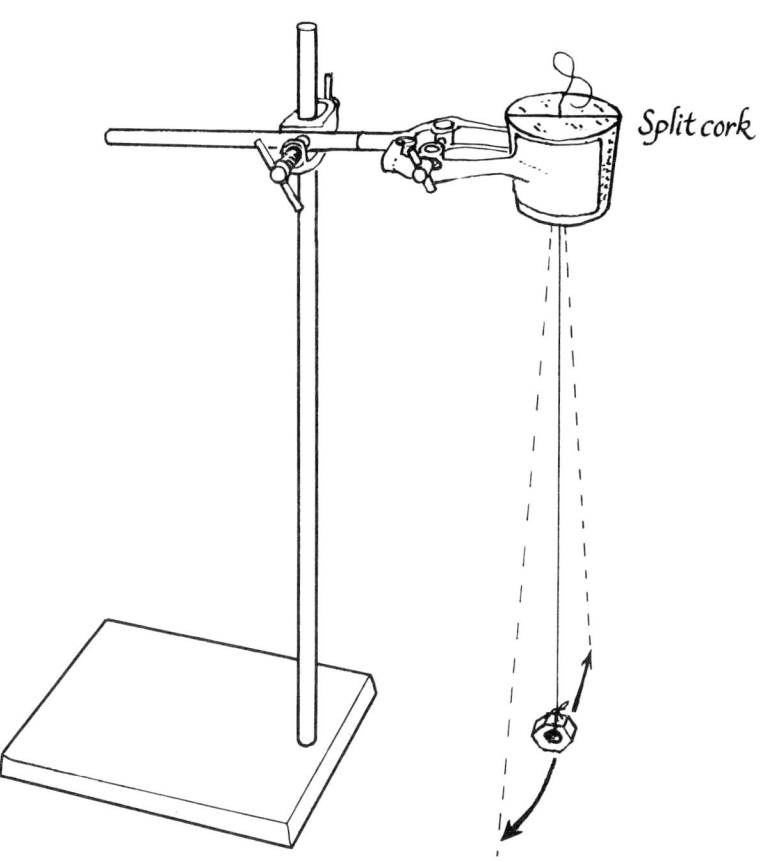

Figure 1.8 The pendulum

TABLE 1.1. RESULTS OF INVESTIGATION 1d

Length of Pendulum (in metres)	1·0	0·75	0·5	0·25
Time for 100 Swings (in seconds)				
Time for 1 Swing (in seconds)				

Adjust the length of your pendulum (measured from the bottom of the split cork to the centre of the nut), so that it is 1 m long. You can do this by easing the jaws of the clamp, pulling the thread through and then re-tightening the clamp. Start your pendulum swinging and, by using a stop-clock, stop-watch or a watch with a sweep seconds hand, measure the time for your pendulum to make one hundred swings. Enter this time in your table and, by dividing this time by

100, calculate the time for one swing. Repeat this process for the lengths of pendulums shown in the table. Can you see any connection between the length of the pendulum and the time taken for one swing?

You will notice that the bottom line of the table has been left blank. For each length of pendulum, square the time taken for one swing (by multiplying the time for one swing by itself) and enter this in the bottom line. Is there any connection between this bottom line and the length of each pendulum?

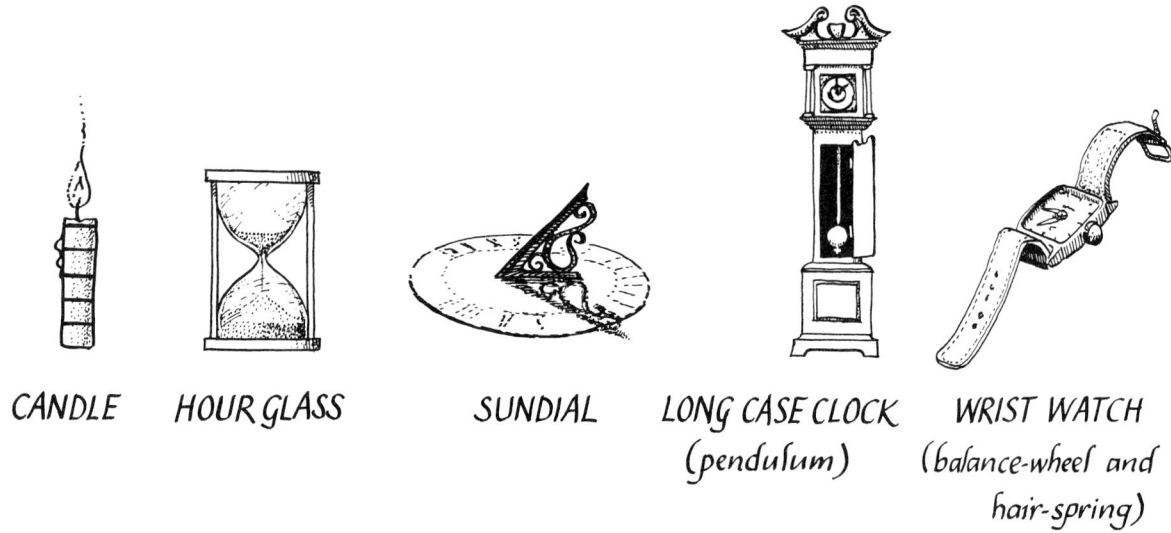

Figure 1.9 Some time measurers

Repeat this investigation, using a large stone instead of the nut. How do your results compare?

Pendulums have been used for many years as a method of measuring time. Most large clocks are regulated by a pendulum, while small clocks and watches have a balance-wheel and hair-spring mechanism, which works on a similar principle. Some time-measuring devices are shown in Figure 1.9.

Investigation 1e. Measurement of mass

Mass is measured on a balance, in which an unknown mass is compared with a known mass. Figure 1.10 shows some types of balance. Use some of them to find the mass of, for example, a stone.

So far, in this chapter, we have been concerned with simple measurement. Science is also concerned with energy and the changes in the forms of energy.

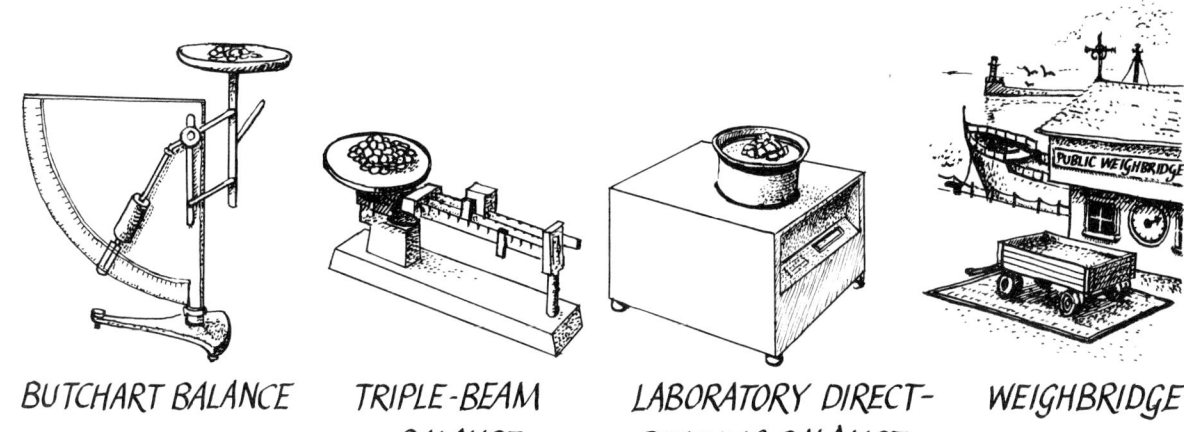

BUTCHART BALANCE   TRIPLE-BEAM BALANCE   LABORATORY DIRECT-READING BALANCE   WEIGHBRIDGE

Figure 1.10 Some mass measurers

ENERGY

1.3. What is force?

An object will always do one of two things:
a. It will stay where it is, or
b. It will keep moving at the same speed in the same direction, unless it is made to change its speed or direction (or both) by a force.

This idea is one of the basic laws of science: Newton's first law of motion. From this idea, we can obtain the meaning of force. **A force is a push or a pull**, which tends to move a stationary object or alters the speed or direction (or both) of a moving object.

When a force causes a change in movement, we say that work has been done. Work is the product of force and the distance moved by the force:

Work = Force × Distance moved by the force

When a force of 1 newton moves through a distance of 1 metre, 1 joule of work is done. If a force of 4 N moves an object through a distance of 7 m, then $4 \times 7 = 28$ joules of work is done.

1.4. Forms of energy

Energy is the capacity to do work, and it exists in many forms, including mechanical energy, heat energy, light energy, electrical energy and chemical energy.

The next investigation will give you some idea of the ways in which one form of energy can be changed into another.

Investigation 1f. Energy changes

1. Saw through a piece of metal with a hacksaw and then feel the blade. What do you notice?
2. Heat a piece of iron (or steel) wire in a bunsen flame. What do you see?
3. Point a photographic exposure meter towards a light. What do you notice?
4. Dissolve some common salt (sodium chloride) in water. Then dissolve some silver nitrate in water and mix the two solutions. Filter the mixed solution through a filter paper and expose the white precipitate (which is silver chloride) collected on the filter paper to a bright light, sunlight if possible, for about ten minutes. Do you notice any change in colour? If you are in any doubt,

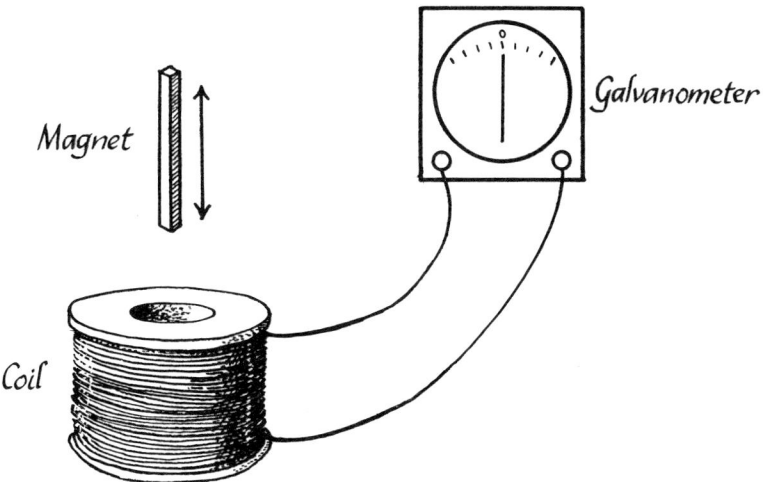

Figure 1.11 Apparatus for Investigation 1f (7)

repeat the experiment but expose only a part of the silver chloride by covering the remainder with black paper.

(*Note.* If you add a drop of photographic developer to the exposed silver chloride, a further change will take place. This is what happens when a film is developed.)

5. Place a thermometer in hot water. What happens to the mercury in the thermometer?
6. Connect an electromagnet through a switch to a suitable low-voltage supply. Suspend the electromagnet about 1 cm above some iron filings and switch it on. What happens to the iron filings? What happens when you switch off?
7. Connect a coil of insulated wire to a sensitive, centre-zero galvanometer. Plunge a bar magnet into the coil, as shown in Figure 1.11. What happens to the reading on the galvanometer when the magnet is plunged into the coil? What happens when the magnet is withdrawn from the coil?
8. Put two match-heads in a hard glass test-tube and fit a cork.

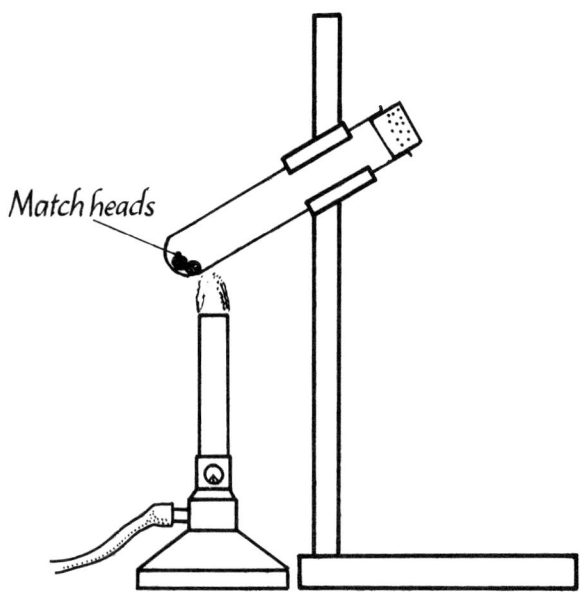

Figure 1.12 Apparatus for Investigation 1f (8)

Support the test-tube in a clamp, as shown in Figure 1.12, so that the cork can do no damage if it shoots out. Place a bunsen burner with a low flame under the match-heads and stand clear. What happens?

★ WARNING. *Stand well clear.*

9. Switch on an electric fire and put your hand *in front of* the element (*DO NOT TOUCH*). Do you notice anything before you see the element glowing?

In each part of this investigation, energy was changed from one form to another. Make a copy of Table 1.2 and complete it by filling in the columns headed 'Energy Put In' and 'Energy Given Out'.

TABLE 1.2. RESULTS OF INVESTIGATION 1f

Part of Investigation	Energy Put In	Energy Given Out
1	Mechanical	Heat
2		
3		
4		
5		
6		
7		
8		
9		

1.5. Energy chains

Everything that happens is due to changes in the forms of energy. In some cases the change is a simple one, but it is more common to find that there are several changes, forming an energy chain. The energy given out in most energy changes is in more than one form.

Let us consider the changes in the forms of energy that take place when a match is struck.

The energy chain starts with radiant energy from the sun. This is changed, by the process of photosynthesis in the leaves of green plants, into starch and sugar (see Chapter 11). Starch and sugar contain chemical energy. This first energy change is of the utmost importance, since without the sun's radiant energy all plant life would die, and if there were no plants, no other form of life (as we know it) would be possible.

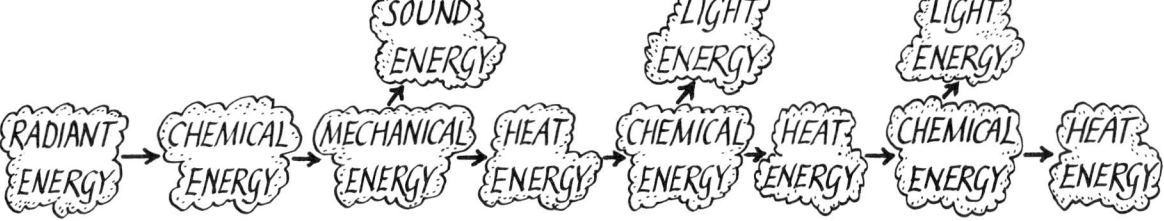

Figure 1.13 An energy chain

The chemical energy stored in the starch and sugar is absorbed by us when we eat and digest foods of vegetable origin or foods of animal origin, since animals eat plants or other animals which have, in turn, eaten plants. The action of rubbing the match-head on the matchbox involves the use of muscles, in which the chemical energy of our food is changed into movement, which is mechanical energy. At the match-head, this mechanical energy produces friction, which releases sound energy and heat energy. The heat energy releases the chemical energy in the match-head, and this chemical energy is changed into light energy and heat energy when the match-head lights. This heat energy now releases the chemical energy stored in the wooden part of the match (originally produced by photosynthesis), and changes it into more light energy and heat energy.

This energy chain can be summarized as shown in Figure 1.13.

1.6. Conservation of energy

Whenever energy is changed from one form to another, the total amount of energy remains the same. Energy cannot be created and cannot be destroyed. This is known as the **law of conservation of energy**.

Unfortunately, the energy output of all machines and devices is never entirely in the form in which we want it.

Suppose we could invent two absolutely perfect devices:

a. A perfect dynamo, converting mechanical energy into electrical energy and nothing else, and

b. A perfect electric motor, converting electrical energy into mechanical energy and nothing else.

By connecting the terminals of these two devices together with pieces of wire, clamping the shafts together and giving the dynamo a spin, the dynamo would produce electrical energy which would drive the motor which would, in turn, drive the dynamo, and so on for ever, as shown in Figure 1.14.

If this were possible, we would have solved the problem of perpetual motion, which has been the vain hope of scientists ever since scientific investigation first began.

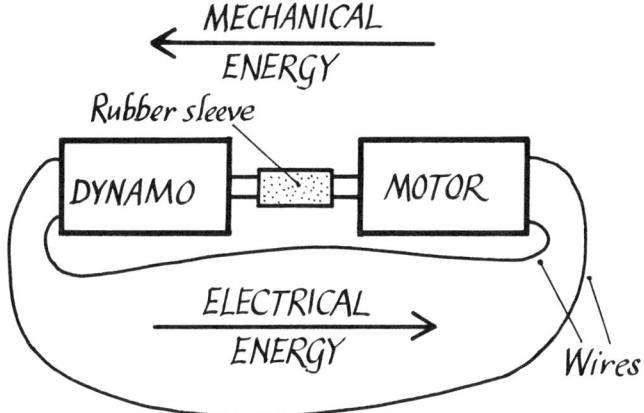

Figure 1.14 Is this perpetual motion?

To discover why perpetual motion is impossible by this method, we must decide where the energy 'loss' occurs. By energy 'loss' we mean any form of energy given out by the dynamo other than electrical energy and any form of energy given out by the motor other than mechanical energy.

Since there must be moving parts in both the dynamo and the motor, there will be friction in the bearings and other working parts, no matter how well they are lubricated. This friction will produce heat energy which, as far as perpetual motion is concerned, is an energy 'loss'.

There will also be some sound energy produced, since no motor or dynamo is perfectly silent, and a certain amount of light and heat energy will be produced by the arcing of the carbon brushes. Even the interconnecting wire, no matter how thick, will become slightly warm and this energy 'loss' by itself would be enough to defeat perpetual motion. These energy 'losses' are shown in Figure 1.15.

Another way of considering the law of conservation of energy is that whenever a known amount of energy 'disappears' in one form, exactly the same amount appears in other forms. (What you lose on the swings, you gain on the roundabouts!) From this, it follows that

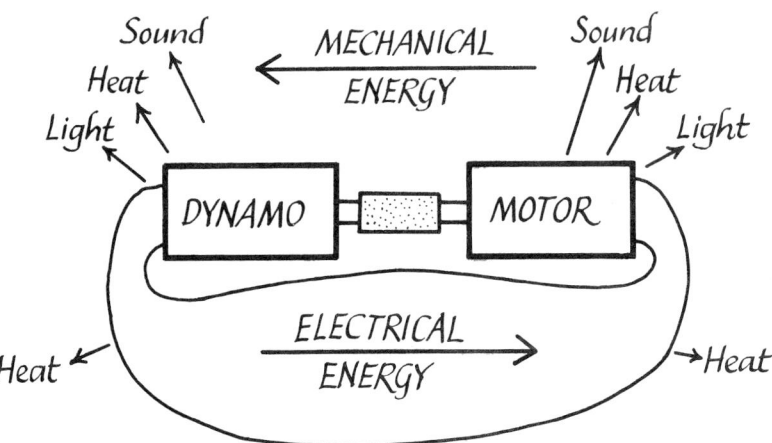

Figure 1.15 Energy 'losses'

the efficiency of any machine or device depends on the proportion of the energy put in that is changed into the form of energy that it is designed to give out. The more energy given out in the 'wrong' form, the less efficient is the machine or device.

A well-lubricated bicycle is more efficient than an unlubricated one, because less energy is changed by friction into heat. A 'cold' electric lamp gives out more light than a 'hot' lamp using the same amount of electricity. No doubt you can think of many more examples. Have you noticed that, in most cases, energy 'losses' are in the form of heat?

In some ways, energy is similar to money. Energy occurs in many forms which can be changed from one form to another. In the same way, money occurs in many currencies, pounds Sterling, French francs, Dutch guilders, American dollars, Spanish pesetas, etc., which are all interchangeable. When we change energy from one form to another, we lose in efficiency because of energy 'losses'. Similarly, when we change one currency into another, we lose on the exchange in the form of the bank's commission.

Test your understanding
1. What is the area of a page in this book?
2. A rectangular container is 6 cm long, 4 cm wide and 10 cm deep. If you poured 144 cm^3 of water into it, how deep would the water be in the container?
3. What length of pendulum would you use to enable you to count in seconds?
4. A boy exerts a steady force of 150 N to move a trolley 40 m. How much work has he done?
5. How much work is done when a steady force of 80 N raises a load through 2·4 m?
6. Make a table of everyday devices. In each case, name the form of energy put in and the form of energy given out.
7. When a man rides a bicycle, the energy put in is in the form of mechanical energy. By drawing an energy chain, trace this energy back to its source, the sun.
8. Make a list of the energy 'losses' which occur when a car is being driven.

Chapter 2

Matter

2.1. What is matter?

In science, when we talk of **matter**, we mean any substance. This may be in the form of a liquid, a solid or a gas.

2.2. All matter possesses heat

All matter has a certain amount of heat energy: no material is absolutely cold. When we describe a substance as cold we mean that it possesses comparatively little heat energy.

2.3. The behaviour of materials when they are heated

Investigation 2a. The expansion of solids

Arrange a strip of aluminium (about 40 cm long, 2 cm wide and 0·5 cm thick), as shown in Figure 2.1, so that one end is securely clamped to hold the strip horizontally about 10 cm above the surface of the bench. A block of wood should be placed under the free end of

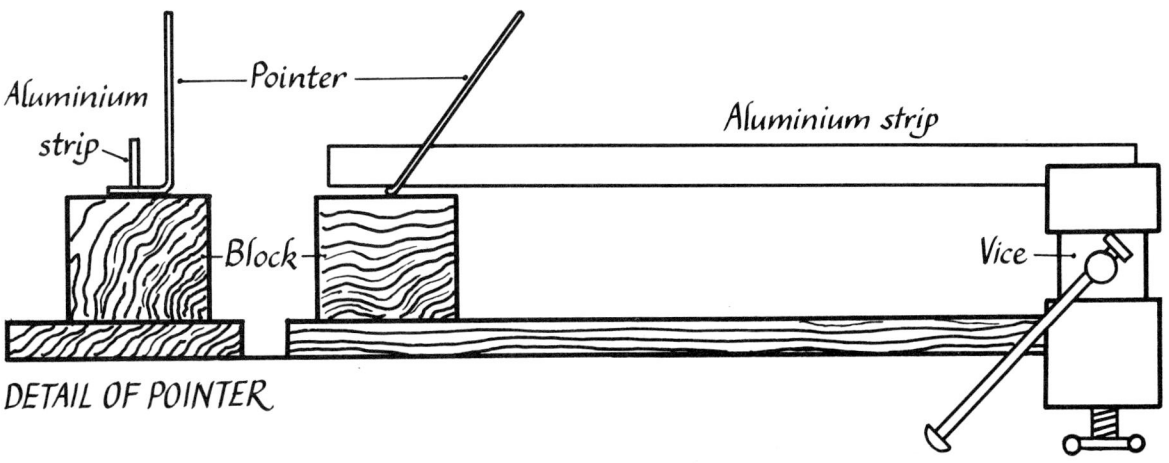

Figure 2.1 Expansion of solids

the strip, so that the pointer is firmly held between the strip and the block. The pointer is made from thick wire.

Heat the middle of the strip with a bunsen burner. Explain what you see to happen.

Investigation 2b. Conduction of heat

Support a steel bar on a tripod. Beginning about 10 cm from one end, place live match-heads at intervals of about 2 cm along the length of the bar (see Figure 2.2).

Heat the end of the bar and describe what happens. Explain this.

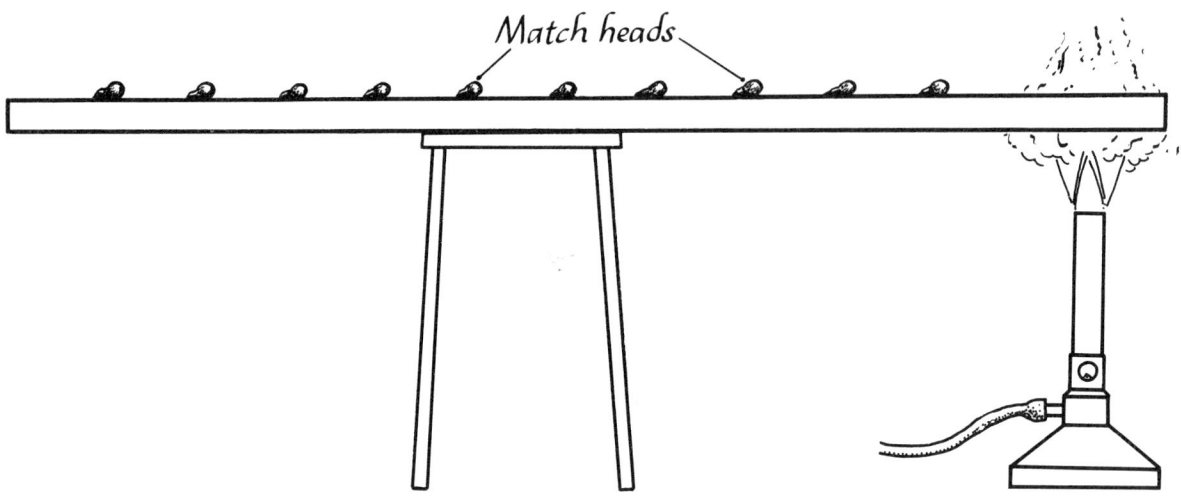

Figure 2.2 Conduction of heat

Investigation 2c. Differential expansion

Obtain a bimetal strip. This consists of two strips of different metals riveted together.

Using a bunsen burner, heat the middle of the strip (see Figure 2.3). Note what happens. What conclusion do you draw from this?

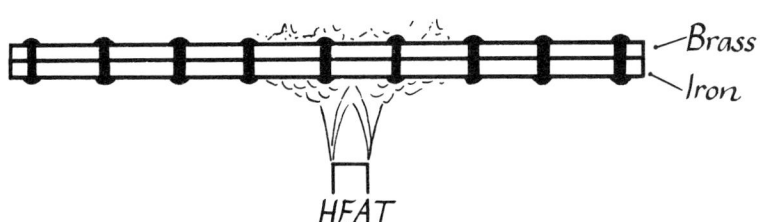

Figure 2.3 The bimetal strip

Investigation 2d. The expansion of liquids

Fill a round-bottomed flask with coloured water (add a little ink to tap water). Fit a bung and a length of glass tubing, as shown in Figure 2.4. Support the flask in a retort stand.

Heat the bottom of the flask with a bunsen burner. Note what happens. Explain these observations.

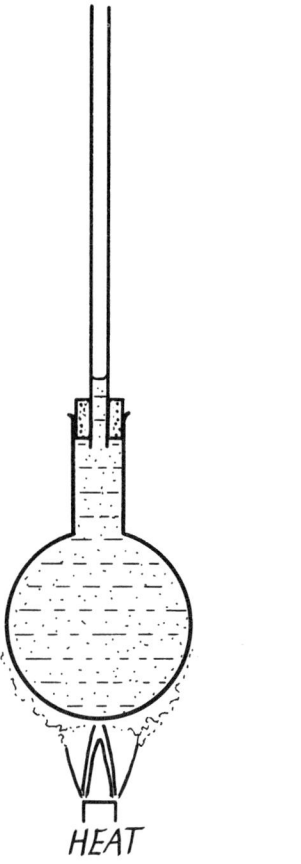

Figure 2.4 Expansion of water

Investigation 2e. Expansion of gases

Bend a piece of small-bore glass tubing (1 to 2 mm inside diameter is very suitable) to a J-shape, as shown in Figure 2.5. Trap a little coloured water in the bend of the tube and then attach the shorter end of the tube to a small flask by means of a rubber bung.

Warm the flask with your hands and observe what happens.

From the investigations you have carried out, you should have noted that solids, liquids and gases all expand when they are heated. You may also have observed that the liquid expanded more than the solid, and that the gas expanded more than the liquid. From the result of Investigation 2c you have learned that brass expands more than iron when they are both heated through the same temperature.

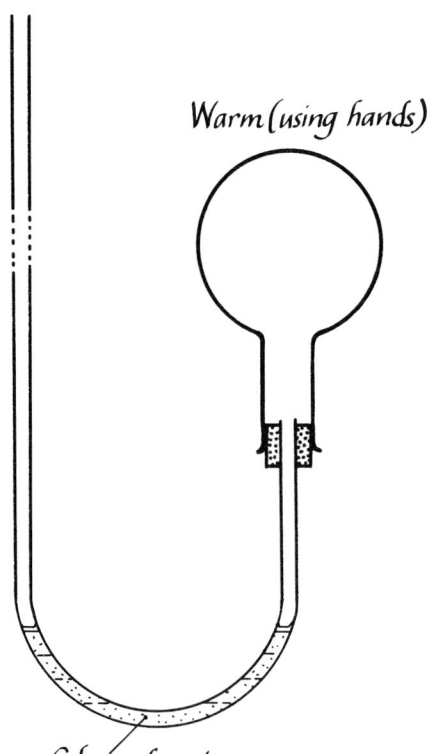

Figure 2.5 The expansion of air

Investigation 2b showed that when a metal rod was heated at one end, the heat travelled along the rod to the other end. The heat is said to be **conducted** along the rod. This is the way in which heat travels in solids. In liquids and gases, heat travels mainly by means of **convection** currents, while heat can only cross empty space by means of **radiation**.

Investigation 2f. Conduction and convection

1. Three-quarters fill a test-tube with water. Heat the water by playing a bunsen burner on the *top* part of the test-tube (*use a test-tube holder*) until the water at the top of the tube boils (see Figure 2.6). Now touch the bottom of the test-tube. What do you notice?

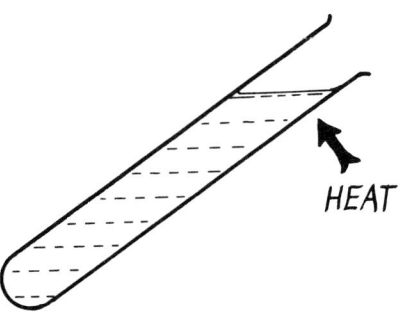

Figure 2.6 Heating water

2. Fill a beaker with water and support it on a tripod without a gauze. Leave this for a few minutes to allow the water to settle and then drop one small crystal of potassium permanganate into the water. Using a very small flame, heat the beaker just below the crystal (see Figure 2.7). Note what happens.

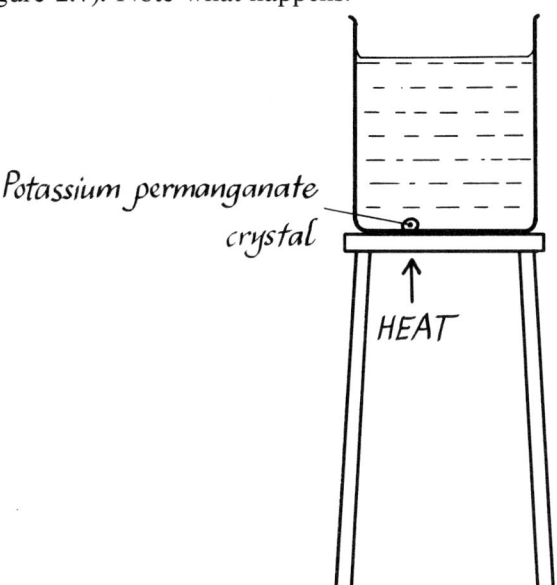

Figure 2.7 Experiment with potassium permanganate

3. Obtain a piece of cartridge paper about 20 cm square and mark it out as shown in Figure 2.8. Then fold it to form a box. Secure the sides of the box by means of paper clips. Place the box on a tripod without a gauze. Fill the box with water and adjust the bunsen burner to give a fairly strong but low flame. Place the bunsen burner under the box. Note what happens.

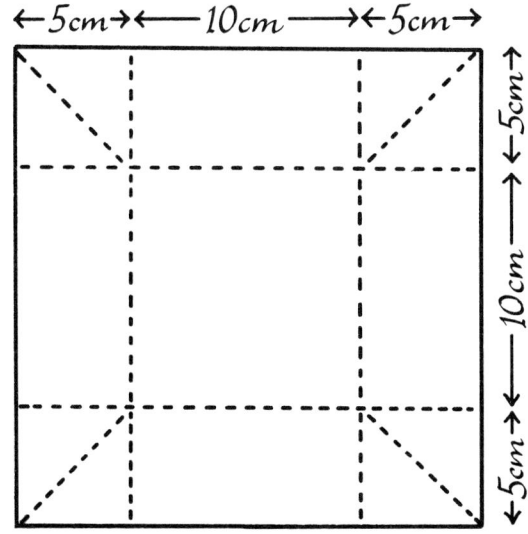

FOLD ALONG THE BROKEN LINES

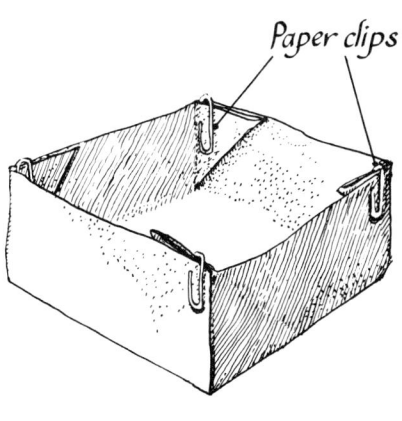

Figure 2.8 Making the box

4. On a piece of cartridge paper, mark out a circle of radius 5 cm. In this, draw concentric circles at intervals of 1 cm. Cut this, as shown in Figure 2.9, to form a spiral. Pass a pin through the centre so that the spiral may be supported. Set the bunsen burner to give a small flame, and hold the spiral *well above the flame*, by means of the pin. Note what happens.

Figure 2.9 Making the spiral

5. Obtain glass and copper rods, each about 20 cm long. Half fill a beaker with water and place the rods in this (see Figure 2.10). Heat the water until it boils. Stop heating and touch the ends of the rods. What do you notice?

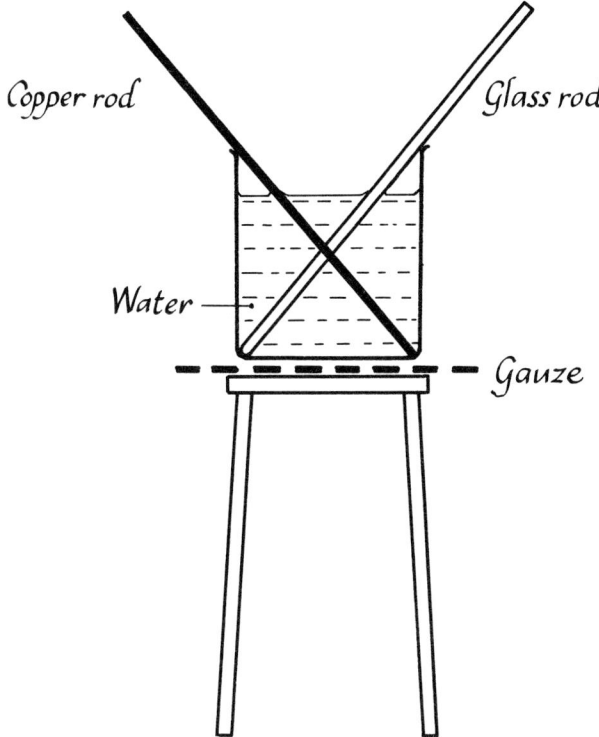

Figure 2.10 Heating rods

2.4. Matter in motion

We have been investigating the ways in which matter is affected by heating. Now we must try to explain this behaviour.

All matter is made up of very small particles which are continually vibrating. If we consider a solid, such as a block of ice, it is really a very large number of vibrating particles. These require space in which to move, so our solid block of ice contains a great deal of empty space. Between each particle and its neighbours are strong forces of attraction, causing the particles to be limited in the space in which they are free to move, and so tending to keep each particle within its appointed place in the solid.

The energy possessed by the moving particles which causes them to vibrate is heat energy. If we give a particle more heat energy it will vibrate more vigorously, while if we take heat from a particle it will vibrate less vigorously. Thus, if we supply heat to the particles in a solid, they will vibrate more vigorously and at the same time will require more space in which to do so, with the result that the solid will expand. Similarly, when heat is removed from a solid, it contracts.

Now, if we heat one end of a metal rod with a bunsen burner, those particles nearest the flame will receive heat and will vibrate more vigorously. This will disturb the nearby particles causing them, in turn, to vibrate faster; before long, particles all along the rod will be vibrating more energetically. In this way, heat is conducted along the rod.

If we continue to heat a solid, the particles will eventually gain so much energy that they will partially overcome the forces of attraction between them, so that they are able to change their position within the substance. At this stage, there will still be attraction between particles, but they will no longer be confined to one appointed position within the substance; the solid has changed into a liquid. You should now understand why it is that a solid has a definite shape, while a liquid will take the shape of any container into which it is placed.

If we supply heat to a liquid, the particles will vibrate more rapidly and will take up more space, and so the liquid will expand. When any substance expands, its volume increases but its mass remains unchanged. This means that the density of the substance decreases, since the density of a material may be stated as being the mass per unit volume of the material. When water in a beaker is heated, the water nearest to the flame becomes heated first and so expands to become less dense. Because less dense liquids float on denser liquids, the hotter water will move to the top of the beaker and its place will be taken by cooler water from the top of the beaker. In this way there will be a continual movement of water within the beaker. This explains how convection currents transfer heat. Such currents can occur only in liquids and gases.

At any given instant, the particles in a substance are not all moving at the same speed nor are they moving in the same direction. In a liquid, some particles are moving at such a high speed that they may break away from the liquid and escape into the surrounding atmosphere. This accounts for the evaporation of liquids. It also explains why a liquid cools when it evaporates. Since it is the faster moving particles which escape, those which are left are slower moving, and therefore possess less heat energy.

When a liquid boils, it is heated until all the particles are given sufficient energy to break away. When all the particles have broken away, the liquid has changed into a gas.

We may think of the three states, or phases, of matter as representing different energy levels, solids being at the lowest level, liquids being at a higher level and gases being at the highest level.

The temperature of a particle is really a measure of its speed. Obviously, it is quite impossible to measure the temperature of a single particle; when we measure the temperature of a substance, we obtain a reading which depends on the average speed of all the particles within that substance.

2.5. Heat and temperature

Heat is not the same thing as temperature. This is obvious when you think about what happens when you boil different quantities of water on the same ring of a gas stove. If we put 2 l of water in a saucepan and boil it, this can reasonably be expected to take about four times as long as it would have taken to boil 0·5 l of water under the same conditions. Both quantities of water would have started at the same temperature and they would both have finished at the same temperature (100 °C). The larger quantity of water would have required more heat because it contained a greater number of particles. Clearly, the amount of heat possessed by a body depends on the number of particles it contains and also on the average speed at which those particles are moving. This is exactly the same as saying that the heat possessed by a body depends on its mass and its temperature. We shall learn (Chapter 5) that another factor called the **specific heat capacity** of the material must also be taken into account, if we are to measure the heat given to, or taken from, a body.

We may now say that **the temperature of a body is a measure of the average speed of the particles within it, while the heat possessed by it is a measure of the total energy possessed by those particles.**

2.6. Heat radiation

We have said that heat can only travel across empty space by means of radiation. Strictly speaking, what travels across space is not heat at all, but electromagnetic waves (having a wavelength longer than the wavelength of light but shorter than radio waves). The particles of matter act as receivers for these so-called heat waves and respond to them by vibrating more rapidly. This means that the matter on which the waves fall becomes hotter. Thus, although these electromagnetic waves are not heat, they produce heat in matter.

2.7. Pressure in gases

When we inflate a tyre, we do so in order to increase the air pressure in the tyre. The air in the tyre is composed of a large number of particles, vibrating at an average speed which depends on the temperature. These particles are continually bombarding the walls of the tyre. The average force exerted per unit area on the walls of the tyre, by this bombardment, is the pressure of the air in the tyre.

If we pump more air into the tyre, the volume does not increase very much, but many more particles of air are forced into the tyre. Thus, the tyre walls are subjected to a greater bombardment which means that the pressure is increased.

When a car is driven on a long journey it is found that the air pressure in the tyres has increased. This is because the force of friction between the tyres and the road converts some of the mechanical energy, produced by the engine, into heat. The air particles within the tyre vibrate more vigorously when they become hotter and bombard the walls more vigorously; in this way the pressure is increased.

If you put your thumb over the air outlet hole of a bicycle pump and force the piston downwards, you will be aware that the pressure exerted by the air increases. At the same time, the barrel of the pump will become warmer. In order to compress the air in the pump, work has to be done on it (by your pushing the piston down), and this means that the particles of gas are given extra energy. This results in their moving more vigorously, giving rise to an increase in temperature.

Summarizing: **when a gas is compressed, its pressure increases and its temperature increases.**

2.8. Expansion of gases

If the temperature of a given volume of gas is reduced, the average speed of the particles is reduced so that their total energy is less. This means that the pressure exerted is less. If the pressure of the gas is to be kept constant while its temperature is reduced, then the volume of the gas container must also be reduced so that the gas particles have a smaller area to bombard. Thus, when the temperature of a gas is reduced, it contracts if its pressure is to be unchanged. Similarly, when the temperature of a gas is increased and its pressure is unchanged, the gas expands.

It has been found that all gases expand at the same rate. If the pressure is unchanged, the volume of any gas will increase by $\frac{1}{273}$ of its volume at 0 °C for each degree Celsius rise in temperature. Thus, a litre of gas at 0 °C will expand to occupy two litres at 273 °C if there is no increase in pressure. Cooling a gas will produce a contraction at the same rate.

The temperature -273 °C is called **absolute zero**. Particularly when dealing with gases, it is often useful to measure temperatures on a scale beginning at absolute zero. This is called the **Kelvin** scale of temperature. To convert a temperature in degrees Celsius to the Kelvin scale, it is only necessary to add 273.
Thus:

$$0 \text{ °C} = 273 \text{ K} \qquad 100 \text{ °C} = 373 \text{ K}$$
$$-60 \text{ °C} = 213 \text{ K} \qquad 245 \text{ °C} = 518 \text{ K}$$

The theory which explains heat in terms of the motion of particles of matter is called the **kinetic theory of heat**.

Test your understanding

1. State the three ways in which heat may be transferred from one place to another.
2. Explain how heat is transferred (a) in a liquid, (b) in a solid. How does heat travel across empty space?
3. Describe how you could show that different metals expand at different rates.
4. What is the difference between heat and temperature?
5. Try to explain all your observations in Investigation 2f in terms of the kinetic theory of heat.
6. Explain what is likely to happen if two litres of air are pumped into a one-litre container.
7. Convert the following temperatures to the Kelvin scale:
   (a) 20 °C, (b) 57 °C, (c) −33 °C.
8. Convert the following temperatures to the Celsius scale:
   (a) 673 K, (b) 350 K, (c) 250 K.

Chapter 3

Chemistry

3.1. Different kinds of matter

All materials are made up of atoms. If the atoms are all of the same kind the material is called an **element**; but if the material is made up of groups of atoms, each group containing two or more different kinds of atom, then the substance is called a **compound**.

There are over a hundred elements, and a very large number of compounds.

Investigation 3a. Chemical changes

Carry out the following investigations, observing what happens and making a note of these observations.
1. Take about 5 cm of magnesium ribbon, hold it in a pair of tongs and burn it. *Do not look directly at the light.*
2. Mix together some powdered sulphur and iron filings. Put the mixture in a small ignition tube. Heat the tube strongly until the mixture at the bottom of the tube glows red hot. Stop heating and observe what happens.
3. Dissolve about 5 g of copper sulphate crystals in water in a beaker. Using a similar amount of water, dissolve about 6 g of sodium carbonate crystals. Mix the two solutions together.
4. Make a strong solution of copper sulphate (about 20 cm^3 will do). Clean a short length of magnesium ribbon, using emery paper. Cut this into small pieces and place them in the solution.
5. Put a little copper carbonate in a test-tube and add a little dilute sulphuric acid. When the reaction has stopped, slowly add ammonia solution.
6. Half fill a test-tube with iodine solution (iodine in potassium iodine). Add a few crystals of sodium thiosulphate. Shake the test-tube.

In each of these reactions you will have noticed that the substance, or substances, left at the end of the reaction were different from those with which you started. These changes were chemical changes.

3.2. Formulae and equations

When you burned magnesium, it combined with some of the oxygen from air to form a new substance called magnesium oxide. Magnesium and oxygen are both elements, but magnesium oxide is a compound. We may write this as a word equation thus:

$$\text{MAGNESIUM} + \text{OXYGEN} \rightarrow \text{MAGNESIUM OXIDE}$$

Similarly, the reaction between iron and sulphur could be written:

$$\text{IRON} + \text{SULPHUR} \rightarrow \text{IRON SULPHIDE}$$

In chemistry, we often write equations in symbols. This we shall find to be of great value, but we must be quite sure of exactly what the symbols mean. Each atom has its own particular symbol, so if we write 'O' we mean *one atom of oxygen*. Similarly, '3S' means *three atoms of sulphur*. If we write 'Fe+S→FeS', we mean that iron and sulphur have combined in the proportion of one atom of iron to one atom of sulphur to form a compound called iron sulphide.

TABLE 3.1. SOME COMMON ELEMENTS AND THEIR SYMBOLS

Element	Symbol	Element	Symbol
Hydrogen	H	Zinc	Zn
Helium	He	Iron	Fe
Oxygen	O	Nickel	Ni
Chlorine	Cl	Copper	Cu
Sulphur	S	Magnesium	Mg
Carbon	C	Mercury	Hg
Phosphorus	P	Lead	Pb
Silicon	Si	Silver	Ag
Calcium	Ca	Nitrogen	N
Sodium	Na	Manganese	Mn
Potassium	K		

Hydrogen and oxygen, and many other gases, are made up of **molecules** composed of two atoms. For this reason, the smallest particle of the gas oxygen would be represented as O_2. If hydrogen is allowed to burn in air, it combines with some of the oxygen in the air to form water. The equation representing this reaction is written:

$$2H_2 + O_2 \rightarrow 2H_2O$$

This equation simply tells us that two molecules of hydrogen combine with one molecule of oxygen to produce two molecules of water. If we carry out the experiment, many millions of hydrogen molecules react with half that number of oxygen molecules to produce the same number of water molecules as there were hydrogen molecules.

3.3. Oxides

When we burn an element in air, or oxygen, the new compound formed is called an **oxide**.

Investigation 3b. Making some oxides

Prepare some oxygen by the following method. Set up the gas generation apparatus, as shown in Figure 3.1, and put a little manganese dioxide in the flask. Pour a little twenty-volume hydrogen peroxide down the thistle funnel and allow a little gas to escape; this

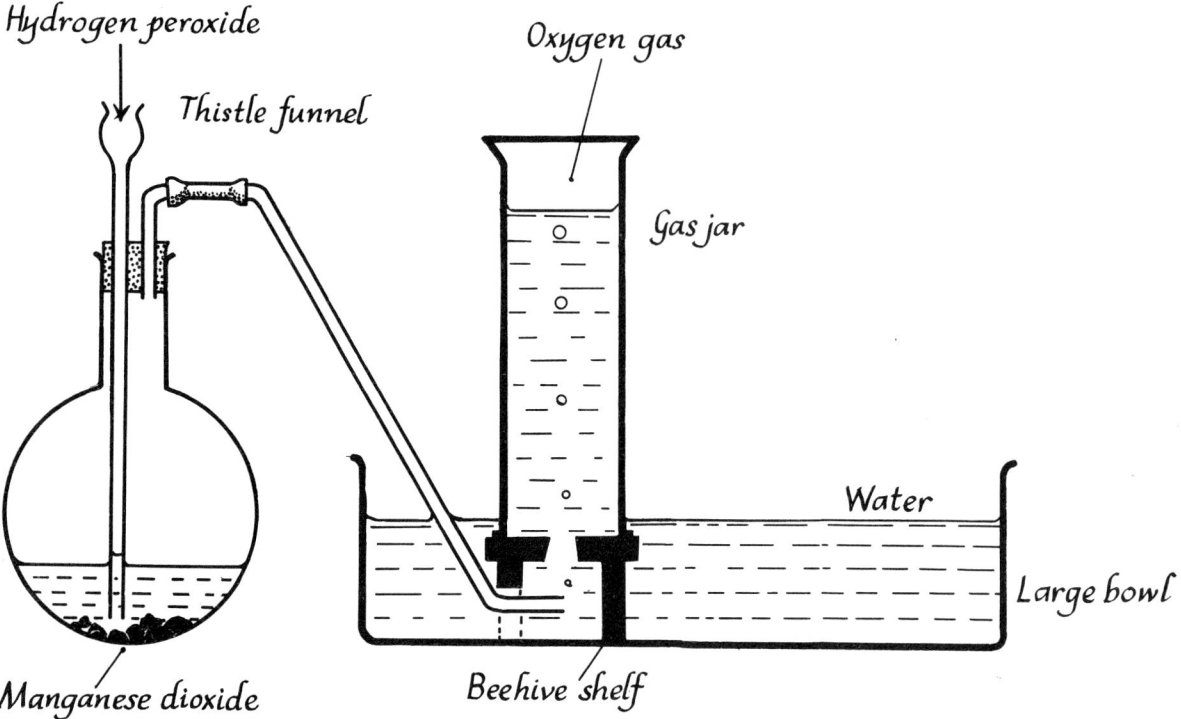

Figure 3.1 Preparation of oxygen

drives the air from the apparatus. Collect five jars of oxygen, adding more hydrogen peroxide as required.

This reaction may be represented by the equation:

$$2H_2O_2 \rightarrow 2H_2O + O_2$$

The manganese dioxide does not change in this reaction, but it enables the hydrogen peroxide to decompose rapidly. When a substance acts in this way, assisting a reaction but not undergoing a chemical change, it is said to be acting as a **catalyst**.

Use the oxygen in the following ways:
1. Heat a small piece of charcoal in a deflagrating spoon until it is red-hot. Place this in a jar of oxygen. When the reaction is complete replace the cover on the jar.

2. Take a little sulphur in a deflagrating spoon and light it in a bunsen burner flame. Hold the burning sulphur in a jar of oxygen and when the reaction is complete, remove the spoon and replace the cover on the jar.
3. Hold a short length of magnesium ribbon in a pair of tongs, light it and hold it in a jar of oxygen. *Do not look directly at the flame.*
4. A small piece of phosphorus is placed in a deflagrating spoon and lighted. This is then placed in a jar of oxygen. When the reaction is complete the cover is replaced.
5. A small piece of sodium is ignited in a deflagrating spoon and placed in a jar or oxygen. When the reaction is complete the cover is replaced.

★ WARNING. *These should be done as demonstrations by the teacher.*

Put a little water into each of the jars, replace the covers and shake the jars. Add a few drops of universal indicator to each jar and shake again.

Take two test-tubes, put a little dilute hydrochloric acid in one and a little dilute sodium hydroxide solution in the other. Add a few drops of universal indicator to each of these.

Make a note of all your observations.

Charcoal is a form of carbon and burns to form carbon dioxide:

$$C + O_2 \rightarrow CO_2$$

When sulphur burns in oxygen it forms sulphur dioxide:

$$S + O_2 \rightarrow SO_2$$

Magnesium burns to form magnesium oxide:

$$2Mg + O_2 \rightarrow 2MgO$$

Phosphorus burns to give phosphoric oxide:

$$4P + 5O_2 \rightarrow 2P_2O_5$$

Sodium burns to yield some sodium peroxide:

$$2Na + O_2 \rightarrow Na_2O_2$$

Can you draw any conclusions about the solutions you have made?

3.4. Acidic and basic oxides

When the oxide of a non-metal dissolves in water, the solution is usually acidic; when the oxides of metals dissolve in water, they give alkaline solutions.

For this reason, the oxides of non-metals are called **acidic oxides** while those of metals are called **basic oxides.**

3.5. Acids, bases and salts

Investigation 3c. Acids and indicators

Put a little lemon juice in a test-tube and dip a piece of neutral litmus paper in it. Then use a piece of universal indicator paper instead of the litmus paper. Finally, add a few drops of methyl orange to the juice. Note your observations. Then repeat the experiment, using the following liquids in place of the lemon juice: tartaric acid solution, vinegar, dilute sulphuric acid, dilute hydrochloric acid, dilute nitric acid and sour milk. Tabulate your observations.

Investigation 3d. The action of acids on metals

For this investigation you will need test-tubes, dilute hydrochloric acid, dilute sulphuric acid, vinegar, splints and specimens of the following metals: iron filings, zinc granules, copper turnings, powdered aluminium, lead shot and a few short lengths of magnesium ribbon.

Put some iron filings in a test-tube and add a little dilute hydrochloric acid. Place the test-tube in the rack and observe what happens. Repeat the experiment, using each of the other acids. Then repeat the series of experiments, using each of the other metals. List the metals and acids which react together.

If there was an effervescence, the gas produced was hydrogen. Some of these reactions are sometimes used to prepare hydrogen in the laboratory.

Investigation 3e. The preparation of hydrogen

★ WARNING. *This should be carried out as a demonstration by a teacher. A safety screen should be used.*

The gas generation apparatus is set up, as shown in Figure 3.2, with zinc granules in the flask. Sufficient dilute sulphuric acid is added to cover the bottom of the thistle funnel. The gas is allowed to escape for a while in order to drive all the air from the apparatus. The gas is then collected by the displacement of water.

A jar of the gas is held mouth downwards, the cover is removed and a lighted taper is held *at the mouth* of the jar. The taper is then held *in* the jar. Does the gas burn? Does it support combustion?

The equation representing the preparation of hydrogen is:

$$Zn + H_2SO_4 \rightarrow ZnSO_4 + H_2$$

Investigation 3f. Burning hydrogen in air

★ WARNING. *This should be carried out as a demonstration by a teacher.*

The gas generation apparatus is set up as in the previous investigation, and the hydrogen is dried by passing it through calcium chloride. The gas is allowed to escape for a few minutes to drive all the air from the apparatus. *This is most important: if air remains in*

Figure 3.2 Preparation of hydrogen

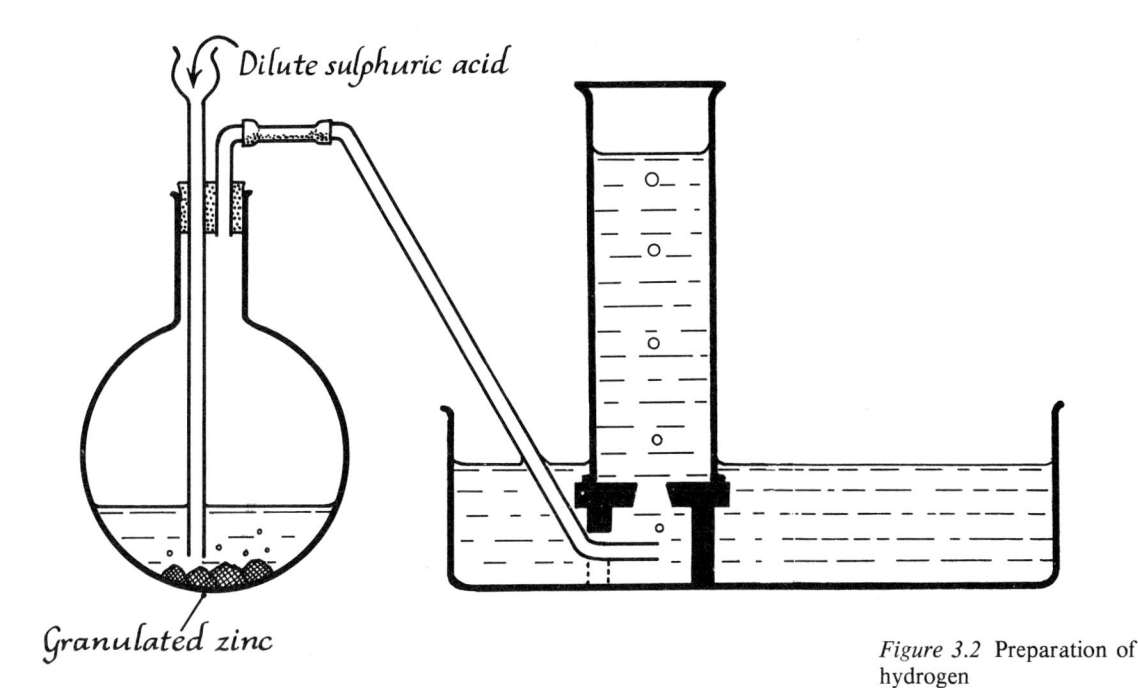

the apparatus a dangerous explosion may result. The gas is then ignited and the flame is played on the cold surface of a glazed tile. After a few minutes the tile is examined.

The equation representing the burning of hydrogen is:

$$2H_2 + O_2 \rightarrow 2H_2O$$

A safety screen should be used.

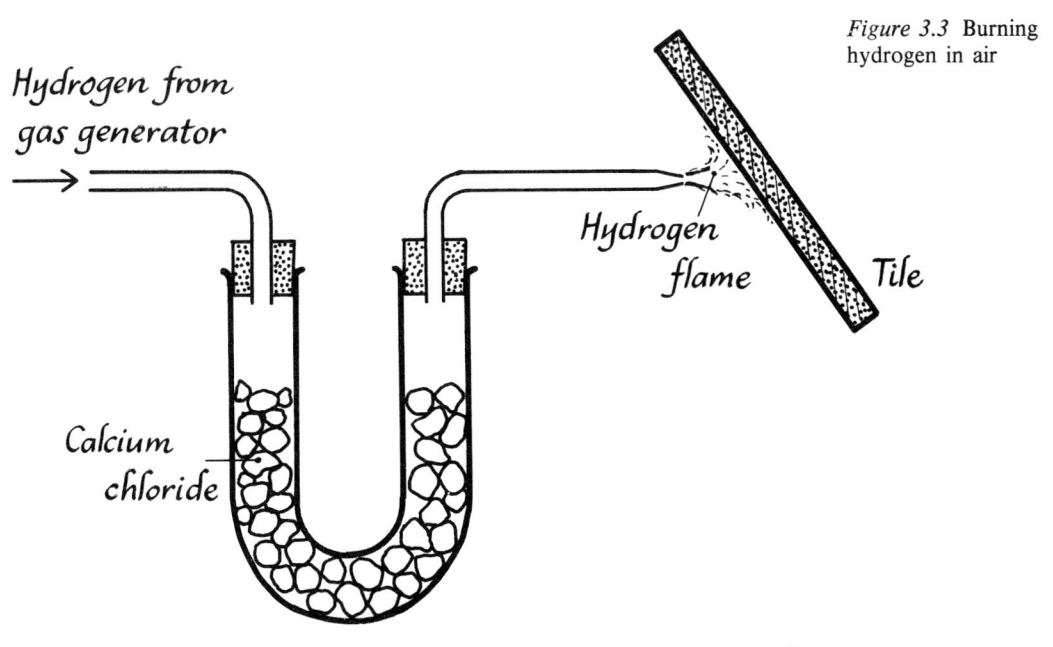

Figure 3.3 Burning hydrogen in air

Investigation 3g. The action of acids on carbonates

For this investigation you will require test-tubes, a teat pipette, some limewater, calcium carbonate, copper carbonate, sodium carbonate and specimens of the following acids: dilute hydrochloric acid, dilute sulphuric acid, dilute nitric acid, vinegar, lemon juice and a solution of tartaric acid.

Put a little calcium carbonate in a test-tube and add a little dilute hydrochloric acid. If there is an effervescence, squeeze the teat of the pipette, insert the pipette in the test-tube and withdraw some of the gas. Put a little limewater in another test-tube and bubble the gas from the pipette through this. If the limewater turns cloudy, the gas is carbon dioxide.

Repeat the investigation, using each of the other acid solutions in turn, in place of the hydrochloric acid. Then repeat the series of investigations, using each of the carbonates in turn. Tabulate your results.

From Investigations 3c, 3d and 3g you should have gained some information about the behaviour of acids.

a. Certain substances change colour in the presence of acids. These substances are called **indicators**. They are often vegetable dyes. Try adding acids to slices of beetroot.

b. Most acids react with certain metals to produce hydrogen. This reaction is often used in the laboratory preparation of hydrogen.

c. Acids react with carbonates to produce carbon dioxide. Carbon dioxide is often prepared in the laboratory by the action of dilute hydrochloric acid on marble chips (a form of calcium carbonate).

At this stage, **an acid may be defined as a compound containing hydrogen, which may be replaced by a metal.** If we write the equations for some of the reactions we have recently carried out, this will explain this definition.

$$Zn + H_2SO_4 \rightarrow ZnSO_4 + H_2$$
(sulphuric acid) (zinc sulphate)

$$Na_2CO_3 + H_2SO_4 \rightarrow Na_2SO_4 + H_2O + CO_2$$
(sodium carbonate) (sodium sulphate)

$$Mg + 2HCl \rightarrow MgCl_2 + H_2$$
(magnesium chloride)

$$Na_2CO_3 + 2HCl \rightarrow 2NaCl + H_2O + CO_2$$
(sodium carbonate) (sodium chloride)

$$CaCO_3 + 2HCl \rightarrow CaCl_2 + H_2O + CO_2$$
(calcium carbonate) (calcium chloride)

When the hydrogen contained in an acid has been replaced by a metal, the compound produced is called a **salt**. The name of the salt is derived from the name of the acid and that of the metal used to make it: the salts obtained from **hydrochloric acid** are **chlorides**; those obtained from **sulphuric acid** are **sulphates**; and those obtained from **nitric acid** are **nitrates**. Thus, when magnesium reacts with sulphuric acid the salt which is formed is called **magnesium sulphate**.

Investigation 3h. Preparation of copper sulphate from copper oxide

For this investigation you will need a beaker, an evaporating dish, a bunsen burner, a tripod, gauze, a stirring rod, some dilute sulphuric acid and some black copper oxide.

Pour about 30 cm^3 of dilute sulphuric acid into the beaker and warm (*do not boil*). Add a little copper oxide and stir until it has all reacted with the acid. Continue adding the oxide, a little at a time, until no further reaction takes place. Then clear the solution by adding a little more acid.

Pour the liquid into the evaporating dish and evaporate. At intervals, dip the stirring rod into the liquid, and allow the liquid on the end of it to cool. When small crystals form on the end of the rod, stop heating the liquid. Allow the evaporating dish and its contents to cool. The solid produced is crystalline copper sulphate.

Copper oxide is called a **base** and will react with an acid to form a salt and water *only*. Bases are metallic oxides or hydroxides. If a basic oxide dissolves in water it forms a **hydroxide**.

Calcium oxide (quicklime) will react with water to produce calcium hydroxide (slaked lime):

$$CaO + H_2O \rightarrow Ca(OH)_2$$
(quicklime) (slaked lime)

In the previous investigation, copper oxide reacted with sulphuric acid to form copper sulphate and water:

$$CuO + H_2SO_4 \rightarrow CuSO_4 + H_2O$$
(copper oxide) (copper sulphate)

Hydroxides react in a similar way:

$$NaOH + HCl \rightarrow NaCl + H_2O$$
(sodium hydroxide) (sodium chloride)

Summary

a. **An acid is a compound containing hydrogen which may be replaced by a metal.**

b. **A salt is formed when the hydrogen in an acid is replaced by a metal.**

c. **A base will react with an acid to form a salt and water only.**

3.6. Preparation of salts

Investigation 3i. Preparation of sodium chloride (common salt)

Set up a burette and fill it to the zero mark with a dilute solution of hydrochloric acid (about 0·1 M). Put 25 cm³ of dilute sodium hydroxide (about 0·1 M) into a beaker. This may be done by means of a pipette fitted with a safety bulb, or by very careful measurement with a measuring cylinder. Put a few drops of litmus solution in the sodium hydroxide solution. This should be seen to turn blue. Run acid from the burette slowly into the sodium hydroxide solution until the litmus changes to a purple–pink colour. While adding the acid, keep swirling the liquid in the beaker. Note the amount of acid required.

Throw away the liquid in the beaker and pour in another 25 cm³ of the same sodium hydroxide solution. To this add the same amount of dilute hydrochloric acid as you used previously. Pour the solution into an evaporating dish and evaporate until crystals begin to form. Allow to cool. The salt is common salt.

In this preparation we made use of the reaction:

$$BASE + ACID \rightarrow SALT + WATER$$

Write the equation for this reaction.

Investigation 3j. Preparation of copper sulphate from copper carbonate

Put about 50 cm³ of dilute sulphuric acid in a large beaker. Add copper carbonate, a little at a time, until there is no further reaction. Decant the resulting liquid carefully into an evaporating dish and evaporate until crystals begin to form. Allow to cool.

In this preparation we have made use of the reaction between an acid and a carbonate:

$$ACID + CARBONATE \rightarrow SALT + WATER + CARBON\ DIOXIDE$$
$$H_2SO_4 + CuCO_3 \rightarrow CuSO_4 + H_2O + CO_2$$

Investigation 3k. Preparation of magnesium sulphate

Put about 25 cm³ of dilute sulphuric acid in a beaker. Clean some magnesium ribbon with emery paper and cut it into short lengths. Drop these one at a time into the acid. When there is no further reaction decant the liquid into an evaporating dish and evaporate until crystals begin to form. Allow to cool.

Here we have used the reaction between an acid and a metal:

$$Mg + H_2SO_4 \rightarrow MgSO_4 + H_2$$

Investigation 3l. Preparation of lead sulphate

Make up separate solutions of lead nitrate and sodium sulphate, using a few grammes of each salt to about 50 cm³ of water. Mix the two solutions together. When the two solutions are mixed a precipitate will form. This may be separated from the liquid by filtering. The precipitate is lead sulphate.

This reaction is known as **double decomposition**, in which the metallic 'parts' of the salts have changed 'partners'. The newly formed salts are sodium nitrate and lead sulphate. The former is in solution, but the latter is insoluble and so falls out as a precipitate.

$$Pb(NO_3)_2 + Na_2SO_4 \rightarrow 2NaNO_3 + PbSO_4$$
(lead nitrate) (sodium sulphate) (sodium nitrate) (lead sulphate)

Investigation 3m. Preparation of sodium chloride

In Investigation 3a (2) you heated together the elements sulphur and iron to form the compound ferrous sulphide. This reaction is **direct combination of elements**.

★ WARNING. *This should be carried out as a demonstration by a teacher.*

The gaseous element chlorine will readily combine with metallic elements to form chlorides. Chlorine is prepared in the following way:

Hydrochloric acid is added, drop by drop, to bleaching powder. The chlorine which is produced is first passed through water to remove any hydrochloric acid vapour which may be present, and then through concentrated sulphuric acid to remove the water vapour. The heavy chlorine may be collected by the displacement of air as shown in Figure 3.4.

A small piece of sodium is placed in a deflagrating spoon which is then lowered into a jar of chlorine. It will burn with a vivid yellow flame, producing sodium chloride.

$$2Na + Cl_2 \rightarrow 2NaCl$$

Sodium chloride has been made by the direct combination of the elements. If more chlorine is available, the following investigations may also be carried out.

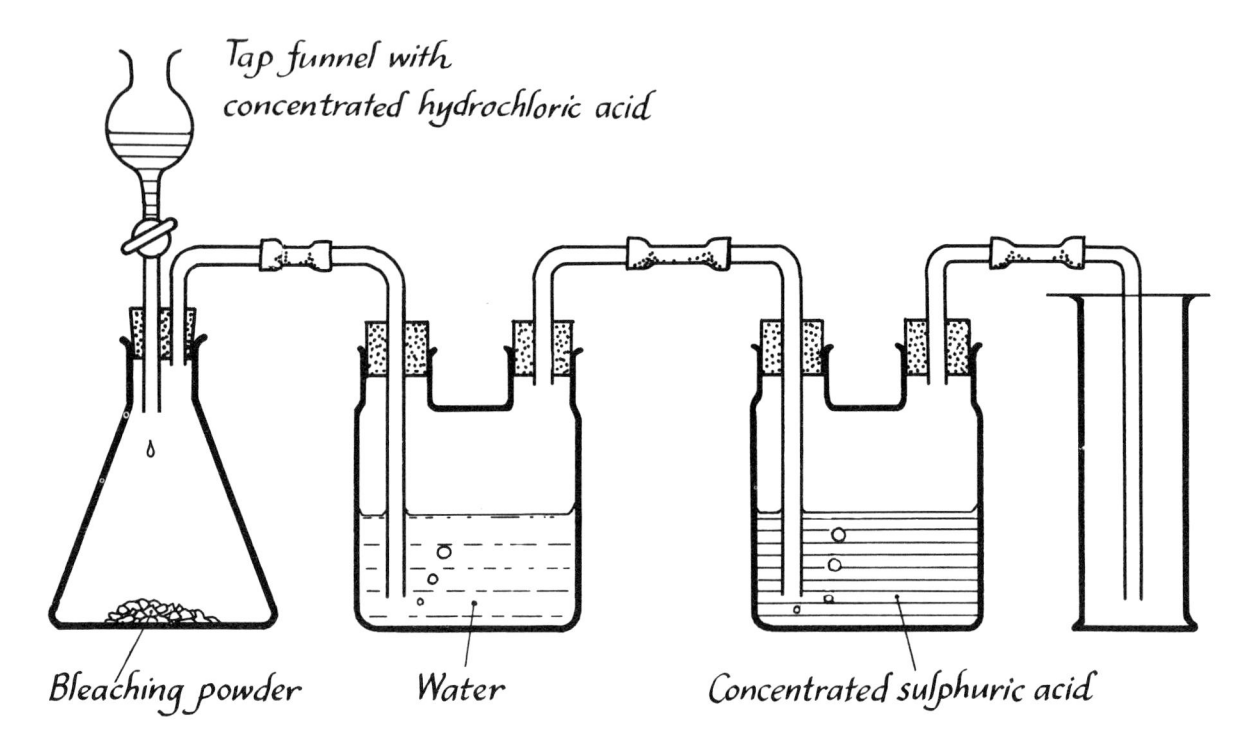

Figure 3.4 Preparation of chlorine

1. A piece of Dutch metal (an alloy containing copper and zinc) is held in a pair of tongs and plunged into a jar of chlorine.
2. Some powdered antimony is sprinkled into a jar of chlorine.

Summary of the methods of preparing salts

　a. **Acid on a base.**
　b. **Acid on a carbonate.**
　c. **Acid on a metal (this is effective with SOME acids and SOME metals).**
　d. **Double decomposition.**
　e. **Direct combination of the elements (useful for preparing pure chlorides).**

3.7. Solutions

If you dissolve some copper sulphate in water the result is called a **solution**. The substance which dissolved is called the **solute** and the liquid in which it dissolved is called the **solvent**.

Investigation 3n. Solvents

For this investigation you will require a few test-tubes with corks; the following liquids: water, alcohol, acetone, xylene; and the following solids: sodium chloride, iodine crystals, celluloid (small pieces),

★ WARNING. *As all of these liquids except water are very inflammable, be sure that there are no naked lights near them.*

perspex (small pieces), beeswax, paraffin wax, potassium permanganate crystals, sealing wax, shellac and barium sulphate.

Put a little sodium chloride in a test-tube, add some water, insert the cork and shake for a few minutes. Does the salt dissolve? Try to dissolve the salt in each of the other liquids in turn.

Repeat the procedure, using each of the other solids in place of the sodium chloride. Tabulate your results, as shown in Table 3.2.

TABLE 3.2. RESULT OF INVESTIGATION 3n

Solid	Water	Alcohol	Acetone	Xylene

Investigation 3o. Heating water

Obtain a round-bottomed flask and fit it with a rubber bung and a delivery tube. You will also need a trough, a beehive shelf, a gas jar and a cover slip.

Set up the apparatus as shown in Figure 3.5, making sure that the

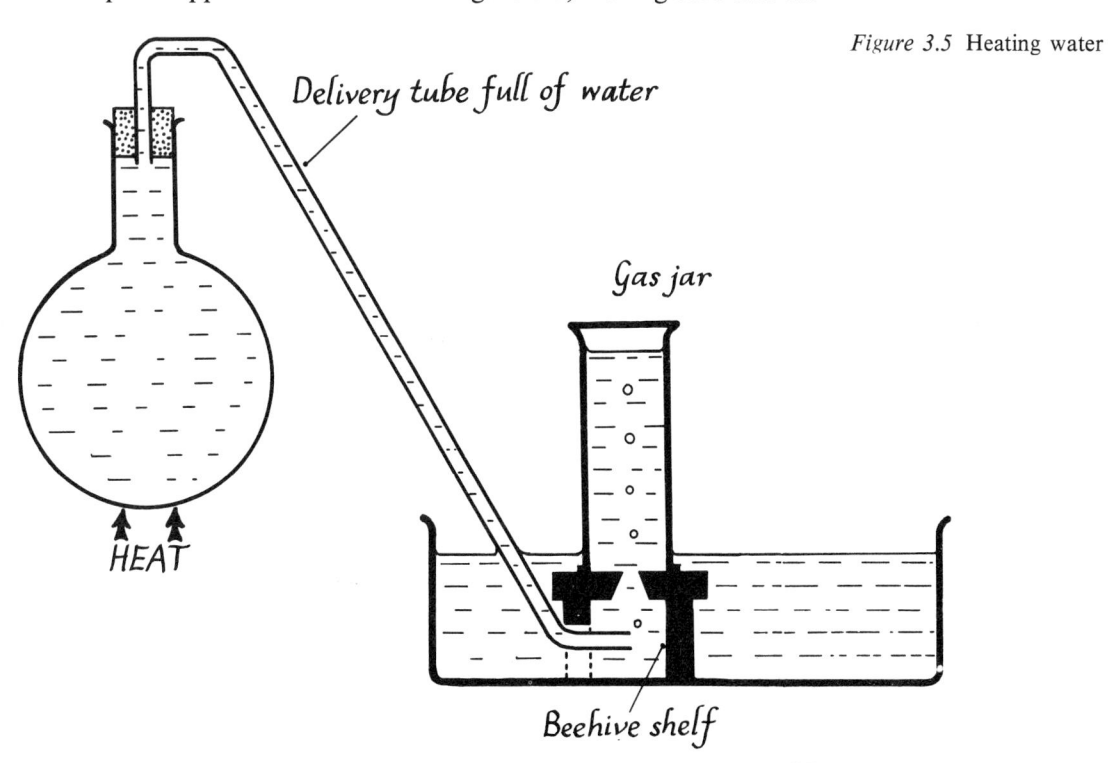

Figure 3.5 Heating water

flask and delivery tube are completely full of tap water. Heat the flask, but do not boil the water. What conclusion can you draw from this investigation?

Investigation 3p. Saturated solutions

Measure 20 cm^3 of water into a small beaker. Put a little lead nitrate into the water and stir until it has dissolved. Add more of the salt and dissolve this. Continue until no more of the lead nitrate will dissolve in the water.

Heat the solution and try to dissolve more salt in it. Allow the solution to cool.

A solution which will allow no more salt to dissolve in it is called a **saturated** solution. Can you draw any conclusions about saturated solutions from Investigations 3o and 3p?

3.8. Crystals

Investigation 3q. Examination of crystals

Obtain specimens of the following substances: sodium chloride, sodium thiosulphate, alum, magnesium sulphate and sodium sulphite. Examine these with the naked eye and then with the aid of a hand lens.

Many chemical compounds may be seen to be of definite shape. For the same compound, this shape is always the same although some particles may be larger than others. These particles are crystals.

Sketch the shapes of the crystals you saw in Investigation 3q.

Investigation 3r. 'Growing' a large crystal

Heat some water in a beaker and in this dissolve as much potash alum as you can. Allow the solution to cool and filter it to remove any undissolved crystals. You now have a cold saturated solution of potash alum.

Dip a thread of cotton in the solution, hang it up and allow it to dry. Some crystals will form on the thread.

Hang the thread in the solution and leave it for a few days. Examine the thread and you will find that the crystals have started to 'grow'. Pick off all but the best crystal and replace the thread in the solution (see Figure 3.6).

Cover the beaker with a filter paper and leave it in a place where the temperature is constant. After several weeks you should have a large crystal.

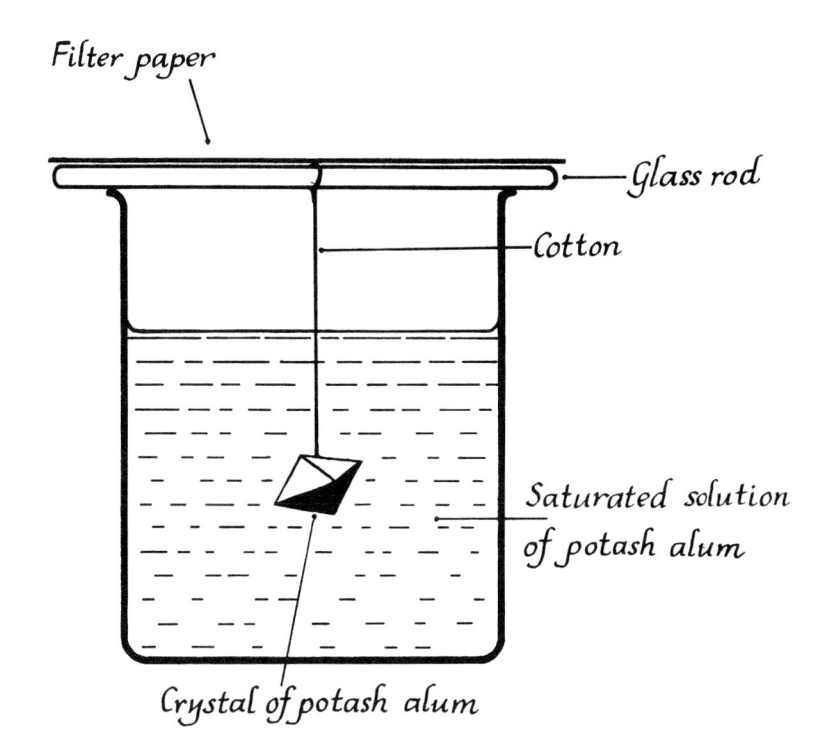

Figure 3.6 'Growing' a crystal

Investigation 3s. Crystal shapes

For this investigation you will need a piece of hardboard about 20 cm square, some 'Plasticine' and about sixty marbles.

Arrange some strips of 'Plasticine' in the form of a square (as shown in Figure 3.7), so that you can put twenty-five marbles in it. On this base, build up a pile of marbles. Does this pile of marbles bear any resemblance to the crystals you have seen?

Try building piles of marbles on bases of different shapes.

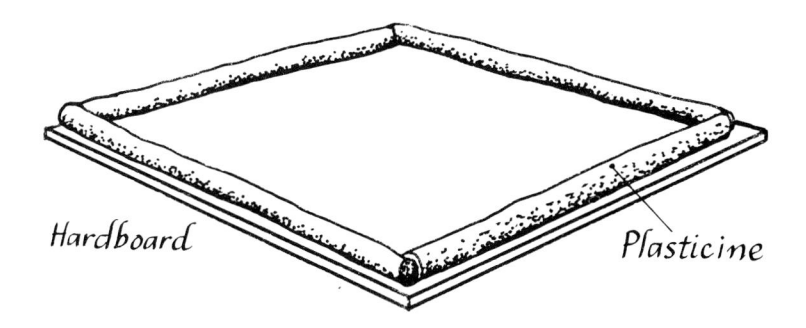

Figure 3.7 The crystal base

3.9. Water of crystallization

Investigation 3t. Heating crystals

Put some crystals of copper sulphate in an evaporating dish and heat them gently. Observe what happens.

When a white powder has been produced, stop heating and allow it to cool. When the powder is cool, add a few drops of water to it.

Crystals of some salts contain water. This is called **water of crystallization**. When the crystals are heated the water of crystallization is driven off as steam and the salt is then said to be **anhydrous**.

Test your understanding

1. Draw a labelled diagram of the apparatus you would use to prepare oxygen.
2. Explain the following terms: acidic oxide, basic oxide, acid, base, salt.
3. What would you expect to be produced when dilute hydrochloric acid is added to sodium carbonate? Write the equation for this reaction (a) in words, (b) using symbols.
4. Describe, with the aid of a diagram, how you could prepare hydrogen. Write an equation for the reaction. What is produced when hydrogen burns?
5. Describe how you would prepare copper sulphate from copper oxide. What happens when a solution of copper sulphate is mixed with a solution of sodium carbonate? What type of reaction is this?
6. What happens when (a) sodium, (b) Dutch metal, is placed in a jar of chlorine? What type of reaction is this?
7. Baking soda is a mixture of tartaric acid (a solid) and sodium hydrogen carbonate. When water is added, baking soda gives off a gas. Explain this.
8. Give a labelled diagram of the apparatus used in the preparation of chlorine.

Chapter 4

The Measurement of Temperature

4.1. 'Hotness'

If you touch a beaker of ice and then a beaker of water, your sense of touch will quickly tell you that one is hotter than the other.

Investigation 4a. Comparing 'hotness'

Fill two large beakers with water, one with water at room temperature and the other with slightly warmer water. Dip one finger of your left hand into one beaker and one finger of your right hand into the other beaker. What difference can you feel?

Figure 4.1 Comparing 'hotness'

Tepid water
Water at room temperature
(a)

Tepid water
Water at room temperature
(b)

Now change your fingers over, as shown in Figure 4.1. Does the difference seem to be more or less? Can you detect, by your sense of touch alone, which beaker contains the warmer water? Check the difference in temperature with a thermometer.

Repeat this investigation in pairs, one partner being blindfolded to act as a 'guinea pig'. Start by having a difference in temperature of about 10 kelvin and reduce the difference until the blindfolded partner cannot detect any difference. What is the smallest temperature difference that you can detect? What is the purpose of blindfolding the 'guinea pig'?

Having found your sensitivity at comparatively low temperatures, find out if you are equally sensitive at higher temperatures by using the same method.

One of the oldest methods of checking the temperature of a baby's bath is to dip your elbow in it. Does this mean that the elbow is more sensitive to temperature difference than the hands? Devise an investigation to find out whether this is so or not.

Investigation 4b

Fill three large beakers with water, one with cold water, one with tepid water and one with hot water, as shown in Figure 4.2. Place

Figure 4.2 Another method of comparing 'hotness'

one finger of one hand in the cold water and one finger of the other hand in the hot water. After about thirty seconds, transfer both fingers to the tepid water. You know that all the water in the tepid beaker is at the same temperature. Does your sense of touch confirm this?

From these two investigations, you will realize that although the human body is sensitive to temperature change, it is unreliable for the measurement of temperature.

4.2. The laboratory thermometer

From Chapter 2 you will know that most substances expand when they are heated. The laboratory thermometer is filled with mercury, which expands and extends along the capillary tube when the bulb is heated. The temperature of a substance is a measure of the average speed of the particles in the substance. If this average speed is higher than the average speed of the particles of mercury within the thermometer, energy is conducted through the glass to the mercury. This increases the average speed of the particles of mercury and, consequently, the mercury expands.

To be of any use, a thermometer must have a marked scale, in the same way as a ruler is marked with a number of equal divisions. The process of dividing a scale into divisions is called **calibration** and, in order to calibrate a thermometer, two **fixed points** are needed.

These fixed points are the temperature at which a pure substance changes its state from solid to liquid and from liquid to gas. The logical substance to use for this purpose is water. The fixed points are **freezing point** and **boiling point**.

The scale in common use is the **Celsius** scale, named after the Swedish scientist, **Anders Celsius**, who first suggested dividing the temperature range between the two fixed points into 100 degrees. On a Celsius thermometer, freezing point is 0° and boiling point is 100°.

Investigation 4c. Checking the lower fixed point

Although the lower fixed point is called freezing point, it is more convenient to place the bulb of the thermometer in melting ice. To check the accuracy of a thermometer's lower fixed point, put some crushed ice into a funnel so that the bulb of your thermometer is completely surrounded by it, as shown in Figure 4.3. As the ice melts, water will drip from the funnel. When this occurs, compare the level of the mercury in the thermometer with the lower fixed point as marked on the thermometer scale. How accurate is your thermometer's lower fixed point?

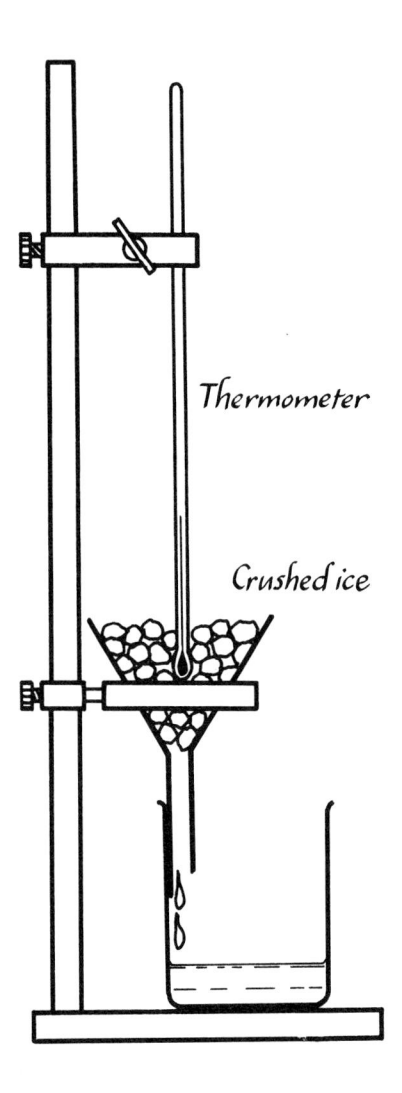

Figure 4.3 Checking the lower fixed point

General-purpose thermometers, such as those used in a school laboratory, are mass-produced. Although this method produces thermometers which may have inaccuracies in their fixed points, the calibration between these points is much more accurate. Since many experiments require a measurement of the rise or fall in temperature, rather than the measurement of an actual temperature, this type of thermometer is perfectly adequate. More accurate thermometers are calibrated individually. Naturally, these thermometers are more costly than the mass-produced type and, as they break just as easily, are not often used in schools.

Investigation 4d. Checking the upper fixed point

To check the upper fixed point of a thermometer, the bulb must be surrounded by steam. Set up the apparatus as shown in Figure 4.4 and, when the water is boiling, compare the level of the mercury in

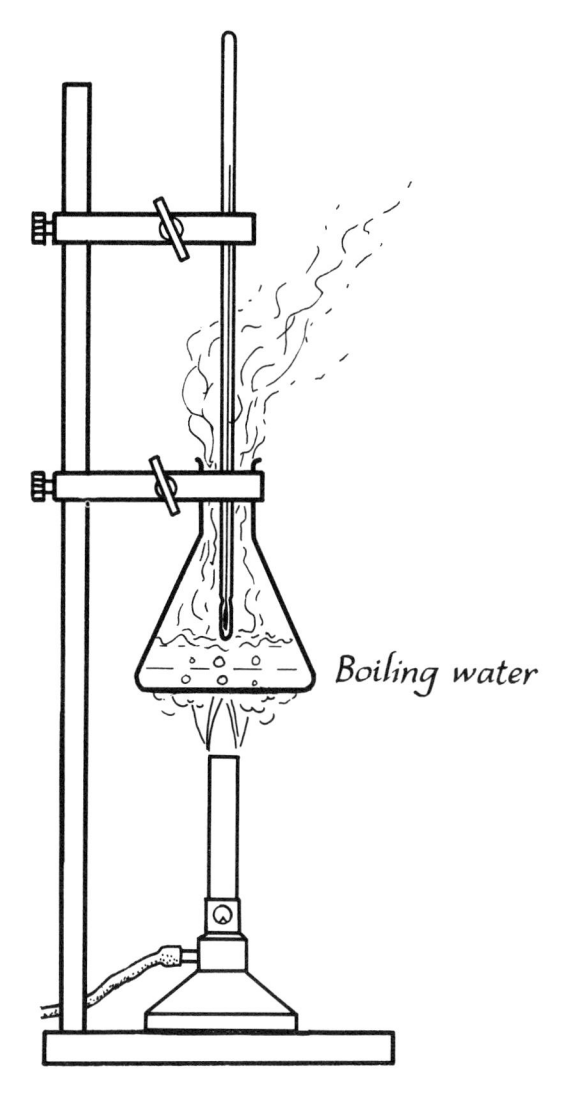

Figure 4.4 Checking the upper fixed point

the thermometer with the upper fixed point as marked on the thermometer scale. How accurate is your thermometer's upper fixed point?

(*Note.* This 'indicated' boiling point will be accurate only if the atmospheric pressure is normal, i.e. if the barometer reading is 760 mm, since increased pressure raises the boiling point and reduced pressure lowers it.)

4.3. Mercury and alcohol as thermometric liquids

When most liquids are poured from a container, a small amount of the liquid remains on the inner surface of the container. Mercury does not wet the inside of the thermometer tube in this way. For this reason, mercury is the liquid most commonly used in general-purpose thermometers. It is also a useful liquid to use for it has a high boiling point (357 °C) and a low freezing point (−39 °C), it expands evenly when heated and it is a good conductor of heat. Another liquid frequently used in thermometers is alcohol, but because alcohol is a colourless liquid, it is coloured by the addition of a dye to make it more easily seen. Alcohol thermometers cannot be used for measuring high temperatures since the boiling point of alcohol is only 78·5 °C, well below the boiling point of water. They are, however, ideal for measuring very low temperatures because the freezing point of alcohol (−117 °C) is much lower than the freezing point of mercury, and well below the coldest temperature generally experienced.

Can you think of any disadvantages of using water as a thermometric liquid?

4.4. The clinical thermometer

Although it would be possible to measure body temperature with a general-purpose laboratory thermometer, there are several disadvantages:

a. Only a very small part of the total range would be used. The normal range is from −10 °C to 110 °C on a laboratory thermometer, compared with a range of about 34 °C to 44 °C on a clinical thermometer.

b. Because of the small distance between one degree mark and the next on a laboratory thermometer, it is not sufficiently accurate for clinical purposes.

c. The thermometer would have to be read while it was still in the patient's mouth.

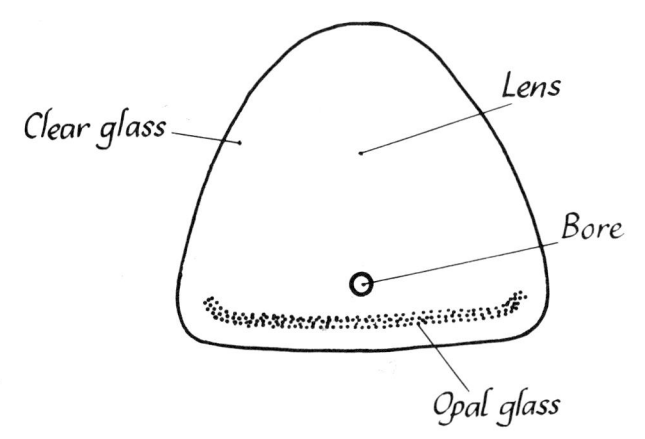

Figure 4.5 Cross-section of a clinical thermometer

Because the range of the clinical thermometer is small, the diameter of the bore through the stem is much smaller than the diameter of the bore through a general-purpose thermometer. This reduction increases the distance moved by the mercury thread for a small rise in temperature, compared with the distance moved by the mercury thread in a general-purpose thermometer caused by the same rise in temperature. This results in the clinical thermometer having a greater accuracy over its limited range.

In order to make the thin thread of mercury in a clinical thermometer more easily seen, the cross-section of the stem is shaped so that the front acts as a lens, which magnifies the thickness of the mercury thread. A layer of opal glass is moulded in the stem behind the bore to act as a background against which the mercury shows more clearly. A cross-section of the stem of a clinical thermometer is shown in Figure 4.5.

Between the bulb and the lower end of the scale of the clinical thermometer is a **constriction** (see Figure 4.6). This constriction is

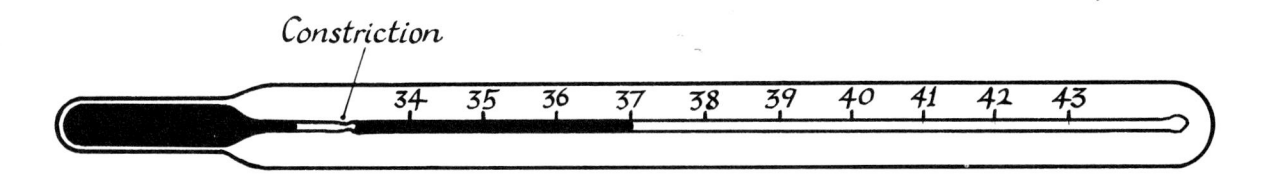

either a narrowing of the bore or a sharp bend in the bore. The effect of the constriction is to resist the flow of the mercury. When the bulb of the thermometer is placed in the patient's mouth, heat is conducted through the glass to the mercury in the bulb. This makes the mercury expand and it is forced past the constriction. When the thermometer is removed from the patient's mouth, the mercury in the bulb contracts but, because of the resistance of the constriction, the mercury above the constriction remains where it is and the thread of mercury breaks below the constriction. The thermometer will continue to record the patient's temperature until it is re-set. In order to re-set a clinical thermometer, it is necessary to re-join the break in the mercury. This is done by shaking the thermometer vigorously to force the mercury down past the constriction.

Figure 4.6 The clinical thermometer

4.5. The Six's maximum–minimum thermometer

In the recording of weather conditions, it is necessary to know the highest and lowest temperatures reached during a definite interval of time. These temperatures can be recorded by a **Six's maximum–minimum thermometer**, as shown in Figure 4.7.

In each limb is an **index** or marker which is shaped like a dumbbell with a steel spring. The thermometric liquid in the left-hand limb with the spherical bulb is alcohol. When the alcohol in this bulb is heated it expands, forcing the mercury in the bottom of the tube upwards in the right-hand limb. This forces the index up the tube. When the temperature falls, the spring on the index in the right-hand

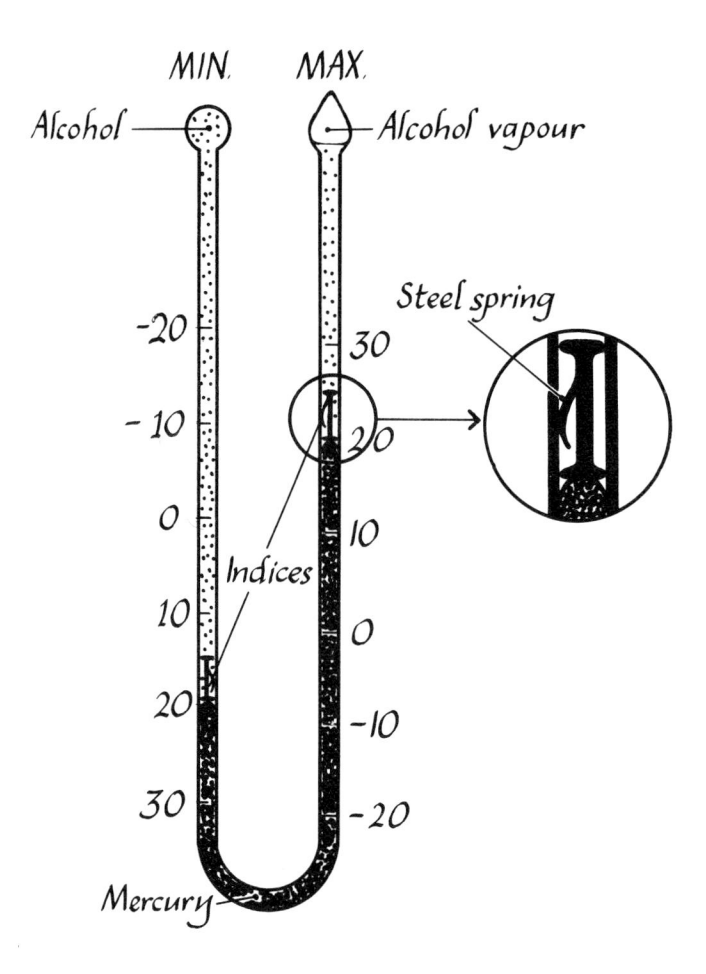

Figure 4.7 The Six's maximum–minimum thermometer

limb keeps the index in position and the maximum temperature is indicated on the scale opposite the bottom of the index.

As the temperature falls, the alcohol contracts and the mercury rises in the left-hand limb. The index is forced up the tube by the mercury and is kept in position by its spring when the temperature rises. The minimum temperature is indicated on the scale opposite the bottom of the index.

When the maximum and minimum temperatures have been read, the indices are moved down to touch the top of the mercury in the limbs by using a magnet.

4.6. The gas thermometer

All of the thermometers in this chapter, so far, have relied upon the expansion of a liquid in a glass tube. As you will know from Chapter 2, a gas expands more than a liquid if its temperature is increased by the same amount.

Investigation 4e. A simple gas thermometer

Take a piece of glass capillary tubing about 30 cm long and connect it to one end of a 10 cm length of rubber tubing. Clamp the open end of the rubber tubing with a Mohr clip, pinch the middle of the rubber tubing between your finger and thumb and put the open end of the capillary tubing in some mercury. By releasing the pressure of your finger and thumb on the middle of the rubber tubing,

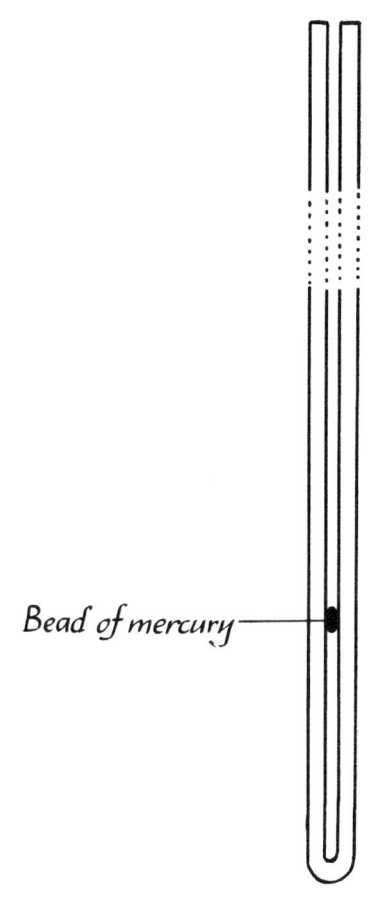

Figure 4.8 A simple gas thermometer

introduce a small bead of mercury into the capillary tube. Lay the tube on a level surface, release the Mohr clip and, by carefully tilting the tube, position the bead of mercury about 5 cm from the open end of the capillary tube when the Mohr clip is replaced. Now seal the open end of the capillary tube by heating it in a bunsen flame. Finally, remove the rubber tubing.

In this thermometer, the thermometric gas is the air trapped below the bead of mercury (see Figure 4.8). Fix your gas thermometer to a laboratory thermometer with rubber bands, so that the bottom of the bead of mercury is level with the top of the mercury in the laboratory thermometer. Place both thermometers in a beaker of cold water and gently heat it. In which of the thermometers does the mercury move faster? Can you account for this? What alteration could be made to the gas thermometer to make it more sensitive?

In SI the gas thermometer is the standard against which all others are calibrated.

4.7. The Bourdon principle

Another type of gas thermometer is the remote-reading type often used in car engine cooling systems, boilers and ovens. This type consists of a sensing unit, placed in the position where the temperature is to be measured, a connecting tube and a dial mechanism on which the temperature is indicated.

The sensing unit is a capsule of gas. At the opposite end of the connecting tube is a **Bourdon tube**. A Bourdon tube is a metal tube in the shape of a spiral with one end sealed. When the gas inside the sensing unit is heated it expands but, since the entire system is sealed, the pressure increases. When the pressure of the gas inside the Bourdon tube increases, the tube tends to straighten out. This movement is transmitted by means of a quadrant and pinion to the needle which indicates the temperature on the dial, as shown in Figure 4.9.

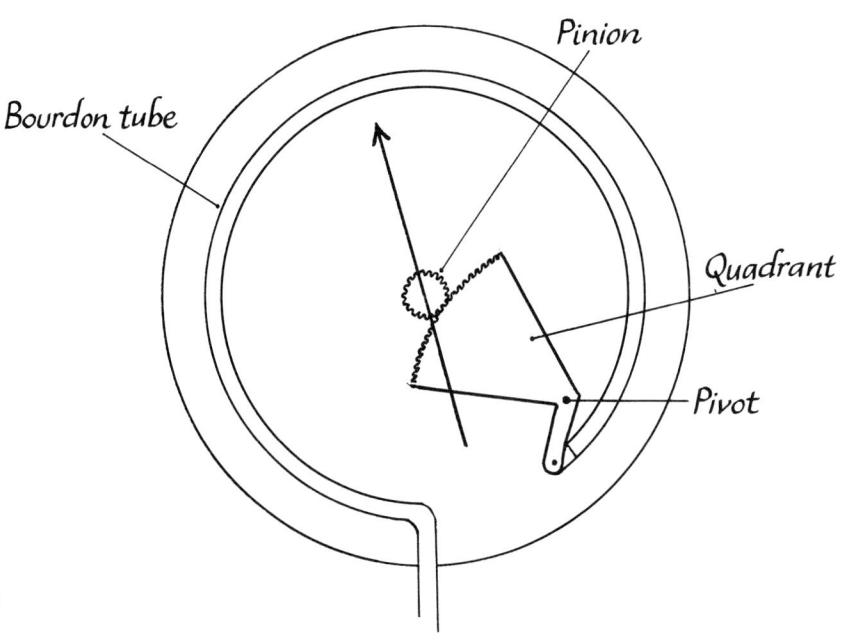

Figure 4.9 The Bourdon tube

Investigation 4f. The Bourdon principle

Close one end of a 60 cm length of rubber tubing with a rubber bung. Arrange the rubber tubing in the shape of a spiral on the bench, with the stoppered end in the centre of the spiral. Put the open end in your mouth and blow as hard as you can. What happens to the spiral? Now use a bicycle pump fitted with a football adaptor to increase the air pressure inside the tubing. Is this more effective?

A popular children's party novelty, consisting of a mouthpiece and a coiled tube of paper with a length of thin steel spring inside it (see Figure 4.10), is an example of the Bourdon principle. Blowing through the mouthpiece causes the paper tube to straighten out and, when blowing stops, the spring returns the paper to its original shape.

Figure 4.10 The Bourdon principle

4.8. The bimetal strip thermometer

Many household dial thermometers for measuring room temperature, and some oven thermostats, rely on the fact that different metals expand by different amounts when heated (see Investigation 2c). In this type of thermometer, the two metals are welded together in the form of a spiral, rather like the hair-spring of a watch or clock. The outer end of the spiral is firmly fixed to the inside of the case and the inner end is fixed to the spindle which carries the needle (see Figure 4.11). When the bimetal strip is heated, one metal expands more than the other. Since they are welded together, the only movement that can take place is an alteration in the shape. If the metal which expands by the greater amount is on the outside, the spiral will tighten. If the metal which expands by the greater amount is on the inside, then the spiral will loosen. Because the outer end

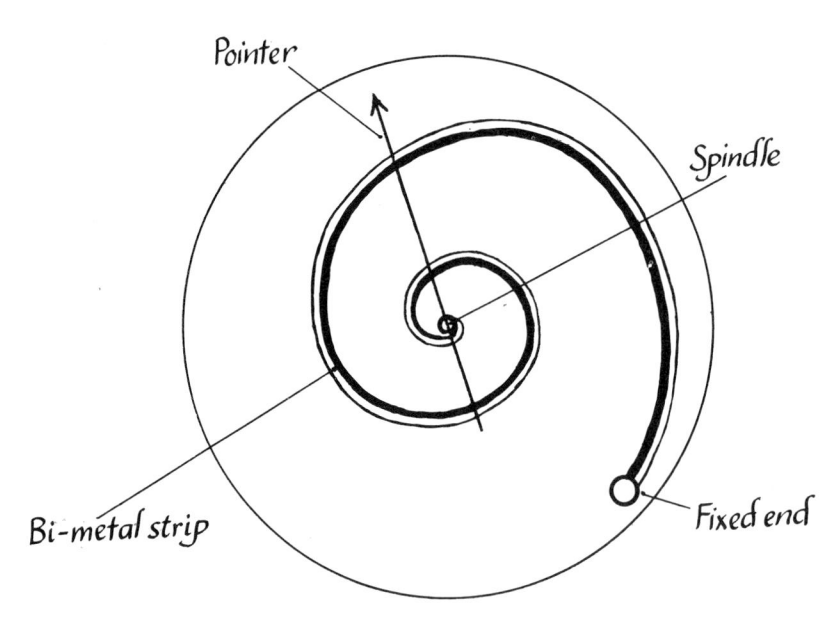

Figure 4.11 The bimetal strip thermometer

of the spiral is fixed, any alteration in the shape of the spiral will move the inner end. This turns the spindle, thus moving the needle.

4.9. The thermocouple

For measuring very high temperatures, such as the temperature of molten steel, the normal laboratory thermometer would be useless as it would melt. For this purpose, an instrument called a **pyrometer** is used. Most pyrometers make use of the principle of the **thermocouple**. When two wires of different metals are connected to form a circuit, and one of the junctions of the two metals is heated while

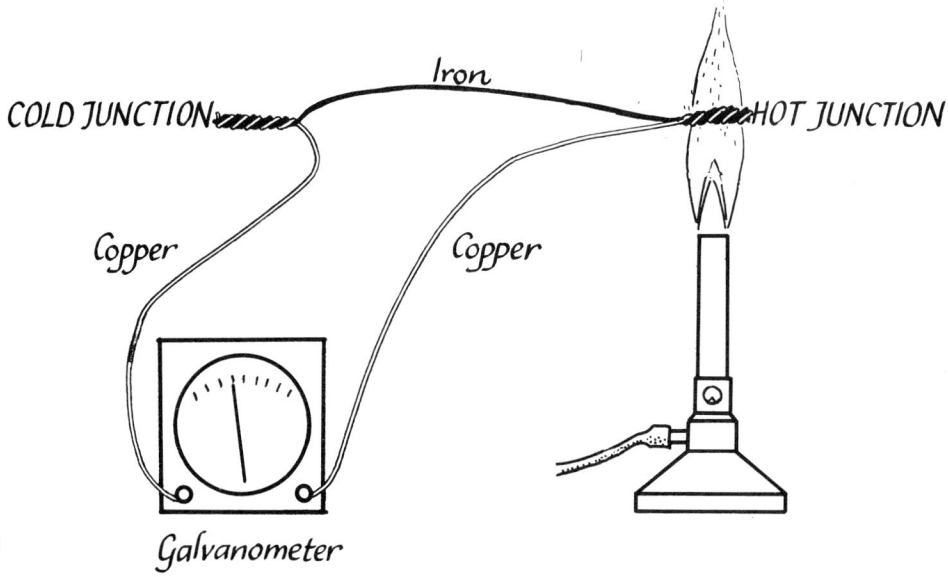

Figure 4.12 The thermocouple

the other junction remains cold, an electric current will flow round the circuit. If a sensitive galvanometer is connected into the circuit, it will measure the current. By placing one junction in a series of known temperatures, the dial of the galvanometer can be calibrated to read temperature instead of current.

Investigation 4g. The principle of the thermocouple

Cut lengths of wire, 20 cm long, of as many different metals as you can obtain. Select wires of two metals and connect them as shown in Figure 4.12 by twisting the ends together. Connect the galvanometer (a centre-zero, 50–0–50 micro-ampere range is suitable) and place one of the junctions in a bunsen flame, first using a luminous flame and then a non-luminous flame. Record the galvanometer reading in each case. Repeat this for each pair of different metals and enter your results in a table similar to that shown in Table 4.1.

TABLE 4.1. RESULTS OF INVESTIGATION 4g

Materials	Galvanometer Reading	
	Luminous Flame	Non-luminous Flame
Copper and iron		
Nichrome and iron		
Constantan and iron		
Copper and nichrome		

Did the use of any of the metals present any problems? To produce a larger reading on the galvanometer, several thermocouples can be connected together in series, as shown in Figure 4.13. This is called a **thermopile**.

4.10. The resistance pyrometer

The electrical resistance to current flowing in a pure metal increases in direct proportion to its temperature. This principle is used in the resistance pyrometer. The metal normally used is platinum; it has a very high melting point and can be used over the range 0 °C to 1 080 °C. The platinum is mounted in a probe which acts as a sensing unit and is connected to an instrument which measures its resistance but is calibrated to read temperature.

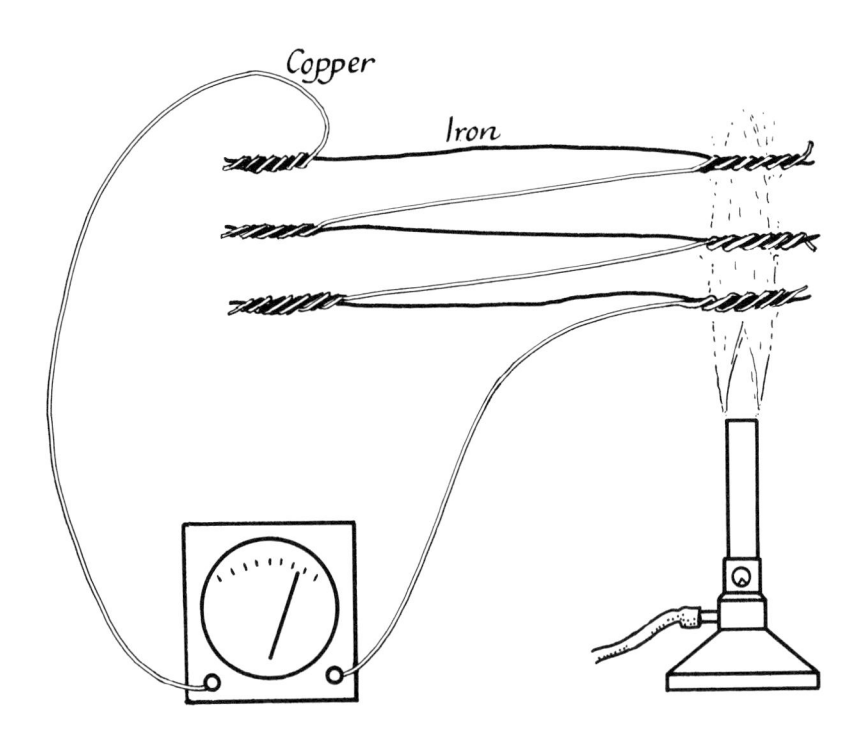

Figure 4.13 The thermopile

4.11. Seger cones

In pottery kilns, the control of temperature is very important because the temperature at which pottery is fired affects the quality of the final product. Although most kilns are fitted with a pyrometer, many potters prefer to rely on a device which indicates when a predetermined temperature has been reached. One such device is a **Seger cone**. A Seger cone is a narrow pyramid which begins to melt and sag at a definite temperature. Seger cones are made to indicate a wide range of temperatures. The potter decides the temperature at which the pottery must be fired, selects the Seger cone that will sag at that temperature and places it in the kiln in the position where that temperature is required. When the cone begins to sag, the potter knows that the predetermined temperature has been reached. Figure 4.14 shows a Seger cone before and after firing.

Test your understanding

1. What type of liquid-in-glass thermometer would be more suitable for measuring the temperature of boiling water, mercury or alcohol? Explain the reason for your answer.
2. What are the main differences between the clinical thermometer and a laboratory thermometer? (Use diagrams to help your explanation.)
3. How would you read and re-set a Six's maximum–minimum thermometer?
4. Explain the action of a thermometer using the Bourdon principle.

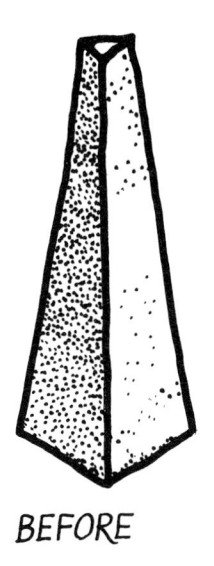

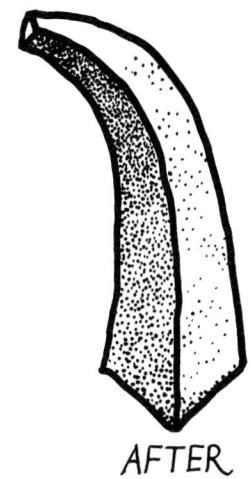

BEFORE   AFTER   *Figure 4.14* A Seger cone

5. Why is mercury a good thermometric liquid?
6. Why must a clinical thermometer never be sterilized in boiling water?
7. In a bimetal strip thermometer, why must the spiral be made of two different metals?
8. Under what conditions would you use a pyrometer instead of a thermometer? Explain your reasons.
9. Clinical thermometers have a time stamped on them. This indicates the length of time that the thermometer must be in the patient's mouth before the reading is taken. What differences would you expect to find in construction between a 'half-minute' thermometer and a 'two-minute' thermometer?
10. When checking the upper fixed point of a thermometer, why should you not put a bung in the flask?
11. What difference would you expect to see on a galvanometer if it was connected to a thermocouple and, instead of one junction being in a bunsen flame, the junction was surrounded with solid carbon dioxide?
12. Why is the spring on the index of a Six's maximum–minimum thermometer made of *steel* and *not brass*?

Chapter 5

Heat Measurement

In Chapter 4 we learned something about temperature and how it may be measured. Heat is not the same thing as temperature and cannot be measured directly with a thermometer.

5.1. The difference between heat and temperature

Investigation 5a. Heating different quantities of water

For this investigation you will need two beakers of the same pattern, two tripods, two gauzes and a bunsen burner. Pour 200 cm^3 of water into one of the beakers and 50 cm^3 of water into the other. Set up each beaker on a tripod and gauze as shown in Figure 5.1.

Light the bunsen burner and adjust it to give a roaring flame. Measure the temperature of the larger quantity of water, then place the bunsen burner under the beaker and measure the time required for the water to reach 80 °C. Repeat this procedure, using the smaller quantity of water. Compare the two times.

Investigation 5b. Heating equal masses of different liquids

★ WARNING. *Note that liquid (medicinal) paraffin is used. On no account should Kerosene (domestic paraffin) be used for this investigation.*

For this investigation you will require two beakers of the same pattern, two tripods, two gauzes, a bunsen burner and supplies of water and liquid paraffin.

Weigh each beaker in turn and then pour water into one of the beakers until the total weight is 100 g more than the weight of the empty beaker. Pour liquid paraffin into the other beaker to obtain the same increase in weight.

Note the temperature of the water, set the bunsen burner to give a roaring flame and heat the water until it reaches a temperature of 80 °C. Note the time required to do this. Repeat this procedure, using the paraffin. It is important that the bunsen burner should not be adjusted between the two parts of this investigation so that it supplies heat at the same rate to each of the liquids. Compare the two times.

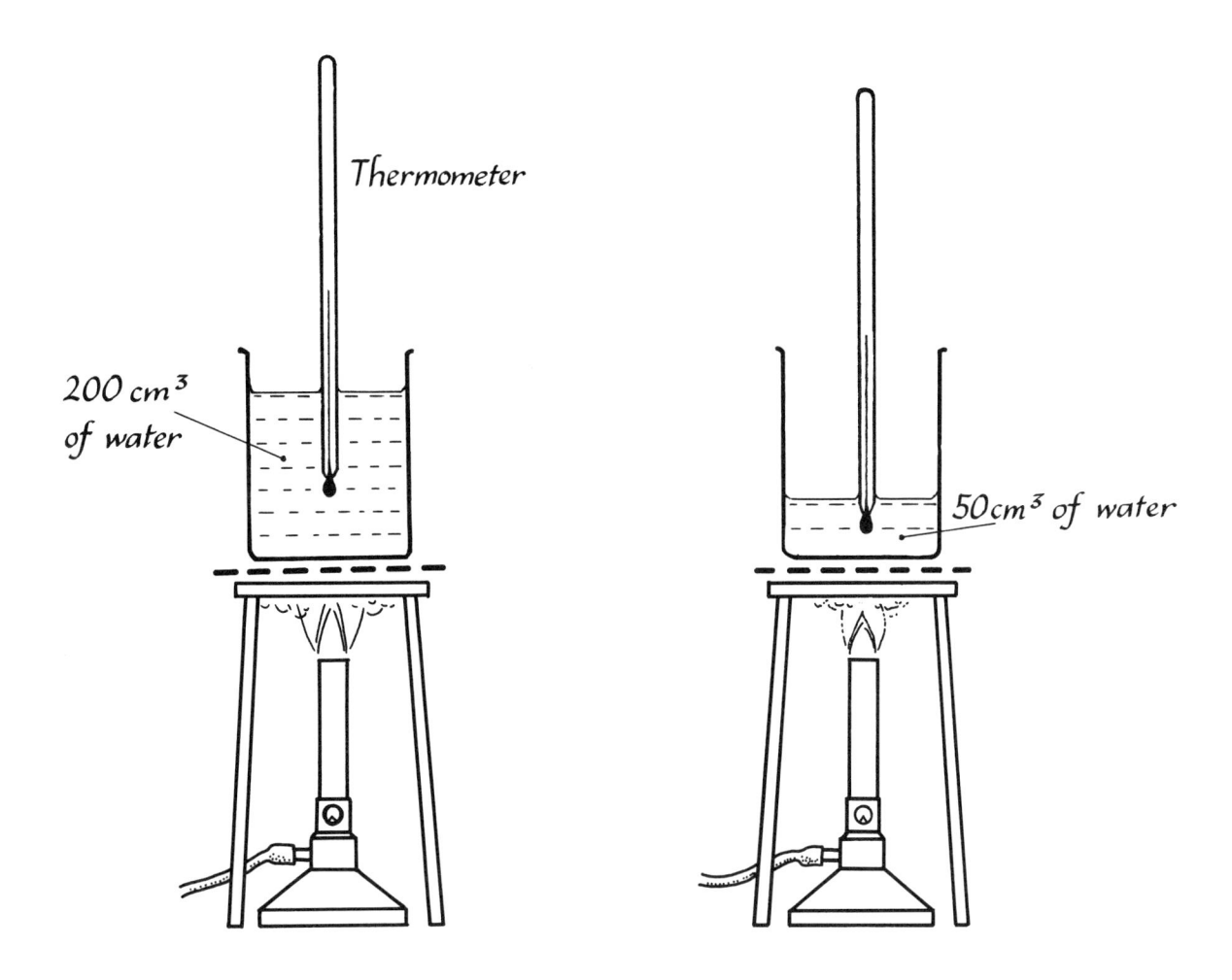

Figure 5.1 Heating different quantities of water

From Investigation 5a you should have learned that the smaller quantity of water required less heat to increase its temperature than the larger quantity, even though the two quantities of water had their temperatures increased by the same amount.

Investigation 5b showed that equal masses of different liquids required different quantities of heat in order to raise their temperatures by the same amount.

From these investigations we may conclude that the amount of heat in a body depends on the mass of the body, the temperature of the body and the substance of the body.

Heat energy is measured in **joules**. In order to raise the temperature of one gramme of water through a temperature interval of one kelvin, 4·2 joules of heat energy are required.

5.2. Specific heat capacity

The **specific heat capacity** of a substance is the number of joules of heat energy required to raise the temperature of one kilogramme of that material through an interval of one kelvin. Table 5.1 lists the specific heat capacities of various materials.

TABLE 5.1. SPECIFIC HEAT CAPACITIES

Material	Specific Heat Capacity (joules per kilogramme per kelvin)	Material	Specific Heat Capacity (joules per kilogramme per kelvin)
Water	4 200	Aluminium	880
Liquid paraffin	2 200	Copper	380
Turpentine	1760	Iron	440
Marble	900	Lead	126
Paraffin wax	2 900	Brass	370

5.3. Heat losses and their prevention

In Chapter 2 we learned that heat may be transferred in three different ways: by conduction through solids, by convection in liquids and gases and by radiation across empty, or nearly empty, space. When a hot body cools, the lost heat is transferred in one or more of these ways.

Investigation 5c. Loss of heat from a metal container

For this investigation you will need a thin metal can with a felt jacket to fit it, a thermometer and a supply of hot water (at about 80 °C).

Pour 100 cm³ of hot water into the container, stir the water and note the temperature of the water. Take the temperature of the water at one-minute intervals for twenty minutes. Stir the water frequently.

Draw a graph, plotting temperature against time (see Figure 5.2).

Repeat the investigation, using the can without its felt jacket. Represent your results on the same graph as before.

The rate at which a hot body loses heat to its surroundings depends on the difference in temperature between the body and its surroundings. Is this statement borne out by your observations in this investigation?

Investigation 5c shows the importance of lagging water pipes and tanks. Materials which are suitable for lagging are poor conductors of heat, such as felt, glass fibre and expanded polystyrene.

A common device for keeping substances hot or cold, by insulating them from their surroundings, is the vacuum flask. It consists of a double-walled glass vessel which has had the air withdrawn from

Figure 5.2 Cooling curves

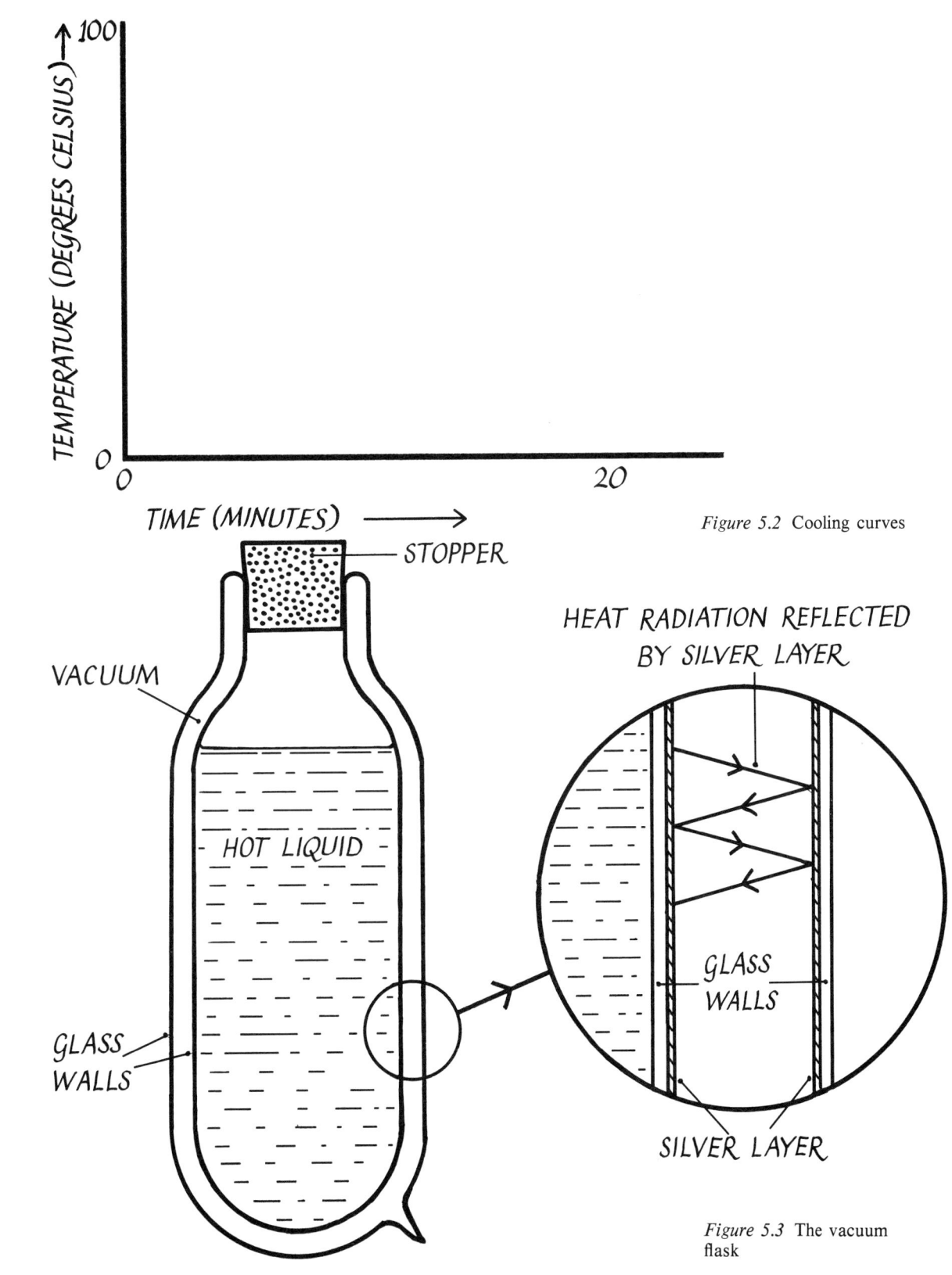

Figure 5.3 The vacuum flask

the space between the walls. The 'vacuum side' of the walls are silvered (see Figure 5.3). Glass is a poor conductor of heat, so heat is lost very slowly by conduction along the wall. Only a little of the heat which is conducted through the inside wall to the silver will be radiated because shiny silver surfaces are poor radiators of heat. Because such surfaces are good reflectors of heat, most of the heat which is radiated will be reflected back and forth in the space between the walls. No conduction or convection can occur in this empty space. The stopper is a poor conductor of heat and, by closing the opening, it also prevents loss of heat by convection in the air above the contents of the flask.

5.4. Latent heat

Investigation 5d. The rate at which a bunsen burner supplies heat

Measure 200 cm^3 of water into a beaker. (Since 1 cm^3 of water has a mass of 1 g, the mass of the water will be 0·2 kg.) Note the temperature of the water.

Place the beaker on a tripod and gauze, and heat the water, using a bunsen burner set to give its hottest flame. Note the temperature of the water at one-minute intervals until its temperature is about 80 °C.

Tabulate your results, as shown in Table 5.2, and calculate the heat supplied per minute.

TABLE 5.2. RESULTS OF INVESTIGATION 5d

Time (minutes)	Mass of Water (kilogrammes)	Temperature (degrees Celsius)	Specific Heat Capacity (joules per kilogramme per kelvin)	Rise in Temperature (kelvin)	Heat Supplied (joules per minute)
0	0·2		4 200	—	—
1	0·2		4 200		
2	0·2		4 200		
3	0·2		4 200		
				Total heat supplied	

Total heat supplied = a joules.
Time to supply heat = b minutes.
Heat supplied per minute = a ÷ b joules.

Investigation 5e. The heat needed to boil away a kilogramme of water

(*Note.* If this investigation is not carried out immediately after Investigation 5d, then that investigation should first be repeated. This is because any variation in gas pressure, or the use of a different bunsen burner setting, will lead to an unsatisfactory result.)

Pour 200 cm^3 of water into a beaker. Weigh the beaker and its contents. Heat the water until it boils. When the water is boiling, start timing. Allow the water to boil for five minutes. Turn off the bunsen burner. When the beaker and its contents are cool, weigh them.

Calculation of result

Mass of beaker and water at start of experiment	$= x$ g
Mass of beaker and water at end of experiment	$= y$ g
Mass of water boiled away (x g $- y$ g)	$= z$ g
Heat supplied in one minute (from previous work)	$= Q$ J
Heat supplied in five minutes	$= 5Q$ J
Heat used to boil away z g of water	$= 5Q$ J
Heat used to boil away 1 g of water	$= 5Q \div z$ J
Heat required to boil away 1 kg of water	$= 1\,000(5Q \div z)$ J

When water boils, the temperature of the water and of the steam remains constant at 100 °C, although heat is still being absorbed by the water. This heat, which does not raise the temperature of the water, is the energy needed to change the water from a liquid to a gas. Heat used in this way is called **latent heat**. Latent heat is also required to change ice at 0 °C into water at 0 °C. When steam condenses to form water, and when water freezes to form ice, latent heat is given out.

When 1 kg of water at 100 °C is converted into steam at 100 °C, 2 267 000 joules (2 267 kilojoules) of heat energy are used. This is called the **specific latent heat of vaporization** of water. When steam condenses, 1 kg of steam will give out this quantity of heat. This explains why a steam scald is so much more serious than a boiling water scald.

To convert 1 kg of ice at 0 °C into water at 0 °C, 396 000 joules (396 kilojoules) of heat energy must be given to the ice. This is called the **specific latent heat of fusion** of ice. If we wish to make a cool drink, we often add ice cubes. Why is this better than adding cold water?

Investigation 5f. The cooling curve of paraffin wax

Put some paraffin wax in a boiling tube. Support the boiling tube in a beaker of water, as shown in Figure 5.4. Heat the beaker until the water is at a temperature of approximately 80 °C. By this time the wax will have melted. Stop heating and note the temperature of the wax at half-minute intervals until it has fallen to less than 30 °C. Tabulate your results and draw a graph, plotting temperature against time.

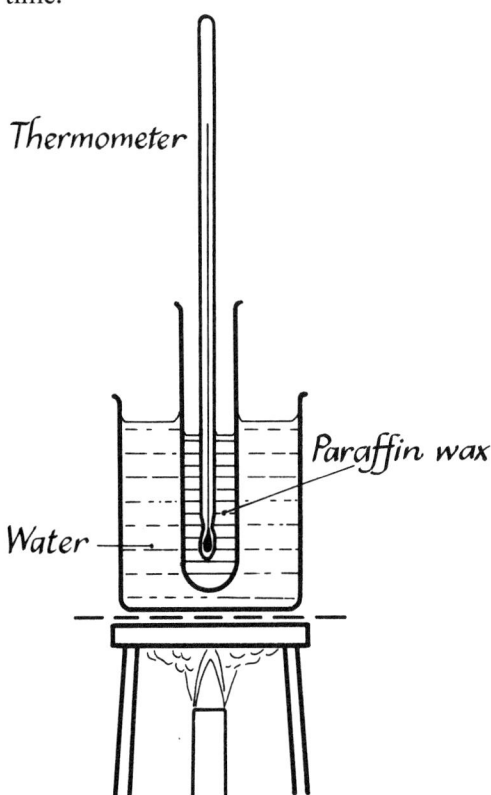

Figure 5.4 Cooling curve of paraffin wax

From your graph, read off the melting temperature of paraffin wax. How does your graph show that latent heat is given out when paraffin wax solidifies?

In order to remove the thermometer it is necessary to melt the wax again.

5.5. Sensible heat and latent heat

When a substance changes its state, heat is given out or heat is absorbed. This is latent heat and no temperature change occurs while the heat is being given out or absorbed.

If a substance loses or gains heat without changing its state, then its temperature must change. Heat causing a temperature change is called **sensible heat**.

Figure 5.5 shows how the temperature of a mass of phenol changes when it is heated steadily. Sections AB and CD show that the temperature is rising as heat is supplied. This heat is sensible

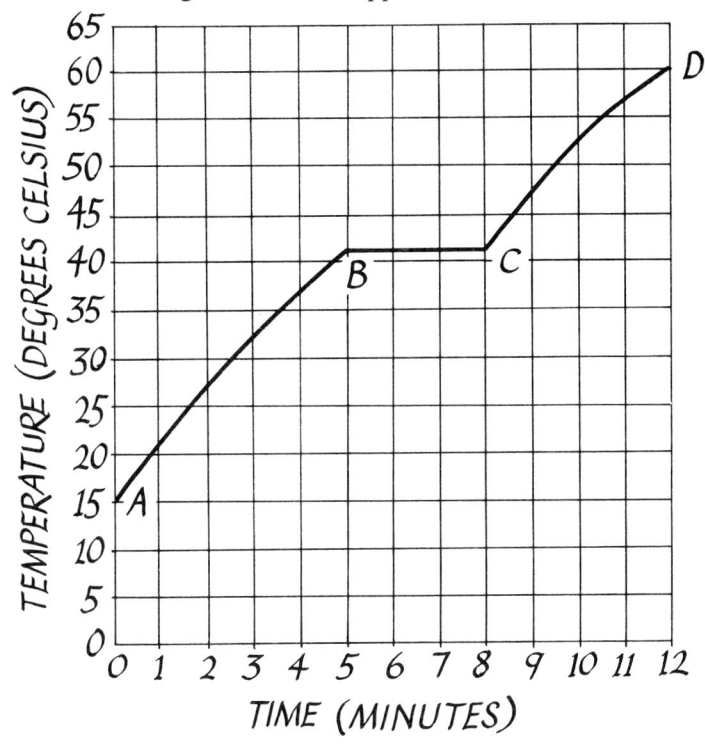

Figure 5.5 Sensible heat and latent heat

heat. Section BC shows that heat is being supplied without a rise in temperature occurring. This is latent heat. Latent heat is absorbed by the phenol in order that it may change its state from solid to liquid.

Test your understanding

1. List the different ways in which heat may be transferred.
2. Draw a labelled diagram of a vacuum flask.
3. What is meant by sensible heat and latent heat?
4. How would you find the rate at which a bunsen burner supplies heat?
5. What is meant by specific heat capacity? In what units is it measured?
6. In a cooling curve experiment using naphthalene, the following results were obtained:

Time (minutes)	0	1	2	3	4	5	6	7	8	9	10
Temperature (degrees Celsius)	95	89	85	81	80	80	80	76	71	66	62

Plot a cooling curve and label the melting point of naphthalene. What was the temperature of the naphthalene 8·5 minutes after the start of the experiment?

Chapter 6

Producing Heat

One of the earliest religions of primitive man was sun worshipping. This was a very logical religion because the sun not only provided man with light but also with heat. Man's ability to produce fire, and to use it and control it, is one of the main reasons for man's supremacy over other animals. The discovery of fire was probably due to natural phenomena, such as lightning or volcanic lava setting fire to trees. Early methods of making fire depended on producing sufficient heat by friction to start some dry tinder smouldering. The heat was produced by rubbing a pointed stick along a groove in a piece of wood or by rotating a pointed stick (between the hands or with a fire-bow) in a dent in a piece of wood, as shown in Figure 6.1.

The tinder, which consisted of dry grass or leaves, was placed along the groove or in the dent. Once the tinder had started to smoulder, it was made to burst into flame by blowing on it. A later method used to start the tinder smouldering was to produce a spark by striking a piece of flint.

Investigation 6a. Making fire

Try to light some fine, dry wood shavings by using one or more of the methods shown in Figure 6.1. Use a selection of different types of wood, both hard and soft woods, for the base (grooved or with a dent) and for the pointed stick. Keep a record of the types of wood that you use and the results you obtain. What combination of types of wood did you find most effective?

6.1. The coke fire

Investigation 6b. Oxidation of carbon

Place a carbon block on a tripod and heat one corner of it with a bunsen flame until the carbon glows. Turn the bunsen out and direct a stream of air on to the glowing part of the carbon block with a blowpipe. Alter the force of the air stream and note any change in

Figure 6.1 Early methods of making fire

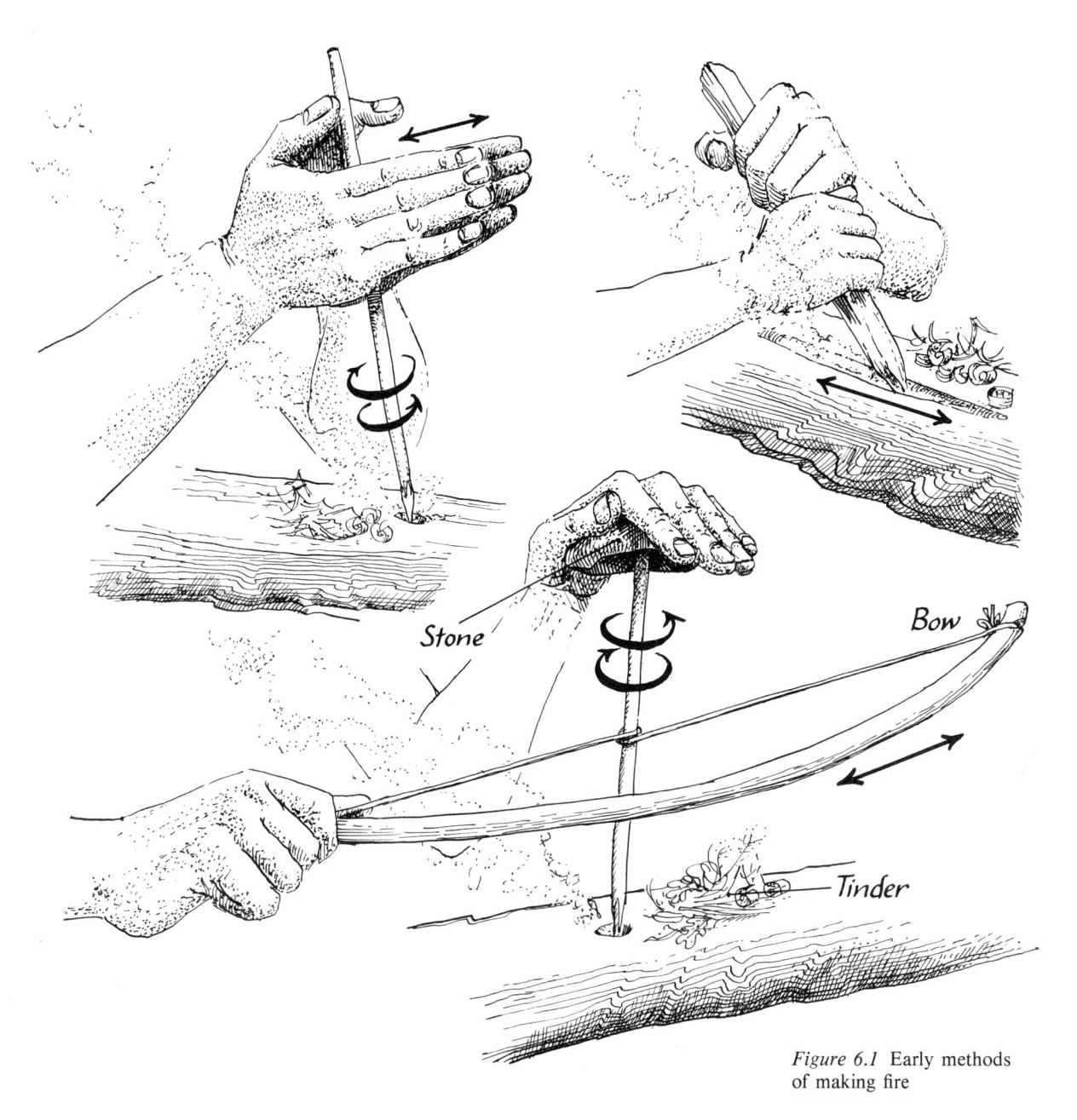

the brilliance of the glow. *Before putting the carbon block away, quench it thoroughly in water.*

Investigation 6c. Oxidation of coke

Insert some small pieces of coke into a length of combustion tube, and fit a single-holed stopper to each end and pieces of glass tubing to each stopper, as shown in Figure 6.2. Place the lower end of the bent piece of glass tubing into a test-tube of limewater and pump air

65

Figure 6.2 Oxidation of coke

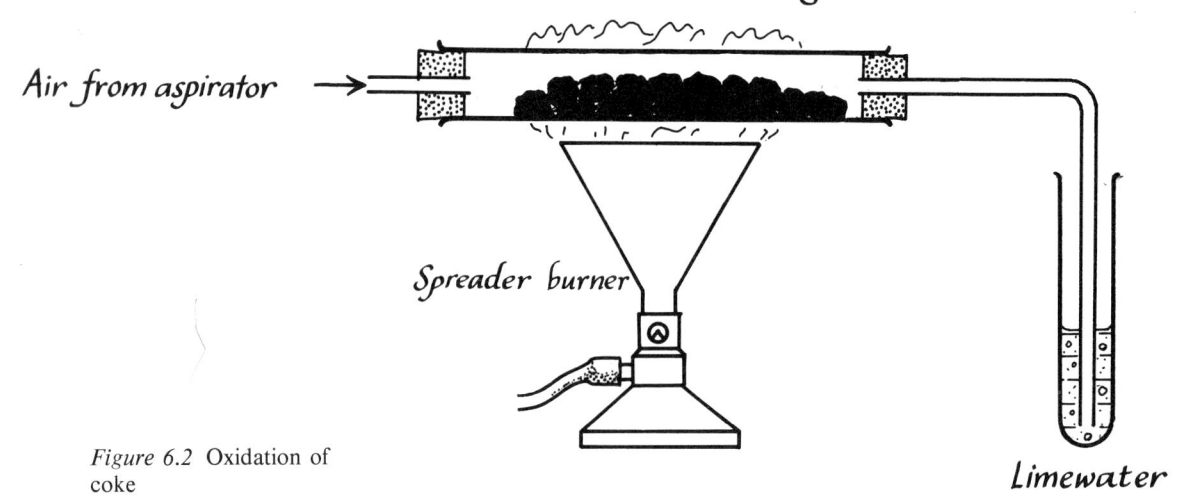

through the apparatus, with an aspirator, while heating the coke. The heat can be spread along the combustion tube by using a Ramsey burner or a bunsen fitted with a fish-tail burner. What effect does the outlet gas have on the limewater? What does this tell you about the outlet gas? If you are not certain, refer to Chapter 3.

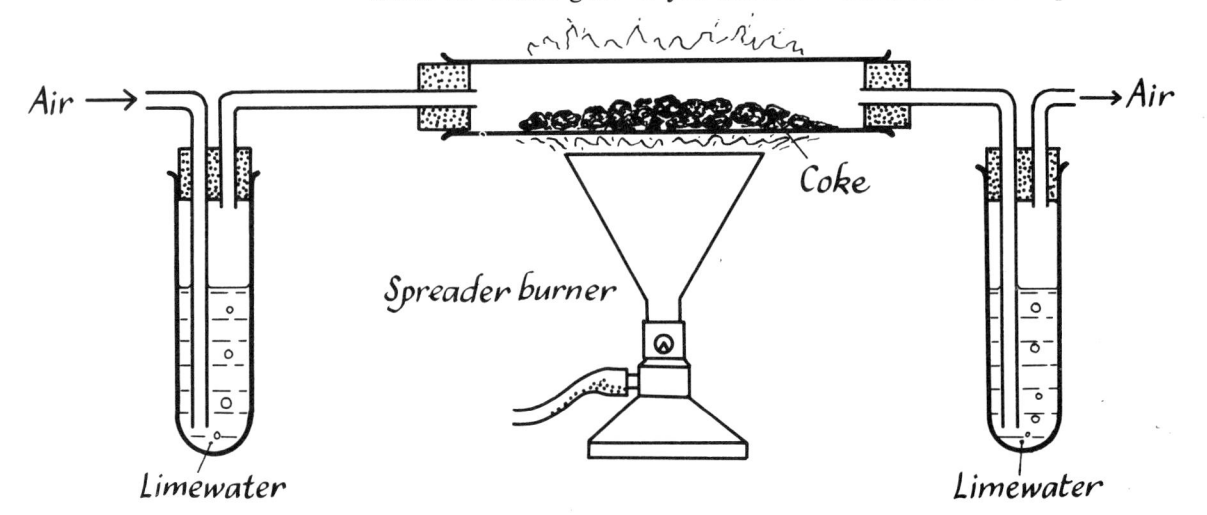

Figure 6.3 Oxidation of coke, with control

From the results of this investigation so far, can we be certain that the changes which took place in the limewater were caused by the effect of the hot coke on the stream of air? In order to ensure that the hot coke is responsible for these changes, the air should be bubbled through limewater before and after passing over the hot coke by arranging the apparatus as shown in Figure 6.3. The air can be made to pass through the apparatus either by being pumped in at the left-hand end or by being sucked out from the right-hand end. In order to suck the air out from the right-hand end, the outlet tube can be connected to an aspirator, a vacuum pump or a bunsen pump. When using this more complicated apparatus, the test-tube of limewater on the left-hand side is acting as a control. In which test-tube does the limewater change first? Does this confirm that the hot coke was responsible for the change?

Repeat this investigation without heating the coke. Do you get the same results?

When any substance combines with oxygen, an oxide is formed and heat energy is released. The speed with which this reaction (oxidation) takes place depends upon three factors: the nature of the material, the temperature and the amount of oxygen available.

A burning fire can be made to burn faster by increasing the supply of air to it. In a domestic fire, this is done by opening a vent below the fire, thus forcing more air through the fire by convection. In a forge, additional supplies of air are forced through the furnace by bellows or by an air pump driven by an electric motor.

In a coke fire, there are three different chemical reactions, each one taking place in a different zone of the fire (see Figure 6.4). In

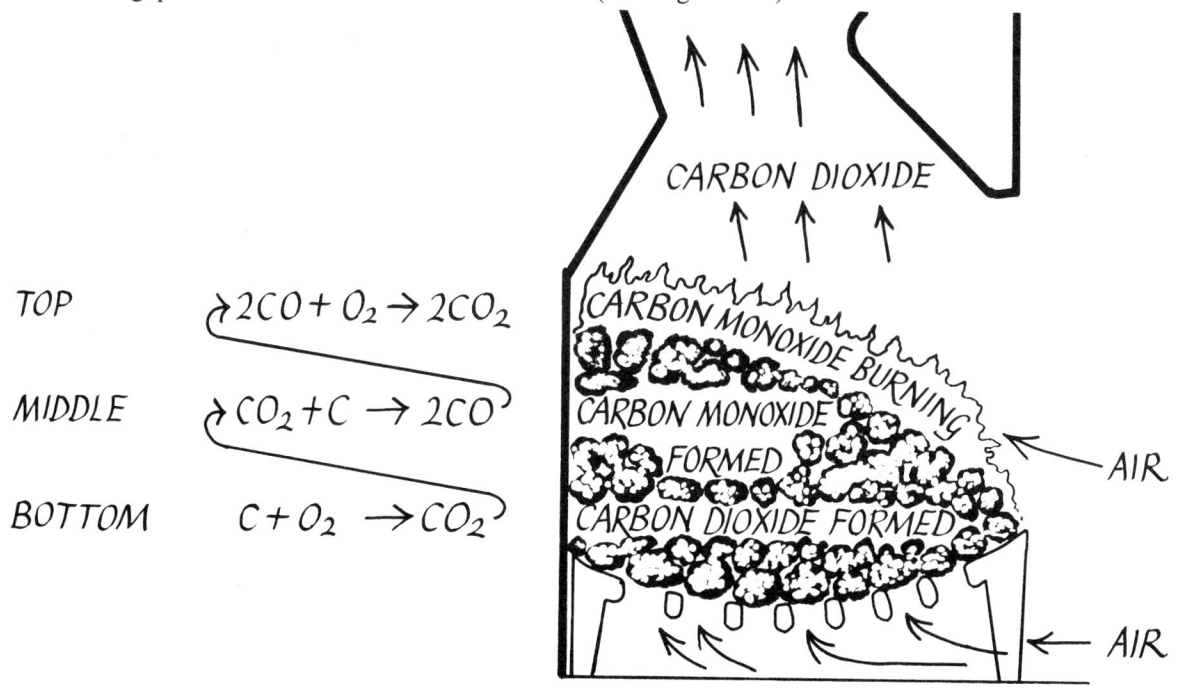

TOP $2CO + O_2 \rightarrow 2CO_2$

MIDDLE $CO_2 + C \rightarrow 2CO$

BOTTOM $C + O_2 \rightarrow CO_2$

Figure 6.4 Zones of combustion in the coke fire

the bottom zone, the reaction is oxidation of the hot coke, which releases heat and produces carbon dioxide.

$$C + O_2 \rightarrow CO_2$$

This reaction relies on an adequate supply of air entering the fire from below. The heat released by this reaction increases the temperature of the coke in the middle zone.

67

In the middle zone, the reaction is the reduction of the carbon dioxide (produced in the bottom zone) to carbon monoxide.

$$CO_2 + C \rightarrow 2CO$$

In this zone, heat energy is absorbed.

In the top zone, the carbon monoxide is oxidized to release more heat and produce carbon dioxide, which is carried up the chimney by convection.

$$2CO + O_2 \rightarrow 2CO_2$$

This reaction produces a mantle of blue flame in the top zone, and relies on an adequate air supply entering the top zone of the fire.

Carbon monoxide is a poisonous gas. When air is inhaled, some of the oxygen combines with the pigment in the red blood cells (**haemoglobin**), forming **oxyhaemoglobin**. When carbon monoxide is inhaled, it combines with the haemoglobin of the blood, forming **carboxyhaemoglobin** which, unlike oxyhaemoglobin, is stable. Red blood cells containing carboxyhaemoglobin are unable to release carbon monoxide and, therefore, become incapable of combining with oxygen. If the body tissues are deprived of oxygen, they cease to function. Because carbon monoxide is colourless and almost odourless, its effects on the body occur without warning. The final result of inhaling carbon monoxide is, of course, death.

Carbon monoxide is produced whenever a carbon compound is incompletely oxidized in a limited supply of oxygen. For this reason, it is essential that a coke fire has an adequate supply of air to ensure the complete oxidation of the carbon monoxide produced in the middle zone. In a properly designed fireplace, stove or boiler, the danger of carbon monoxide poisoning is negligible. Because petrol is a carbon compound and because the air supply to the engine of a car is limited, the exhaust gases from a car contain carbon monoxide. For this reason, a car engine should never be allowed to run in a closed garage.

As a fuel, coke has two main advantages over coal: it leaves a fine powdery ash (as opposed to clinker which is often formed in the bottom of a coal fire, limiting the air flow through the fire) and it produces no smoke, an important factor in the reduction of air pollution.

6.2. The coal fire

The heat given out by a coal fire is caused partly by the same reactions that occur in the coke fire and partly by the oxidation of gases which are released when coal is heated.

Investigation 6d. Destructive distillation

Put a depth of about 2 cm of coal chips in the bottom of a hard glass tube. Insert a cork and delivery tube and support the hard glass tube so that it is almost horizontal but with the delivery tube pointing slightly upwards, as shown in Figure 6.5. Heat the coal chips with a bunsen burner and, when smoke is issuing freely from the delivery tube (*AND NOT BEFORE*), apply a lighted splint to the smoke. Move a piece of paper from side to side just above the flame and then examine the paper. From the colour of the deposit, what do you think it might be? Examine the liquid residue in the top of the hard glass tube. From its colour and smell, what do you think it might be? What is the colour of the solid residue at the bottom of the hard glass tube? What do you think it is?

★ WARNING. *Do not use coal dust as this could cause an explosion.*

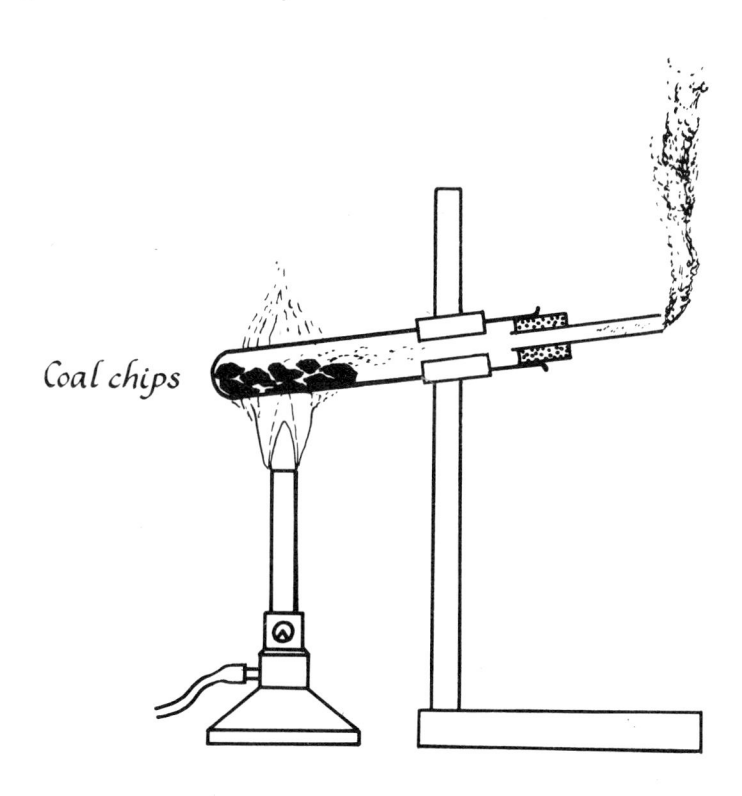

Figure 6.5 Destructive distillation

When you were heating the coal, you may have noticed some droplets of water forming near the top of the hard glass tube. Can you explain the reason for setting the apparatus up in an almost horizontal position?

Repeat this investigation, using sawdust, cheese, paper, soap flakes, wool, cotton wool, leaves or any other animal or vegetable product in place of the coal chips and enter your results in a table, as shown in Table 6.1.

TABLE 6.1. RESULTS OF INVESTIGATION 6d

Substance	Does the Smoke Burn?	Colour of Solid Residue
Sawdust		
Cheese		

Coal was once wood, mainly ferns. If you examine some pieces of coal, you may find leaf prints. All substances which are of animal or vegetable origin are called organic compounds and are always compounds of carbon. The solid residue of each of the substances used in Investigation 6d was carbon. The gas produced by **destructive distillation** of coal was a crude form of coal gas, which used to be supplied for domestic heating and cooking. In a coal fire, the coal is heated and the products of destructive distillation are burned as soon as they are formed. The process is called *destructive* distillation because it breaks down complicated molecules into simpler molecules.

6.3. Heat from gas

Investigation 6e. Gas flames

Cut a short length (about 10 cm) of 1·2 mm diameter (18 S.W.G.) copper wire. With the air holes closed, light a bunsen burner. Hold one end of the wire in a pair of crucible tongs and put the other end in different parts of the flame. Note how long it takes for the end of the copper wire to become red-hot in each part of the flame. Gently blow across the middle of the flame. Does the flame move? What colour is the flame? Is this type of flame quiet or noisy?

Now, slowly turn the air collar round and note the effect on the flame as you do so. Continue turning the air collar until the air holes are completely uncovered. Put the end of the copper wire in different parts of the flame, as you did before, and note how long it takes for it to get red-hot. Which is the hotter flame? Which part of the flame is the hottest and which part is the coldest?

★ WARNING. *Before attempting this experiment place a bowl of water nearby and if necessary plunge the burning paper into it.*

Bring a piece of paper quickly down on to the flame at an angle of about 45° so that the paper is touching the top of the bunsen burner, with the flame on the far side of the paper. As soon as you see the paper beginning to char, remove it quickly from the flame and blow on it to prevent it from catching alight. The secret of this part of the investigation is in knowing the exact moment to remove the paper from the flame—too early, and you will still have a piece of unmarked paper; too late, and you will have a piece of burning paper. After a few attempts, you will soon learn the knack. If you do happen to leave the paper in the flame too long and it bursts into flame, do not panic; be bold and quickly screw up the burning piece of paper in your hands. This will put out the flames and, so long as you are very quick, you will not get burned. Can you explain why this method of extinguishing a burning piece of paper is effective?

When you have succeeded in charring a piece of paper in the bunsen flame, examine the shape of the charred pattern. Does this confirm your previous result regarding the hottest and coldest parts of the flame?

When an inflammable gas is burned, it produces the type of flame shown in Figure 6.6. Immediately above the outlet is a zone of unburned gas; this is so because there is no air to enable the gas to

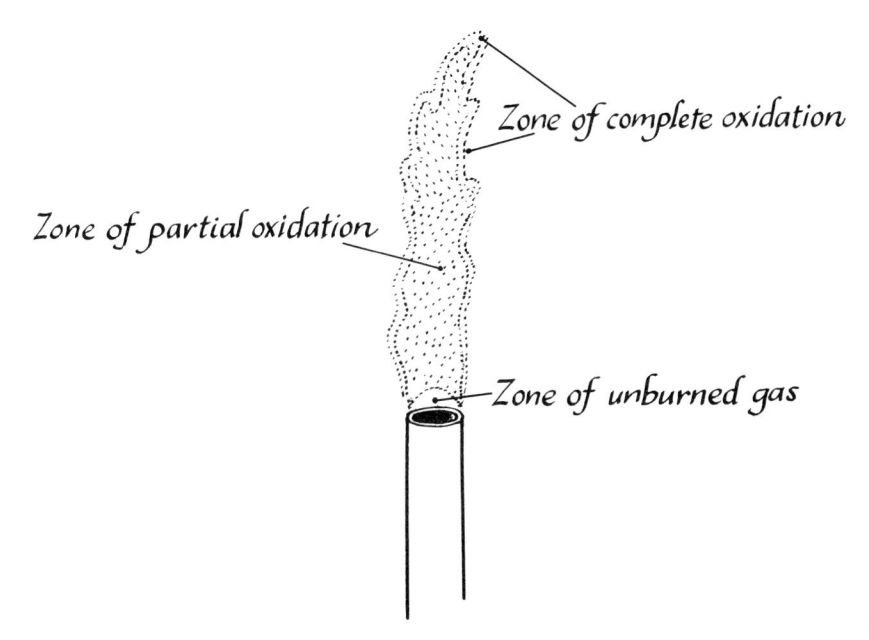

Figure 6.6 The gas flame

oxidize. The zone of partial oxidation is yellow. This colour is caused by thousands of minute particles of glowing carbon. Immediately outside this is the zone of complete oxidation. This outer zone is colourless, and the gas and carbon particles oxidize completely because they are in direct contact with the air.

The flame produced by burning pure gas is not very powerful because the zone of complete oxidation is limited to the outside fringe. In order to increase the rate of oxidation of the gas, it must be mixed with air. A convenient method of doing this is by the principle illustrated by the bunsen burner, the parts of which are shown in Figure 6.7.

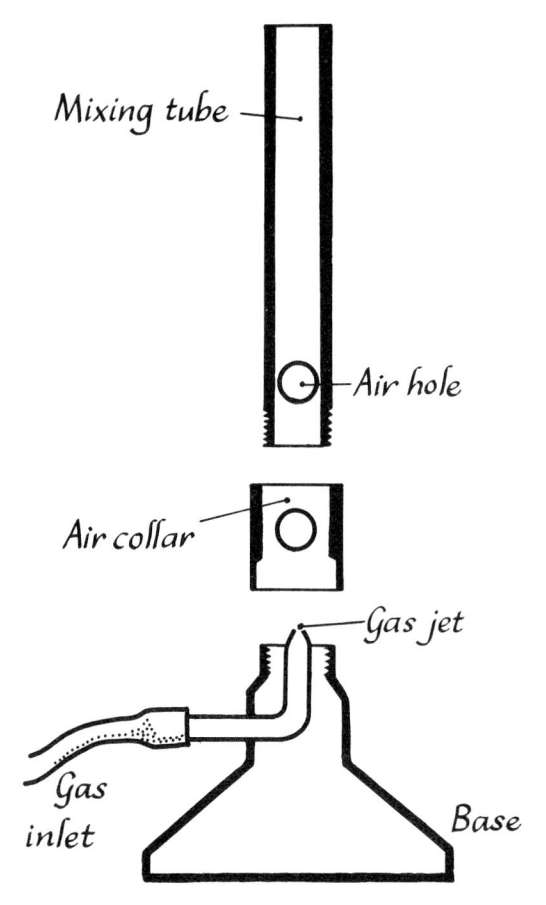

Figure 6.7 The bunsen burner

With the air holes covered by the air collar, the bunsen burner produces a normal luminous gas flame. By rotating the air collar, the holes are progressively uncovered and an increasing volume of air enters and mixes with the gas in the mixing tube. This mixture

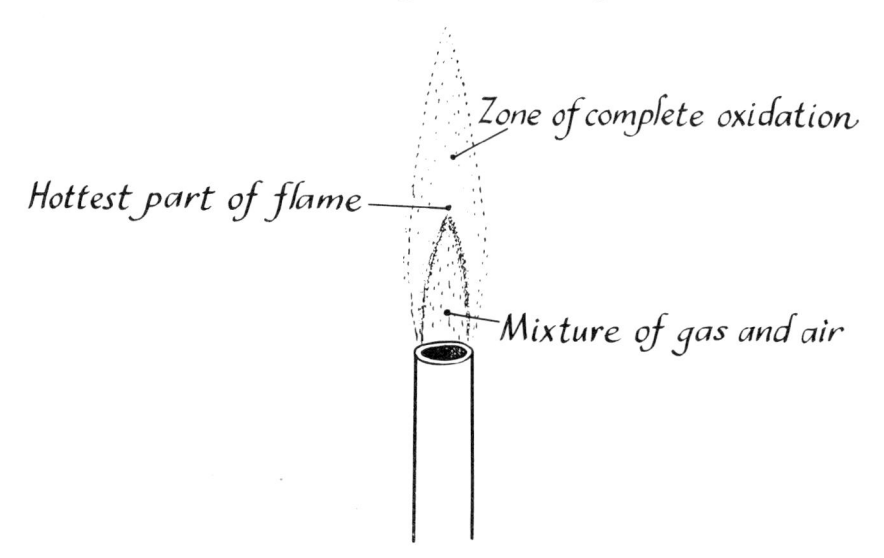

Figure 6.8 The non-luminous bunsen flame

burns at the top of the tube. Air enters through the air holes because the stream of gas issuing from the jet causes a reduction of pressure inside the mixing tube. The non-luminous flame produced by burning a mixture of gas and air (see Figure 6.8) is much hotter than the luminous gas flame.

Most gas appliances in common use employ the bunsen principle in order to obtain the maximum heat output from the gas. In the gas fire, the gas ring and the gas cooker, the size of the air inlet is pre-set by a knurled ring threaded on to the jet tube. The amount of air is reduced as the knurled ring is screwed towards the jet (see Figure 6.9).

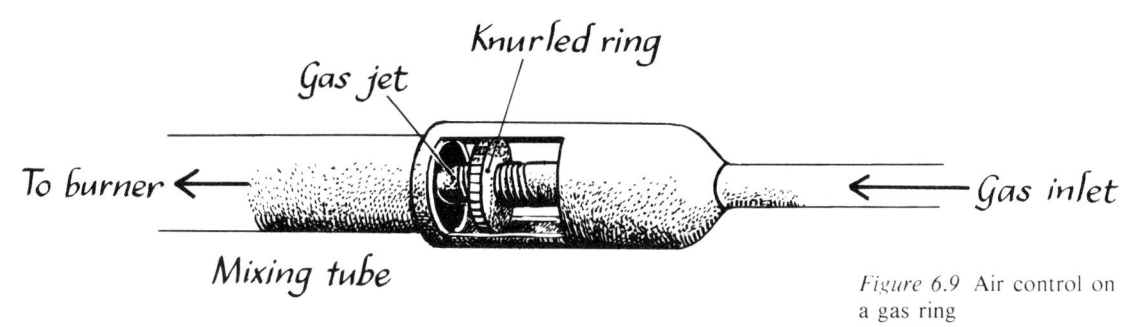

Figure 6.9 Air control on a gas ring

The gas poker, as shown in Figure 6.10, has no such adjustment, the air holes being drilled round the circumference of the tube adjacent to the jet.

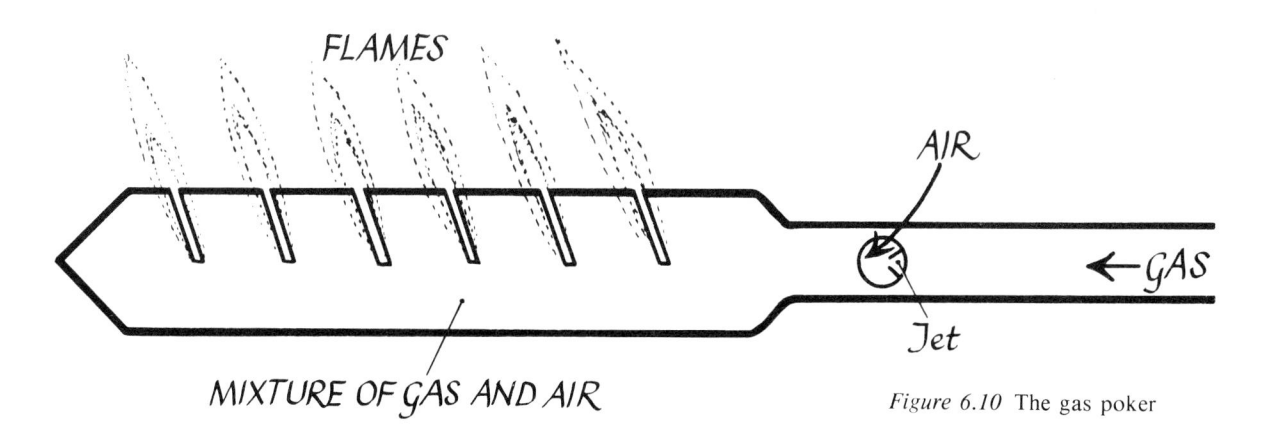

Figure 6.10 The gas poker

In a gas fire the mixture of gas and air is burned as it emerges from a series of jets. Above each jet is a fireclay radiant which glows when heated (see Figure 6.11). The radiants are lattice-work tubes which enclose the flames. They have a large surface area and, when hot, radiate heat. To avoid breakage due to thermal expansion, the radiants are loosely supported.

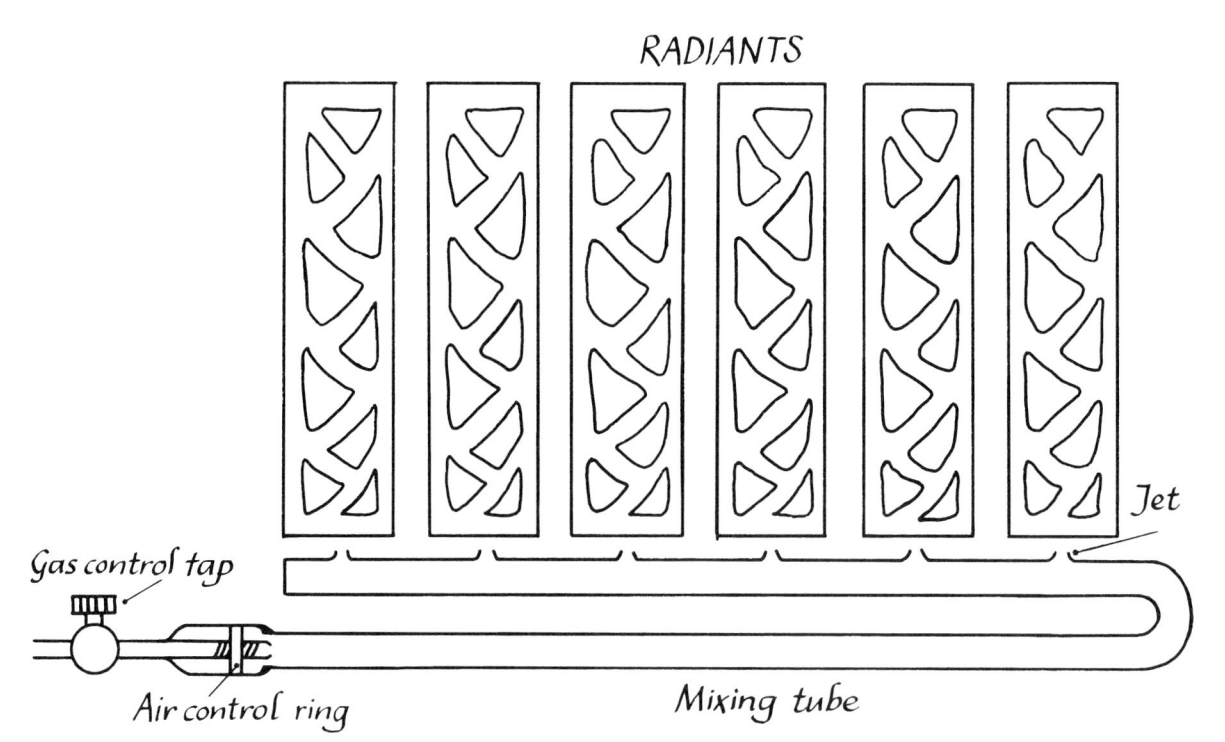

Figure 6.11 The gas fire

6.4. The electric fire

Investigation 6f. Low-voltage heating elements

Cut off about 45 cm of 0·37 mm diameter (28 S.W.G.) nichrome wire and wind it in a tight coil, using a thin piece of glass tubing or rod as a former. Remove the former and connect the ends of your coil to lengths of copper wire by twisting the wires tightly together or by using barrel connectors. Connect your coil in series with an ammeter, a switch and a 12-volt battery (or low-voltage bench supply), as shown in Figure 6.12.

Figure 6.12 A low-voltage heating element

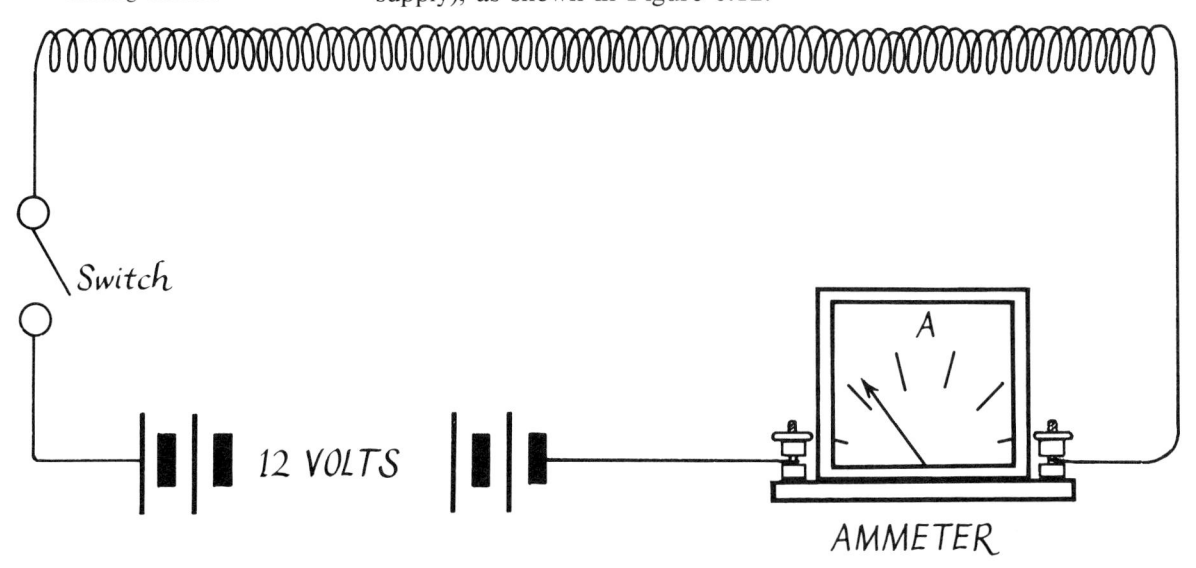

Secure the copper wire near one end of your coil with a clamp and hold the other end of the coil with a pair of pliers or crucible tongs, so that the coil is not touching anything. Switch on and note the current indicated by the ammeter. The coil should now be glowing. If it is not glowing, switch off, shorten the coil by about 2 cm and switch on again. Repeat this process until the coil glows when you switch on.

When the coil is glowing, slowly pull on the coil until it is almost straight. Note the current indicated by the ammeter. Is it greater, smaller or the same as it was when the element was in the form of a coil? Is the element glowing brighter, dimmer or the same as it was before?

The element of a simple radiant type of electric fire is made of wire which has a comparatively high resistance, much higher than copper wire of the same thickness. Elements are usually made of **nichrome**, which is an alloy containing nickel and chromium. The wire is wound round a helical groove, rather like a screw thread, on the outside of a ceramic tube. The ends of the element wire are connected to metal caps at the ends of the ceramic tube, and the

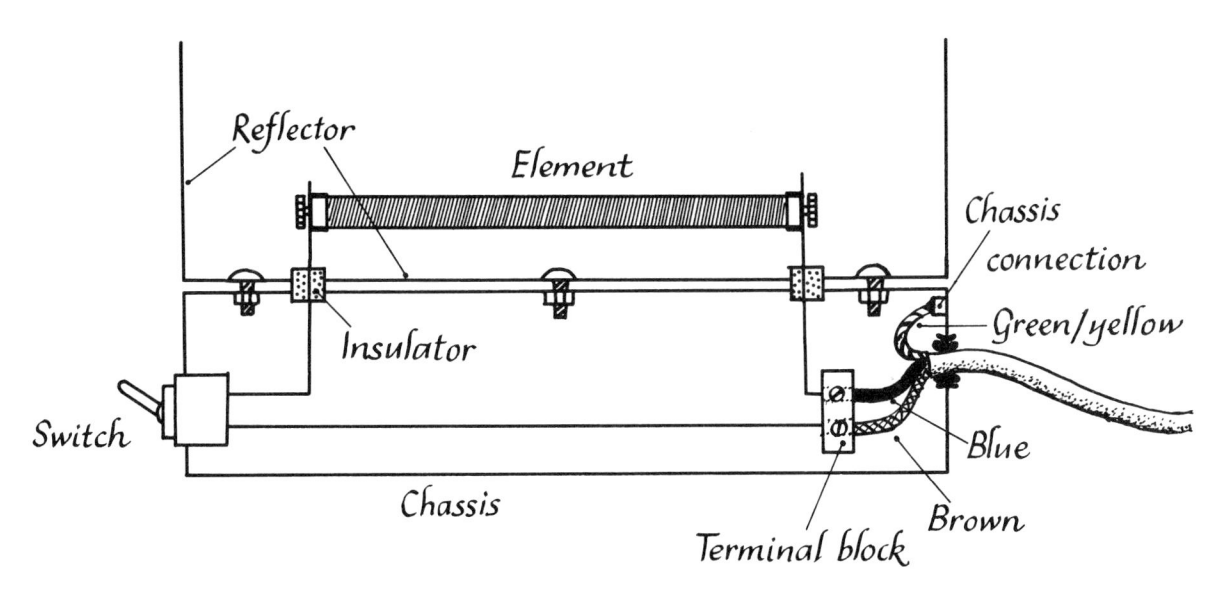

Figure 6.13 Connections to an electric fire

entire element is supported by these caps, which are fixed to metal supports. The metal supports are completely insulated from the chassis (bodywork) of the fire. Because of the heat conducted through these metal supports, the wires carrying the current to the element are not insulated with rubber or plastic along their whole length. Inside the chassis of the fire, the electrical connections from the metal supports to a terminal block are made with wires, which are thick enough to maintain their shape, and bent so that they are well separated from each other and the chassis. Sometimes, this wire is insulated with asbestos sleeving or threaded with ceramic beads. The flex from the mains plug to the fire passes through a rubber grommet in the chassis to the terminal block. The rubber grommet protects the flex from the sharp edges of the hole in the chassis. To ensure electrical safety, the wire from the earth pin (a green and yellow striped wire) of the three-pin plug is firmly connected to the chassis of the fire. When there is a switch on the fire itself, it must be connected in the brown lead, as shown in Figure 6.13.

In the radiant type of electric fire, the element wire gets hot; since the wire is in contact with the ceramic tube, the ceramic tube also gets hot. Because the air is in contact with a relatively small surface area of the element (the outer surface of the wire and the ceramic ridges between the coils), heat loss by convection is reduced and the temperature of the whole element rises until it is glowing. The glowing element transmits heat in all directions by radiation, and this heat is concentrated in the required direction by placing a suitable reflector behind the element. In the interests of safety, a wire guard is fitted in front of the element and reflector.

The radiant type of electric fire does not heat the air in a room directly. Any solid object in the direct path of the rays is heated, and this heat is transferred to the air in the room by convection.

A brightly glowing element transmits heat energy by radiation more efficiently than an element that is only dimly glowing, or not glowing at all, because it is at a higher temperature. An element operating at a lower temperature is a more efficient energy exchanging device, because little or none of the energy available is given out as light; consequently, a greater amount of the energy available is given out in the form of heat. Since a non-glowing element gives out comparatively little radiated heat, it follows that it will give out heat by conduction and convection.

You will know, from Chapter 2, that when a fluid (liquid or gas) is heated, it rises by convection. This occurs because the heated fluid expands, thus making it relatively less dense than its colder surroundings. The denser fluid tends to sink, forcing the heated fluid upwards and thus producing convection currents within the fluid.

For heating the air in a room, a convector type of fire may be used. The element in a convector fire does not glow and is shaped to form a grid across the open base of the chassis, thus presenting a large surface area which heats the air in contact with it. This warm air rises by convection and is deflected through a grille at the top of the front of the heater by a deflector plate, which is set at an angle of 45° inside the top of the heater. The heated air rising inside the convector causes colder air to enter the bottom of the convector and to pass the element, thus producing a circulation of air round the room. A typical convector heater is shown in Figure 6.14.

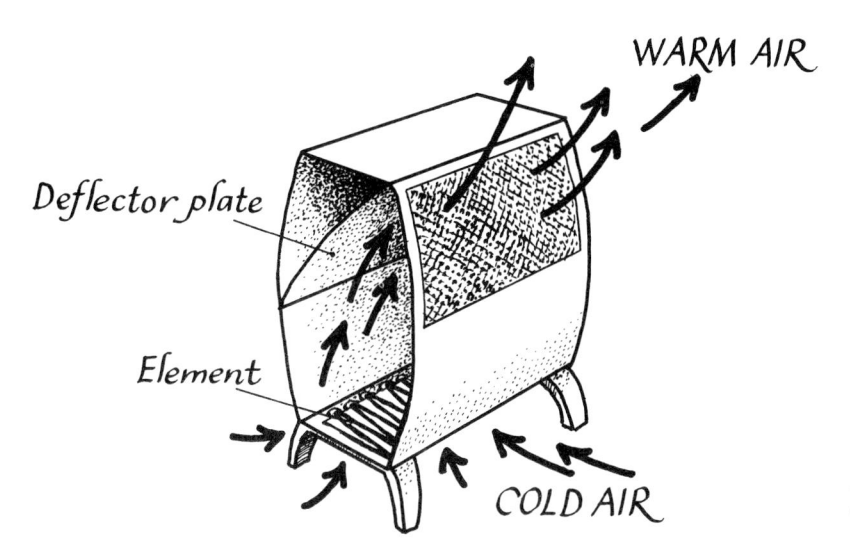

Figure 6.14 The electric convector heater

An efficient method of heating the air in a room quickly is to use a forced-draught heater. In this type of heater, the element is in the form of a grid and is placed vertically in the front of the heater. Air is forced through the element by an electric fan which is placed just behind the element, thus producing a stream of warmed air. By switching on the fan alone, a stream of cool air is produced.

6.5. The domestic hot water system

Investigation 6g. Making a model domestic hot water system

Obtain three tins, two with tight-fitting lids (coffee tins are quite suitable) and the other with no lid. Make holes in the tins, as shown in Figure 6.15, large enough to take single-holed stoppers. These holes can be made with a brace and centre-bit. By using three T-pieces, suitable shaped pieces of glass tubing and short sleeves of rubber tubing, connect the tins and support them so that a bunsen burner can be placed under the bottom tin. The open-ended tube leading to the top of the top tin can be kept in position with a rubber band.

In this model, the piece of rubber tubing with the Mohr clip represents the pipe leading to the hot water tap. The tube which connects the top of the boiler (bottom tin) to the top of the hot water storage tank (middle tin) is called the **flow pipe**. The tube which connects the bottom of the boiler to the bottom of the hot water storage tank is called the **return pipe**. The open-ended tube leading to the cold water cistern (top tin) is called the **expansion pipe**.

Fill the system by pouring water into the cold water cistern, with

Figure 6.15 A model domestic hot water system

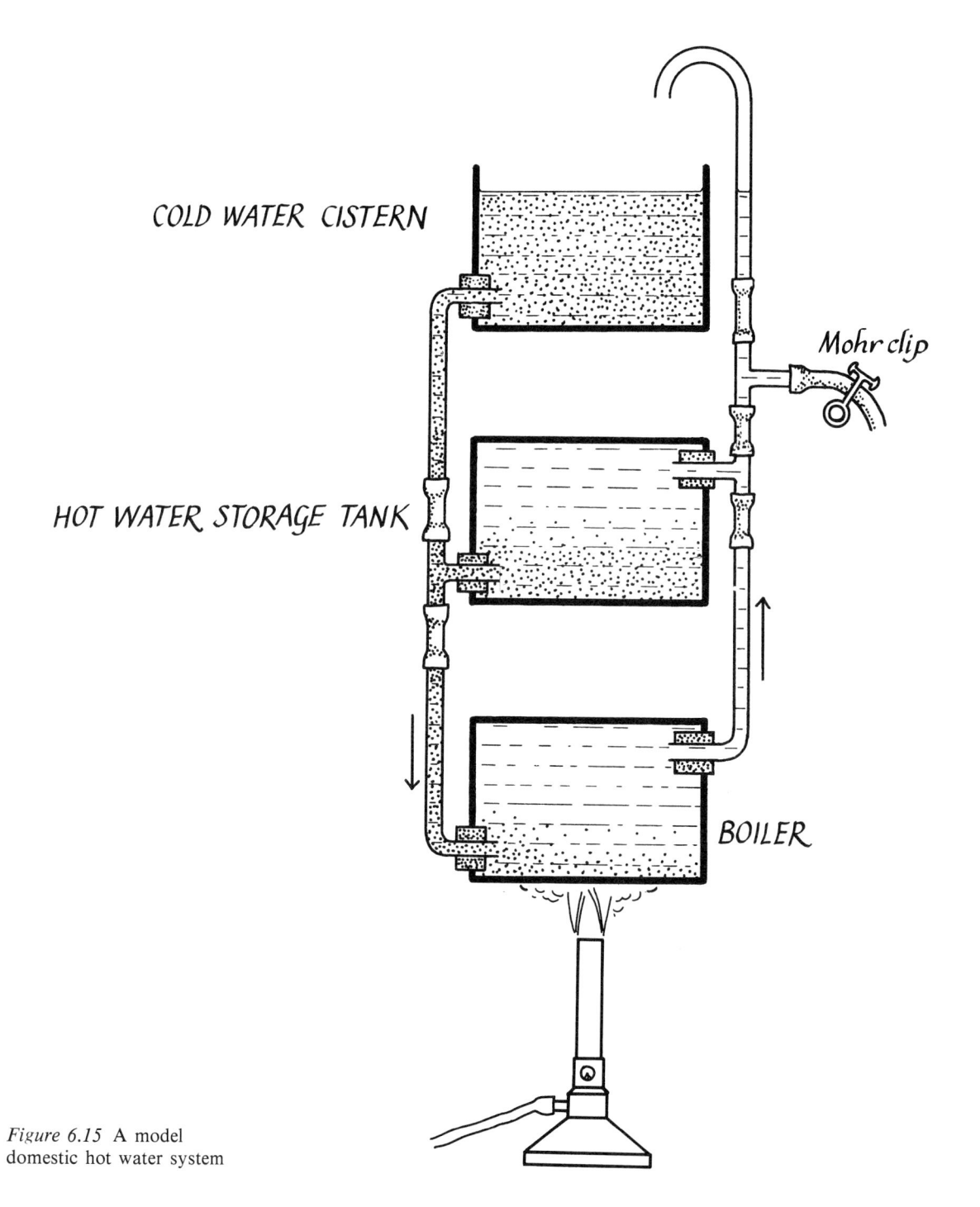

the Mohr clip closed, until the cold water cistern is about three-quarters full. As you do this, notice the flow of water through the glass tubing to the hot water storage tank and the boiler. When the system is full, notice the level of water in the expansion pipe.

Now light the bunsen burner and place it under the boiler. Feel the

78

flow and return pipes with your fingers from time to time. Which pipe is hotter? In which direction is the water flowing through these pipes? Add a few crystals of potassium permanganate or some ink to the cold water cistern and run some hot water from the system, by opening the Mohr clip, maintaining the level in the cold water cistern by pouring cold water into it. This will enable you to see the flow of water through the system quite clearly.

In the directly heated domestic hot water system, as shown in Figure 6.16, the circulation of water from the boiler to the hot water

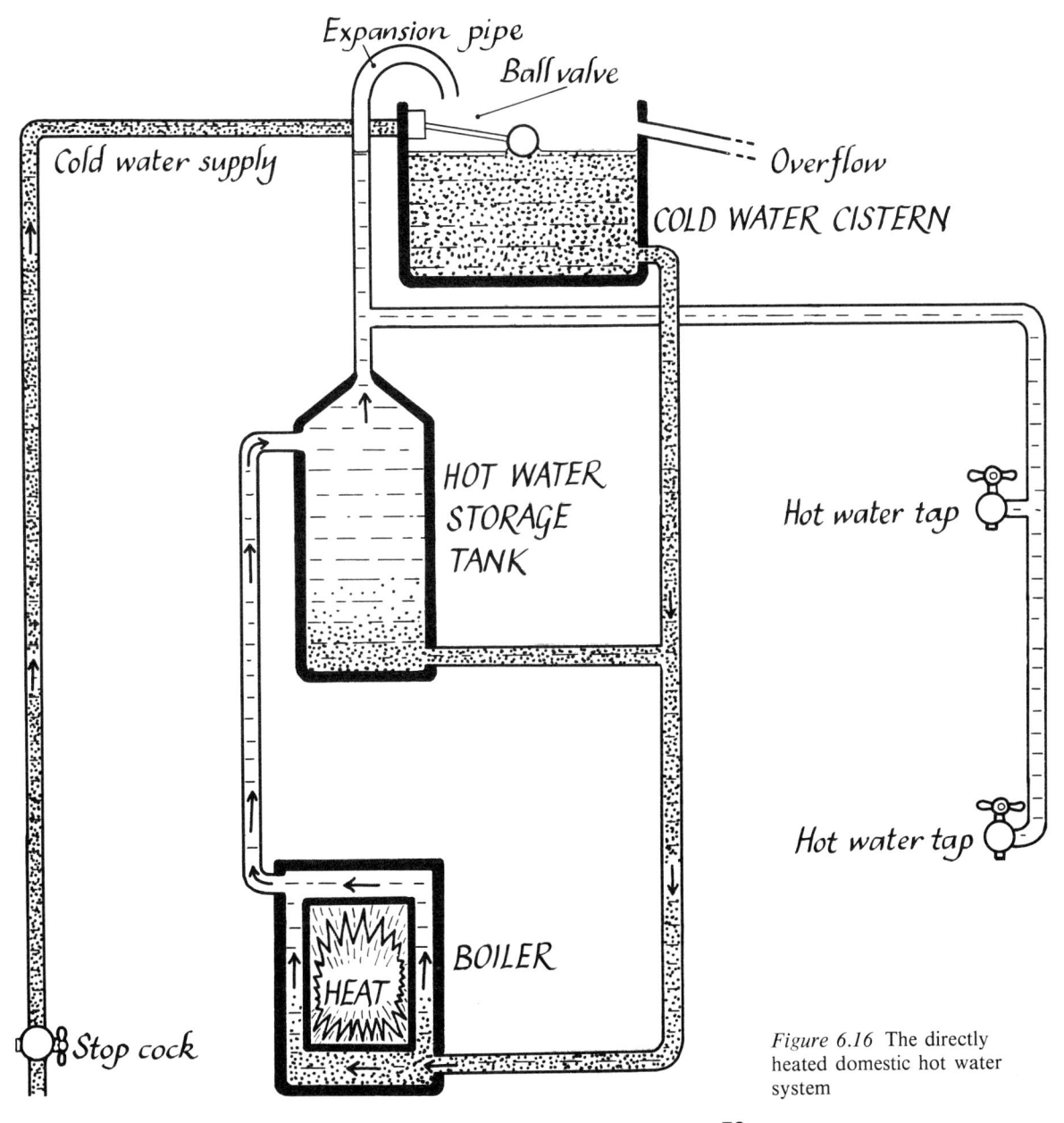

Figure 6.16 The directly heated domestic hot water system

storage tank and back to the boiler is caused by convection. Water in the boiler is heated and expands, thus making it less dense than the colder water in the rest of the system. The colder water, being denser, flows downwards into the boiler, forcing the warm water up into the hot water storage tank. As this constant circulation continues, the temperature of the water increases.

When hot water is taken from the system by turning on one of the hot taps, it is replaced by cold water flowing from the cold water cistern into the bottom of the boiler. The cold water cistern is replenished from the cold water supply pipe, the level being controlled by a ball valve. If the ball fails to rise for any reason, water will continue to enter the cistern, so an overflow pipe is fitted as a safety measure.

As in the model, the open-ended pipe leading from the top of the hot water storage tank to the cold water cistern is called the expansion pipe, although its function has nothing to do with expansion. Any expansion of the water in the boiler and the hot water storage tank will cause the excess volume of water to rise up the pipe leading from the hot water storage tank to the cold water cistern. The function of the expansion pipe is to allow the escape of steam if the water is allowed to boil.

In an area where the water supply is hard, the directly heated system has the disadvantage that the boiler, flow pipe and, to a smaller extent, the return pipe become 'furred up'. When water is run off from the system, fresh hard water is fed back to the system and this, in turn, causes further furring up. To reduce the amount of furring in a hard water area, an indirectly heated system can be used. In this system, heated water from the boiler is fed by convection through a coiled pipe in the hot water storage tank. This circulation from the boiler through the coiled pipe and back to the boiler is sometimes called the **primary circuit**. You will notice that no matter how much hot water is run off from the taps, there is no fresh supply to the primary circuit, thus limiting the amount of furring up. This system is shown in Figure 6.17.

6.6. Central heating

A simple form of central heating can be obtained by connecting an additional flow-and-return circuit to the boiler of a domestic hot water system, as shown in Figure 6.18. When the valves of the radiators are closed, there will be no flow round the radiator circuit. When the radiator valves are opened, hot water from the boiler will circulate through the radiators from top to bottom and then back to the boiler. As the hot water passes through the radiators, heat energy is transferred from the water to the metal of the radiators by con-

Figure 6.17 The indirectly heated domestic hot water system

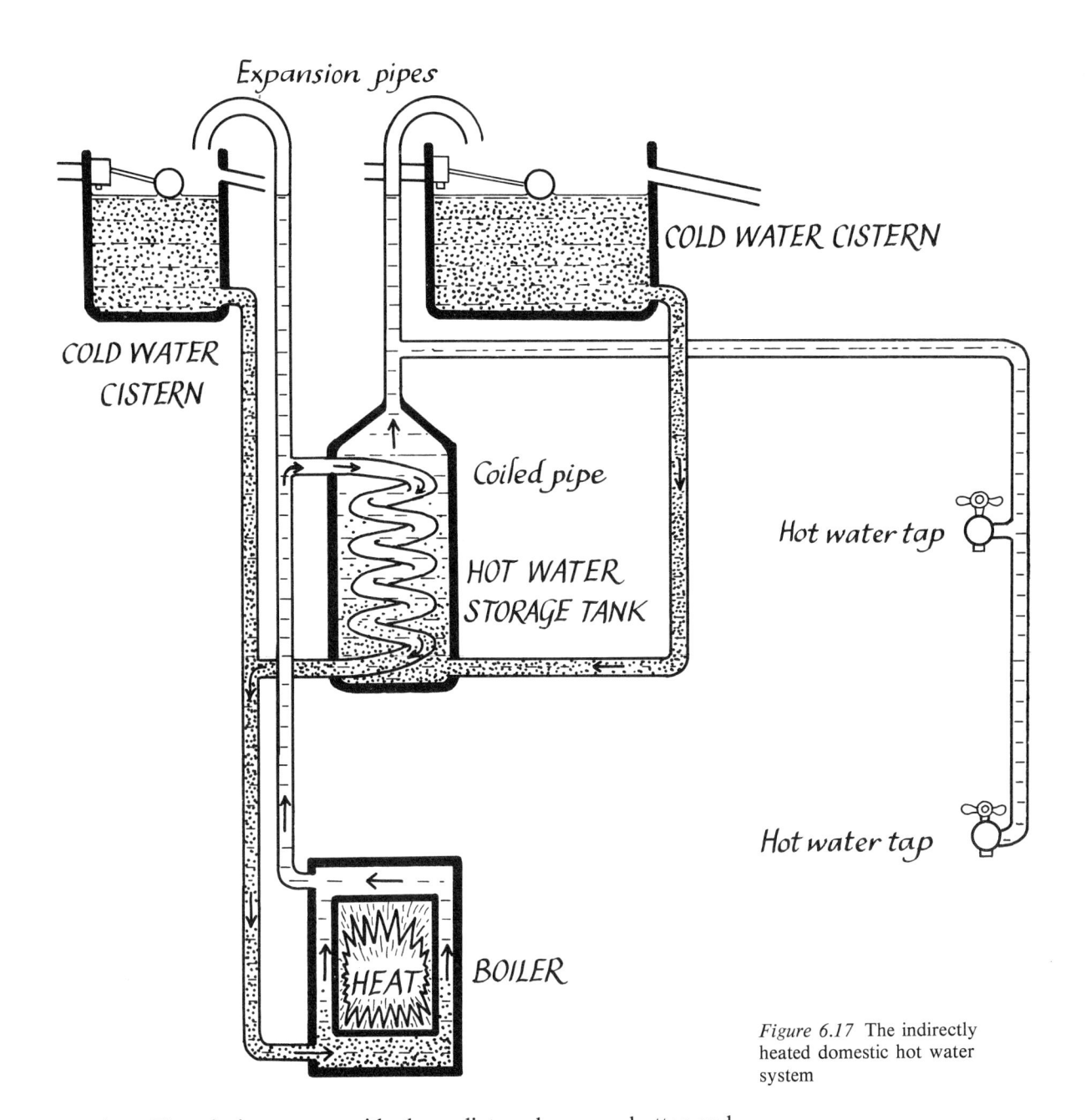

duction. The air in contact with the radiators becomes hotter and rises by convection, thus heating the air in the room. Although this type of heating appliance is called a radiator, a comparatively small proportion of its heat output is by radiation.

When a domestic hot water system is extended to provide central heating in this way, the temperature of the hot water supply will be reduced because of the heat loss from the radiators. This disadvantage can be overcome by installing a boiler with a greater heat output.

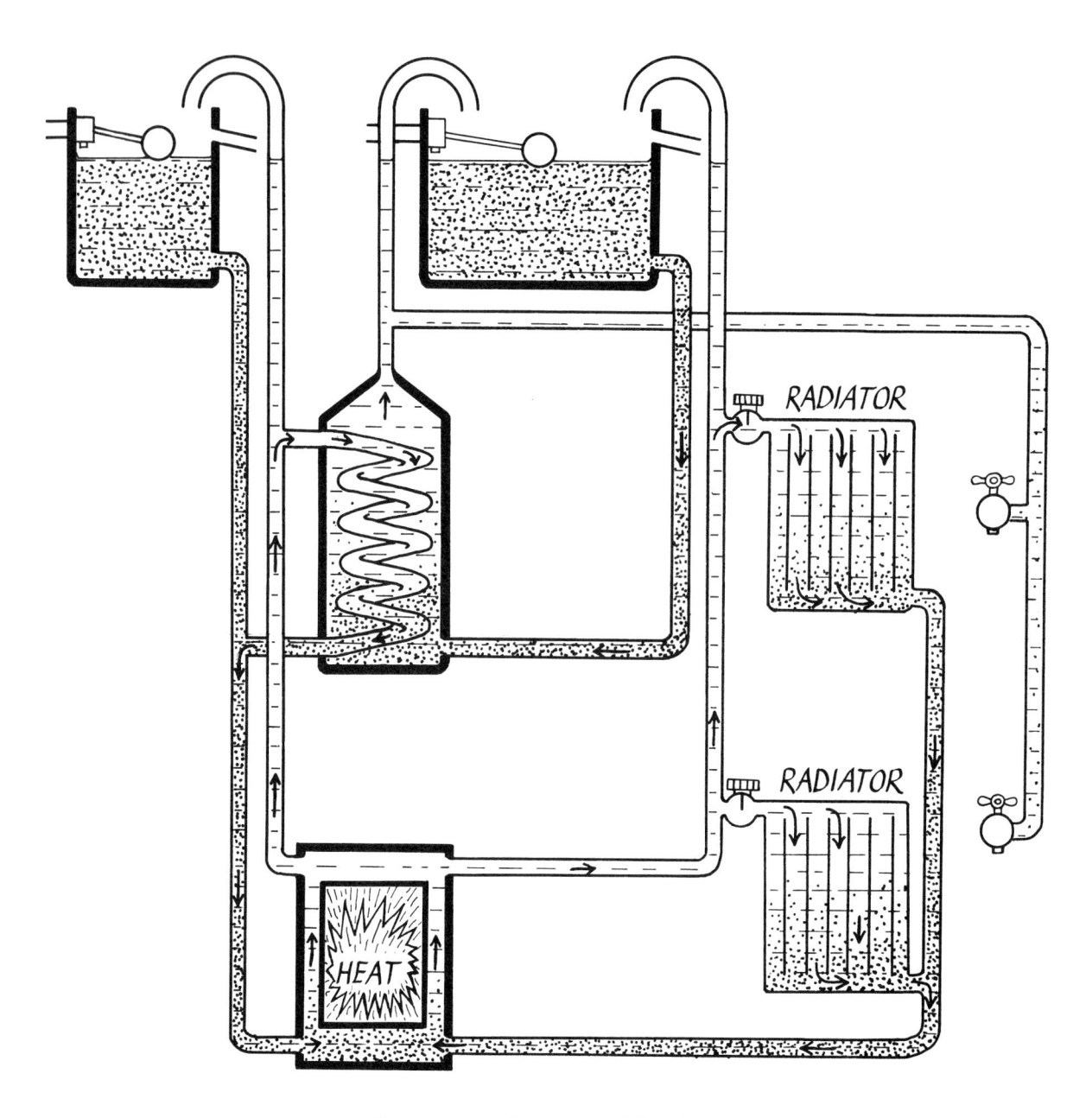

Figure 6.18 A simple central heating system

In many modern central heating systems, the radiators are connected to the boiler by narrow-bore pipes. If this system relied on convection alone to circulate the water, the rate of flow would be insufficient to give an even distribution of hot water to all of the radiators. In order to ensure an adequate rate of flow, a pump is connected into the circuit to draw the water through the narrow-bore pipes.

The use of central heating is not restricted to modern civilizations. The floors of many ancient Roman villas were supported on short piers or pillars so that heated air from a furnace could pass through the space below the floor, thus heating the rooms above. In some modern houses, heated air is pumped through rectangular pipes or ducts which are concealed under the floors and in the walls. The heated air enters the rooms through grilles in the walls just above floor level. This method is called the **ducted air system.**

The introduction of cheaper electricity for heating purposes during off-peak periods has made possible other heating systems, which would otherwise have been too expensive to use to heat a house. Off-peak electricity is generally available for about eight hours of darkness and three hours during the afternoon. These times may vary from one area to another. Because of the lower cost (approximately half price) and the limited times of off-peak electricity, a separate fuse box, meter and time-switch are installed. Cables carrying the off-peak supply are connected directly to the heating system and not to sockets, as is the case with the normal supply cables. This prevents off-peak electricity from being used for purposes other than heating.

Underfloor heating consists of heating cables embedded in the floor. The cables are well insulated and their resistance and length are calculated so that they do not glow. These precautions are necessary to avoid the risk of fire. The cables heat the entire floor area which, in turn, heats the air in the room. A thermostatic switch, which can be set to a predetermined temperature, is usually included in the circuit. When the air in the room reaches this temperature, the current is automatically switched off.

Storage heaters consist of heating elements embedded in large blocks of fire-brick enclosed in a metal case. When the current is switched on by the time-switch, the elements heat the blocks until they reach a predetermined temperature. A thermostatic switch in the circuit then maintains this temperature until the current is switched off by the time-switch. When the temperature of the heated blocks is higher than the temperature of the surrounding air, convection currents are set up inside the heater case and the warm air escapes through slots or a grille near the top of the heater. To ensure a free flow of air, the heater is supported by short legs and should be kept clear of the wall.

The thermostatic switches for underfloor and storage heating can only control the heat input. The heat output from these systems depends on two factors:

a. The higher the thermostatic setting during the heating period, the higher will be the output. In other words, the hotter the floors or blocks are, the more heat they will give out.

b. The colder the air in the room, the higher will be the output. In other words, if the room is cold, the floors or blocks will give out heat faster and become cold faster than if the air in the room is warm.

Because it is impossible in these systems to control the heat output, it is necessary to forecast the amount of heat needed in order to set the thermostat at the correct level. If a warm day is forecast, the thermostat should be set to a low setting. If a cold day is forecast, the thermostat should be set to a higher setting. An attempt has been made in some types of storage heaters to increase the rate at which heat is given out by the inclusion of a fan. The fan is connected to the normal mains supply so that it can be used at any time. Once the blocks are warmer than the surrounding air, the heat output of the heater can be boosted by switching on the fan. This will, of course, reduce the temperature of the blocks more quickly and, once their temperature drops to the temperature of the surrounding air, no further heating is possible until the current is switched on again by the time-switch.

In some countries where the days are hot and the nights are cold, a **solar heating** (heating by the sun) system is used. For this system, a number of straight lengths of pipe are mounted parallel to each other above the roof of the house, with alternate ends connected to form a continuous pipe, as shown in Figure 6.19. Behind each

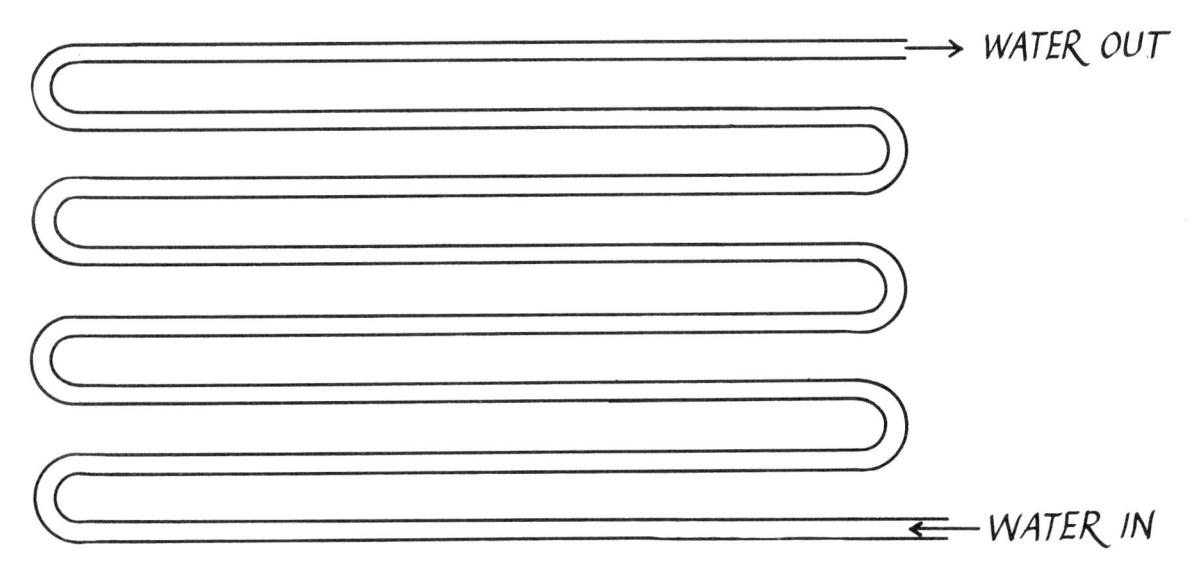

Figure 6.19 Solar heating pipes (plan)

straight length of pipe is a reflector, similar to the reflector of a radiant electric fire. The reflectors are placed so that the radiant heat from the sun is concentrated on the pipes, as shown in Figure 6.20.

The ends of the solar heating pipes are connected to a large underground tank of water, the lower end to the bottom of the tank and the upper end to the top of the tank. During the daytime, a pump circulates water round the circuit. The colder water from the bottom of the tank is heated as it passes through the solar heating pipes on the roof and is then returned to the top of the tank. So long as the sun

Figure 6.20 Solar heating (side view)

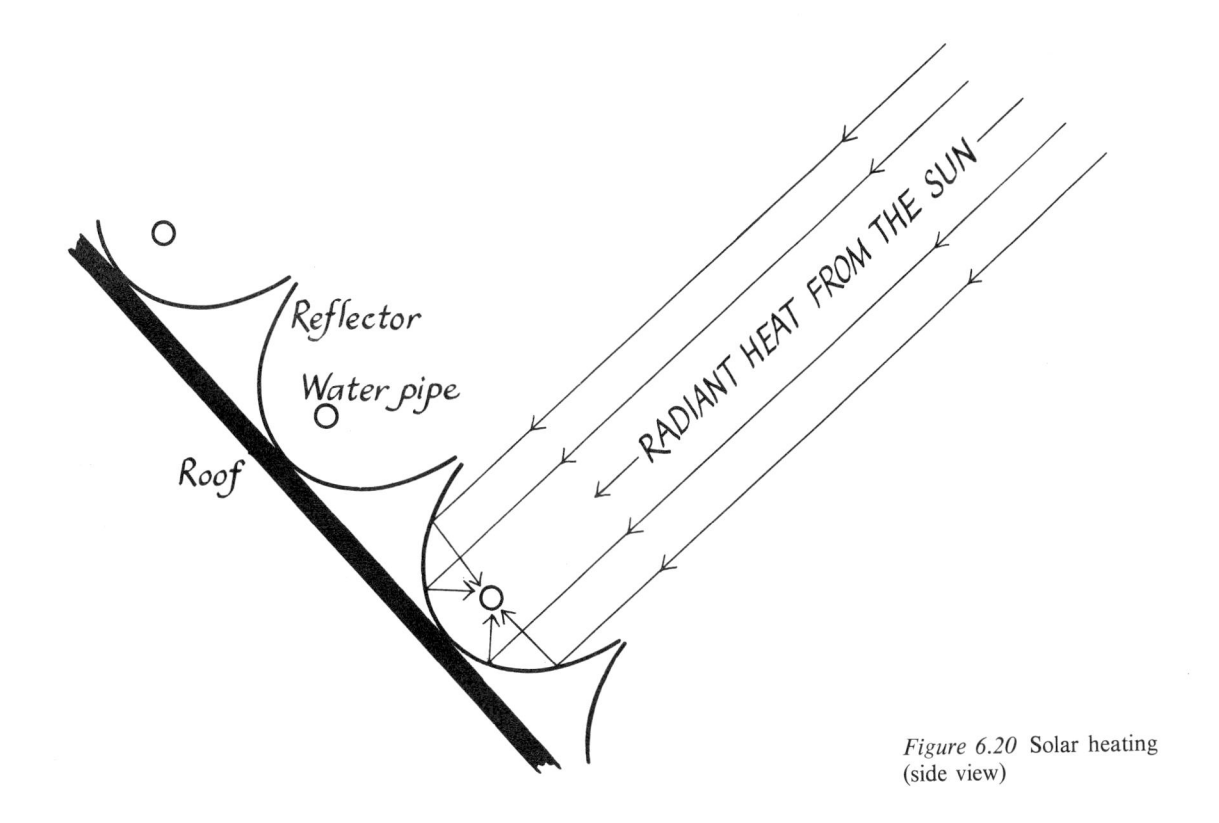

shines, the average temperature of the water in the tank will rise. At night, when the air temperature drops and heating is required in the house, the heated water in the tank is pumped through a series of radiators which, in turn, heat the air in the rooms. While this is happening, the flow of water through the solar heating pipes is cut off, the warmer water from the top of the tank being pumped through the radiators and returned to the bottom of the tank, as shown in Figure 6.21.

Investigation 6h. Solar heating

Blacken the outside of the bottom of two boiling tubes of equal size by holding them over a burning candle or a little burning naphthalene. Put 10 cm³ of cold water and a thermometer into each boiling tube and support both tubes in direct sunlight. Position a plane mirror about 20 cm behind and slightly to one side of one of the tubes, so that the reflected rays of the sun strike the bottom of the tube. Note the temperatures in the two boiling tubes every minute for about ten minutes. Does the plane mirror affect the rate at which the temperature of the water increases?

Mould some modelling clay into the shape of a shallow bowl and press small pieces of plane mirror on to the concave side. Adjust the pieces of plane mirror so that the reflected rays more or less meet at

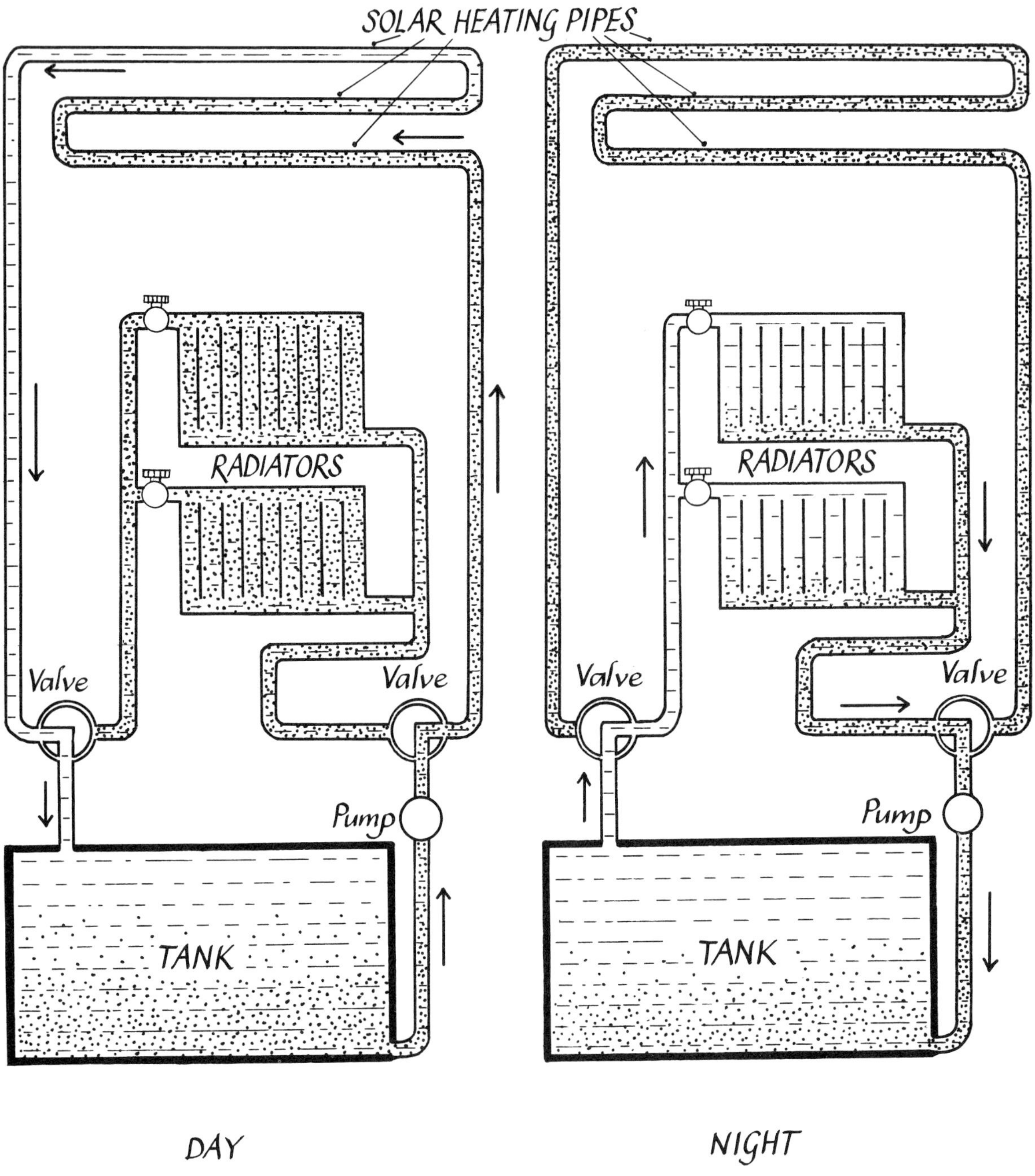

Figure 6.21 The solar heating system

one point. Using fresh cold water (10 cm^3 in each tube), arrange your reflector so that the reflected rays meet on the bottom part of one of the tubes. Take temperature readings every minute for about ten minutes. Is this arrangement more effective than the single plane mirror? What is the purpose of blackening the two tubes? If you do not know, refer to Chapter 7.

Test your understanding

1. What chemical reactions take place in a charcoal fire?
2. Why is carbon monoxide a poisonous gas?
3. What is the function of the air collar on a bunsen burner?
4. Why is a non-luminous flame hotter than a pure gas flame?
5. What is the function of the radiants in a gas fire?
6. What is the difference between the element in a radiant electric fire and the element in an electrical convector?
7. What are the advantages and disadvantages of electrical storage heaters?
8. In a domestic hot water system, which part of the hot water storage tank would you feel to determine whether or not there is enough hot water for a bath? Explain the reasons for your choice.
9. In a solar heating system, what is the function of the underground tank?

Chapter 7

Maintaining Heat

Heating a house is a costly item in a household budget. Therefore, it is wise to ensure that, in cold weather, the heat produced by the heating system remains, as far as possible, inside the house and is not wasted by heating the air outside the house. This can be achieved by insulating the house. A well-insulated house has the added advantage of remaining relatively cool in very hot weather.

7.1. Heat insulation

When two masses at different temperatures are in contact with each other, heat energy will be transmitted from the warmer mass to the colder mass by conduction. This transfer of energy will continue until both masses reach the same temperature. The rate at which the heat transfer takes place depends on three factors:
   a. **The difference in temperature between the two masses.**
   b. **The area of contact between the two masses.**
   c. **The nature of the materials of the two masses.**

In Chapter 5 we found that the heat loss from water in a metal calorimeter is faster during the first few minutes, when the temperature of the water is relatively high, than it is when the temperature of the water is approaching the temperature of its surroundings. In the same way, when a heating device is switched off, the air in a room will cool more rapidly on a cold day than it will on a hot day.

The rate of cooling can be compared with the rate of flow of water. If two tins are connected together with rubber tubing and one of them is filled with water, the rate at which the water flows from one tin to the other will depend on the difference between the levels of water in the two tins. Raising the higher tin will increase the rate of flow; lowering the higher tin will reduce the rate of flow. This comparison between water flow and heat flow is shown in Figure 7.1.

Investigation 7a. Heat loss and surface area

Put 200 cm³ of water into each of two beakers. Place the beakers on tripods and heat them until the water is boiling. Remove one of

WATER FLOW

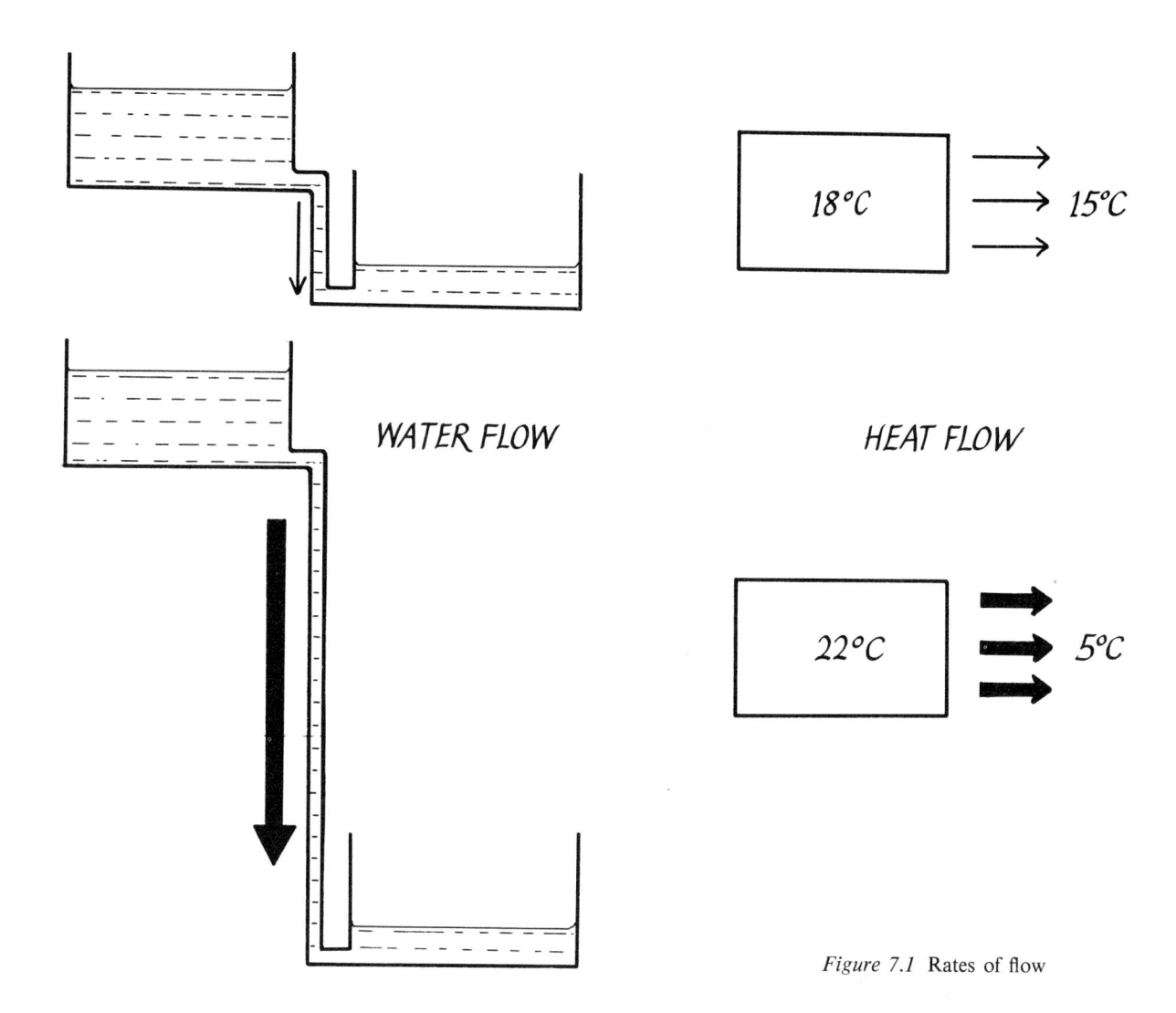

HEAT FLOW

Figure 7.1 Rates of flow

the beakers from the tripod and stand it on an asbestos mat (to protect the bench). Pour the boiling water from the other beaker into a shallow dish, also on an asbestos mat. Take temperature readings of the water in the beaker and in the dish every half minute and compare the results. Which 200 cm^3 has the greater surface area? Which 200 cm^3 of water loses heat faster?

In Chapter 5 we learned that some materials are good conductors of heat and other materials are poor conductors of heat. Materials which are poor conductors of heat are good insulating materials. A house built entirely of good insulating materials would be cool in hot weather and easy to keep warm in cold weather. Unfortunately, most good insulating materials do not possess sufficient physical strength, and a house built entirely of such materials would be liable to collapse. A house built entirely of metals and other relatively dense materials would be in no danger of collapsing, but, because such materials are good conductors, the house would be uncomfortably hot in hot weather and very expensive to keep warm in cold weather.

Of the three states of matter (solids, liquids and gases), solids are the best conductors and gases are the best insulators. The amount of heat loss from a house can be reduced by the use of insulating materials in the following ways:

a. *Floors.* In many older houses, the ground floor consists of wooden floorboards on joists supported by brickwork. The air-space below the floor acts as an insulating layer. In many modern houses, the ground floor consists of a solid concrete raft, surfaced with wooden blocks or tiles made from plastic, rubber or cork, all of which are relatively good insulators. The air-space between the ceiling of downstairs rooms and the floor of the rooms above acts as an insulating layer. Additional insulation can be achieved by laying linoleum or vinyl floor covering with a suitable underlay. The underlay can be made of thin foam rubber or plastic, felt, thin cork sheeting or compressed paper, all of which are good insulators. Carpets, of course, provide additional insulation.

b. *Walls.* The outer walls of a bungalow or a house are normally solid brickwork, about 25 cm thick, or cavity brickwork. A cavity wall consists of two thin walls, each about 12 cm thick, separated by a 5 cm space. To prevent the two walls from collapsing, they are held together by metal tie-bars, which span the cavity at intervals of about 1 metre. A certain amount of heat will be conducted through these tie-bars, but, because they are widely spaced, this heat loss is very small.

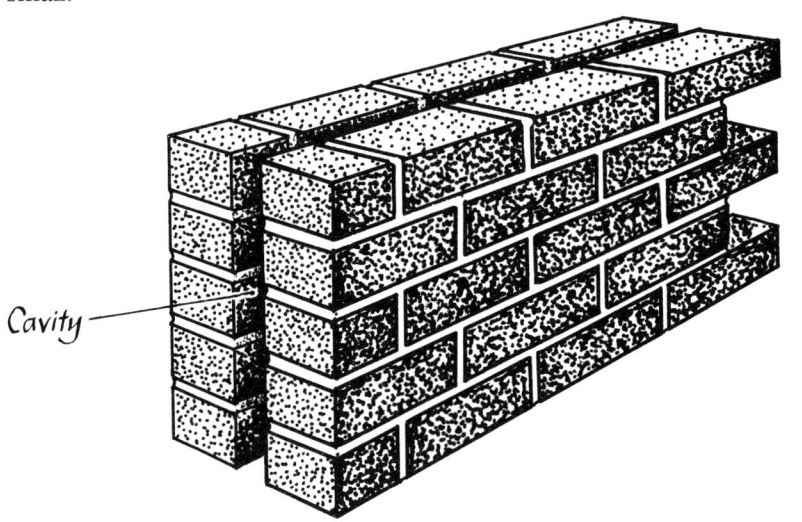

Figure 7.2 A cavity wall

The outer wall of a cavity wall is usually built of bricks or concrete blocks, in order to support the upper part of the house and the roof. The inner wall may also be built of bricks or concrete blocks (which will give the wall extra strength), or it may be built of breeze blocks or thermal bricks, which will improve insulation. The construction of a cavity wall is shown in Figure 7.2.

Compared with a solid brick wall, a cavity wall reduces heat loss by 28 per cent. This reduction in heat loss occurs because the air in the cavity is a poor conductor of heat. A certain amount of heat is transferred across the cavity from one wall to the other by convection currents which are set up within the cavity. These convection currents are set up because the layer of air in contact with the warmer of the two walls will be less dense than the layer of air in contact with the cooler wall.

The heat loss through the cavity by convection can be reduced by restricting the free flow of air within the cavity. This can be done by either filling the cavity with an insulating material such as glass fibre when the wall is being built or, in the case of an existing wall, forcing foam plastic into the cavity. Heat loss through the walls of a house can be further reduced by papering the inside surface. Even better insulation can be achieved by lining the wall with a thin sheet of expanded polystyrene before papering.

c. *Windows*. Although glass is a comparatively poor conductor of heat, window glass is so much thinner than a brick wall that the heat loss through an ordinary glass window is 2·5 times as great as the heat loss through solid brickwork of the same area. The windows of

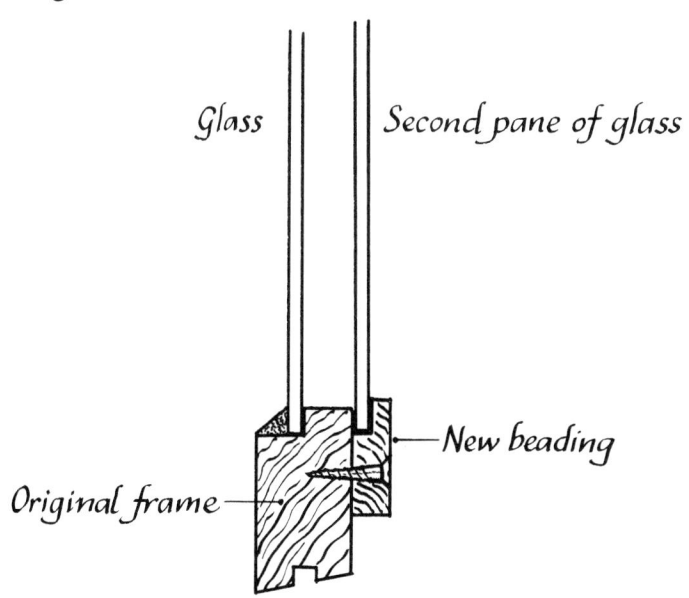

Figure 7.3 Double glazing

modern houses are larger than the windows of houses built some years ago. This tendency towards larger windows creates an insulation problem.

Partial insulation can be obtained by drawing curtains across the window, thus enclosing a relatively still layer of air between the window and the curtain. This method has obvious disadvantages during the day. A more effective method uses the same principle as the cavity wall and is called **double glazing**. This can be achieved by putting a spacer round the edges of an existing pane of glass and then fitting a second pane, as shown in Figure 7.3. This is best done on a hot, dry day, otherwise any moisture in the air between the two layers of glass will condense on the inside surface of the outer pane in cold weather. This surface cannot be cleaned without removing the pane. Another, though more expensive, method is to use factory-made double glazing units. Compared with a single glazed window, double glazing with a gap of 1 cm will reduce heat loss by 44 per cent.

d. *Ceilings.* Heat loss through a ceiling can be reduced by covering the ceiling with expanded polystyrene tiles. The heat loss from a ground floor room through the ceiling to the room above is not a serious matter for three reasons:
1. The difference in temperature between the two rooms is likely to be less than the difference in temperature between the ground floor room and the outside of the house.
2. The layer of air trapped between the ceiling of the ground floor room and the floor of the room above acts as an insulator.
3. Any heat that is transmitted through the ceiling is not wasted, as it is heating the room above.

e. *Roofs.* A more serious problem is the heat loss from a room into a roof space. This can be reduced by insulating the ceiling with ceiling tiles and filling the spaces between the rafters with an insulating material. Suitable materials are glass fibre, in the form of matting which is unrolled between the rafters, or a 'loose fill' such as vermiculite (expanded mica), which looks like some types of breakfast cereal. Because these materials are themselves poor conductors and trap a large volume of air, they reduce heat loss by conduction and convection. The roof space can be further insulated by fixing metal foil immediately below the roof tiles. This reduces heat transfer by radiation and, although the heat loss by radiation from the roof during cold weather is very limited, it is a major factor during hot weather in keeping the inside of the house cool. The insulation of a roof space is shown in Figure 7.4.

If we assume that the heat loss through single window glass is one hundred units, the comparative heat loss through other structures having the same area and under the same conditions is shown in Table 7.1.

TABLE 7.1. HEAT LOSSES

Material	Comparative Heat Loss
Single glass	100
Double glass (1 cm gap)	56
Wooden floor	42
Solid brickwork (2 bricks thick)	40
Cavity brickwork (2 bricks thick with 5 cm cavity)	28
Plasterboard ceiling with wooden floor above	16
Corrugated iron roof	168

The application of an insulating material (lagging) always has the same purpose—to reduce heat transfer. Cold water pipes are lagged to prevent the water inside them from losing heat, freezing and, consequently, bursting the pipes. The hot water storage cylinder in a domestic hot water supply system is lagged to reduce the heat loss from the water inside it, thereby maintaining its comparatively high temperature. Steam pipes in the boiler-room of a ship have a very thick layer of lagging for the same purpose.

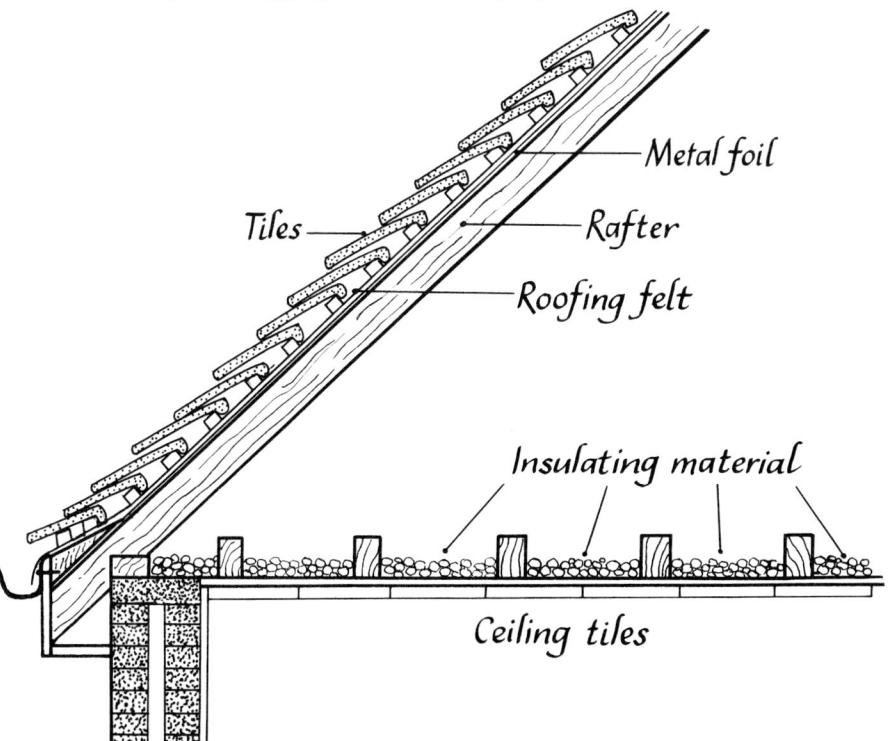

Figure 7.4 Roof space insulation

Investigation 7b. Insulating property of expanded polystyrene

Make a box, using expanded polystyrene ceiling tiles. Reinforce the corners with strips cut from another tile and be certain that you use the correct adhesive. If you use the wrong type of adhesive, the results may be disastrous, as some adhesives contain solvents which will dissolve the tiles.

Cut a hole in the centre of one of the tiles, large enough to allow a 100-watt bulb to pass through it. On one of the other sides, stick a small square of expanded polystyrene near one corner as a reinforcement. Through this reinforced section, cut a hole to take a cork with a thermometer through it. The simplest way to cut these holes is to melt the material with a heated rod of metal (a steel knitting needle is ideal); other methods tend to leave jagged edges.

Mount a batten-type bulb-holder in the centre of a piece of wood, with the flex running along a groove underneath. Put a 100-watt bulb in the holder, place the box over it and insert the thermometer, as shown in Figure 7.5.

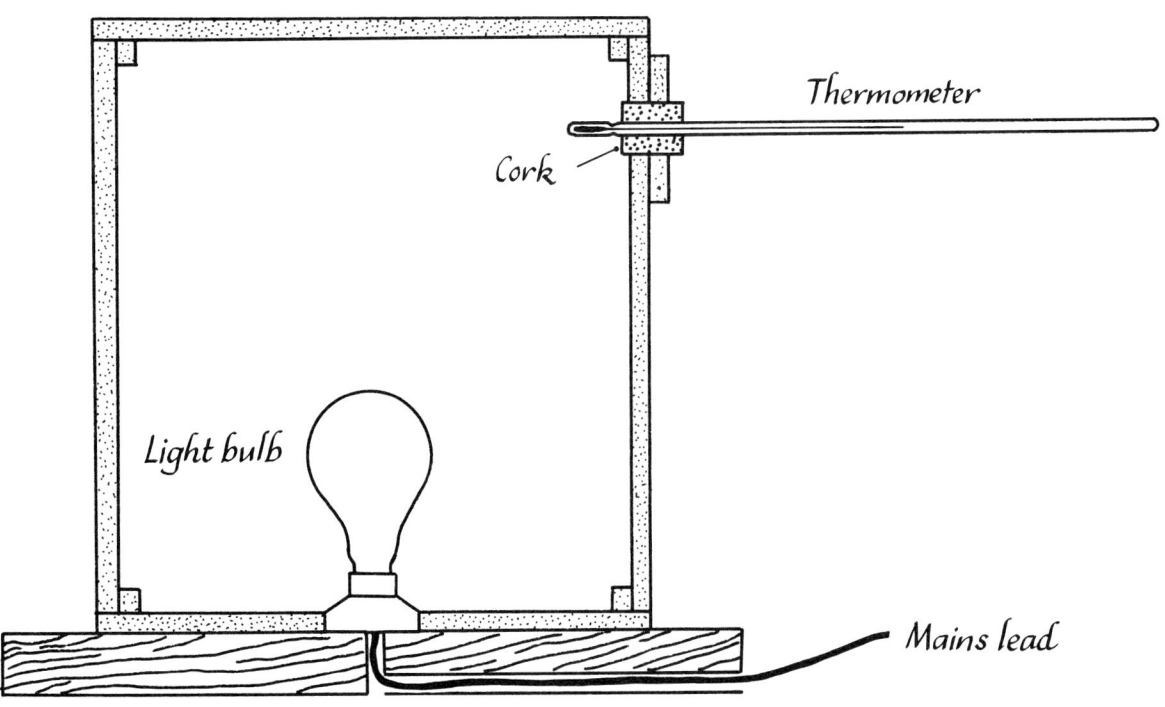

Figure 7.5 Apparatus for Investigation 7b

Note the temperature indicated by the thermometer and switch the bulb on. Take temperature readings every half minute until there is no further rise in temperature, and then plot a graph of these readings. Assuming that the heat output of the bulb remains constant, can you explain why the graph flattens out as the time increases? At what point on your graph does the heat loss through the expanded polystyrene equal the heat output of the bulb?

Repeat this investigation using boxes of the same size made up as follows:
1. All sides made of hardboard.
2. All sides made of hardboard, but with a large window cut in one side which is covered with a single thickness of glass.
3. As above, but with two thicknesses of glass for the window, one sheet on the inside of the box and the other on the outside so that they are separated by the thickness of the hardboard.

Plot graphs of your results for each of these boxes and compare them. In which box did the air reach the highest temperature? In which box was the greatest heat loss?

7.2. Heat and colour

When light strikes an object, it will be transmitted, reflected or absorbed, depending partly on the nature of the material and partly on the nature of its surface. If the object is transparent, the light will be transmitted through it. If the object is opaque, the light will not be transmitted through it but may be absorbed. If the surface of the object is bright and shiny, the light will be reflected from it. In practice, no substance will transmit all of the available light, or reflect all of it, or absorb all of it. When we talk of a transparent substance, we really mean a substance which will transmit more light than it reflects or absorbs. Similarly, a reflecting surface reflects more light than it transmits or absorbs and an opaque substance absorbs more light than it transmits or reflects.

Since light and radiated heat are both forms of radiant energy, it would seem logical that, to a certain extent, they should behave in the same way.

Investigation 7c. Radiating surfaces

Take two water-tight tins of equal size and shape (small coffee tins are ideal), polish the outer surface of one and paint the other with matt black paint. Make a hole in each of the lids, large enough to admit a thermometer. Place each tin on a tripod and gauze, pour equal amounts of water into each of the tins (two-thirds full is a suitable amount, though this will need to be measured) and fit the lids on the tins. Place a thermometer through the hole in each tin lid and clamp it so that the bulb of the thermometer is about halfway down between the surface of the water and the bottom of the tin, as shown in Figure 7.6. Heat the water in the two tins with bunsen burners so that they both reach boiling point at the same time. This can be done by constant observation of the thermometers and adjustment of the bunsen flames.

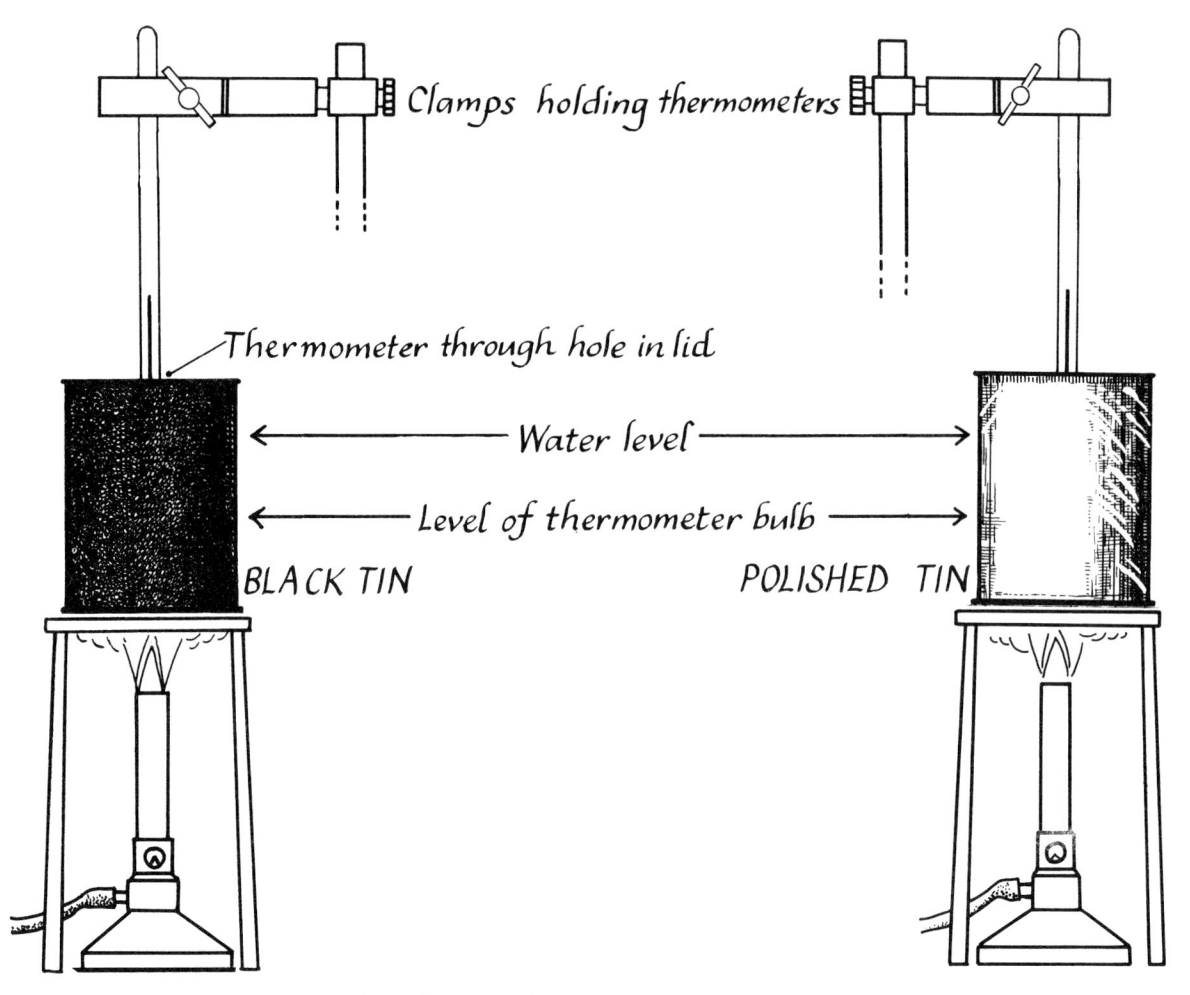

Figure 7.6 Apparatus for Investigation 7c

When the water in both of the tins is boiling, turn out the bunsen burners and note the temperatures indicated by the thermometers every half minute for about fifteen minutes. Make a copy of Table 7.2 and enter your readings in it.

TABLE 7.2. RESULTS OF INVESTIGATION 7c

Time (minutes)	0	$\frac{1}{2}$	1	$1\frac{1}{2}$	2	$2\frac{1}{2}$	3	$3\frac{1}{2}$	4	$4\frac{1}{2}$	
Temperature of water in black tin											
Temperature of water in polished tin											

From your results, can you determine which surface radiated more heat within the time limit of your investigation?

In this investigation, heat is obviously lost from the water by methods other than radiation, i.e. convection and conduction. Heat loss will occur by convection in the air surrounding the tins and, although some heat will have been conducted from the hot tripod and gauze into the water after the bunsen burners are turned out, a little heat may be conducted from the water through the tins to the tripod and gauze during the latter part of the investigation, since the metal of the tripod and gauze will cool faster than the water in the tins. These heat losses by convection and conduction may be ignored, since they will be almost the same for both tins.

The purpose of this investigation is to compare heat transmission from different surfaces and, since other factors would have an effect on the results, it is important to ensure that these other factors are the same for both tins so that their effects cancel out. The following conditions are the same for both tins in this investigation: size of tin, shape of tin, material of tin, liquid and the amount of liquid in the tin, initial temperature, time interval between readings, area in contact with the tripod and gauze, temperature of the surrounding air, type of thermometer and length of thermometer immersed in the water. The only difference between the two tins is the surface and, since this is the factor that is under investigation, we may assume that any difference in the readings is caused by the difference between the surfaces.

Investigation 7d. Absorbing and reflecting surfaces

Cut out two 15 cm squares of sheet metal. Paint *one* side of one of the sheets with matt black paint and polish one side of the other sheet. In the centre of the *other* side of each sheet, stick a penny with wax and support the two sheets in a vertical position, 15 cm from a source of radiant heat (for example, a bunsen burner with a gas radiant), as shown in Figure 7.7. An alternative source of heat for this investigation is the element from an electric fire, suitably mounted and connected to the mains. Light the bunsen burner (or switch on the element) and note the time it takes for the pennies to drop off the sheets. Since both sheets are the same size, material, thickness and distance from the radiant heat source, and both pennies are the same weight and are stuck with the same type of wax with the same melting point, it follows that, if there is any difference in the time, it must be due to the difference in the surfaces. Which surface is the better reflector of heat? Which surface absorbs heat more rapidly?

You will notice from the results of your investigations that surfaces which are good reflectors of light are also good reflectors of heat, and that surfaces which are good absorbers of light (i.e. dark in colour) are also good absorbers of heat.

Figure 7.7 Apparatus for Investigation 7d

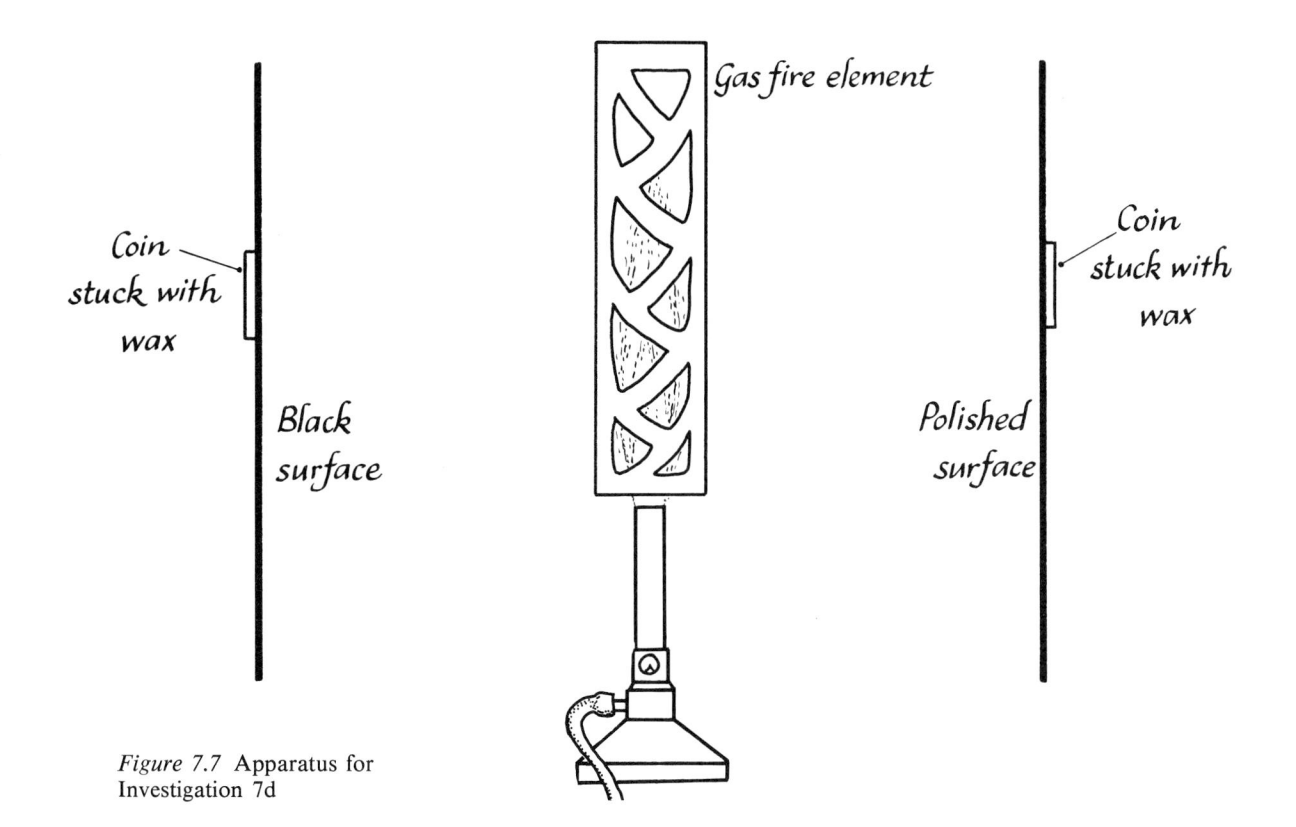

If you ask somebody who is constantly concerned with colour (such as an interior decorator, artist or fashion designer) what they mean by 'warm' colours and 'cold' colours, they will probably tell you that the warm colours are the reds, oranges and yellows and that the cold colours are the blues and greens. You will notice that the warm colours are those that occur at one end of the visible spectrum and that the cold colours are those that occur at the other end. Can you suggest any scientific justification for calling some colours 'warm' and other colours 'cold', or is this just an 'old wives' tale'?

7.3. Cooling by evaporation

You will know from Chapter 2 that, when a liquid evaporates, the faster moving particles escape from the surface of the liquid, leaving the slower particles behind. Since the temperature of a liquid depends on the average speed of the particles, it follows that the temperature of an evaporating liquid will fall because the faster moving particles have escaped.

Investigation 7e. The cooling effect of different liquids

Dip one finger into cold water and another finger into cold methylated spirit. Remove your fingers from the liquids and note which finger feels colder. Which liquid evaporated faster? Using the same method, compare the cooling effect of water with that of lubricating oil, petrol and ether. Which liquid produces the greatest cooling effect?

★ WARNING. *When using petrol and ether ensure there is good ventilation and there are no naked flames.*

Wrap the bulbs of four similar thermometers with small pieces of absorbent lint. Tie the lint with thin cotton. Take four test-tubes and pour into them a depth of about 3 cm of the following liquids:

>   Tube A: water (from a cold tap)
>   Tube B: petrol
>   Tube C: lubricating oil
>   Tube D: ether

Place the wrapped bulbs of the thermometers in the liquids, wait until the temperature indicated by each thermometer is steady and note this temperature. Remove each thermometer in turn and note the lowest temperature indicated. Which liquid evaporated fastest, producing the greatest cooling effect?

Liquids which evaporate readily are called **volatile** liquids. You will notice that most volatile liquids can be easily smelled and are thin 'runny' liquids. Can you suggest any reason why this should be so?

Investigation 7f. Comparison of evaporation rate and surface area

Place a flat-bottomed specimen tube on one scale pan of a beam balance and one half of a petri dish (or an evaporating dish) on the other scale pan. Balance the two sides by adding weights or dry sand to the lighter pan. Lower the balance and, by using two pipettes, put 20 cm^3 of ether into each of the containers simultaneously. Raise the balance and note any movement. Any decrease in weight must have been caused by evaporation. From which container did the ether evaporate faster? In which container did the ether have the larger surface area?

★ WARNING. *Do not breathe any ether vapour. Ensure that there is good ventilation and there are no naked flames.*

Pour a depth of about 2 cm of ether into a flat-bottomed flask with a double-holed bung and glass tubing, as shown in Figure 7.8. Rest the flask in a puddle of cold water on a tile or a piece of hardboard. Connect the shorter glass tube to a bunsen pump and ensure that the other glass tube dips below the surface of the ether. Turn on the water operating the bunsen pump and, after a short time, note what happens to the puddle of water. You may find that you can lift the tile or piece of hardboard by lifting the flask.

This method increases the rate of evaporation in two ways:

a. The air bubbling through the ether increases the surface area.

b. The bunsen pump removes the ether vapour as it escapes from the surface, enabling more ether to evaporate.

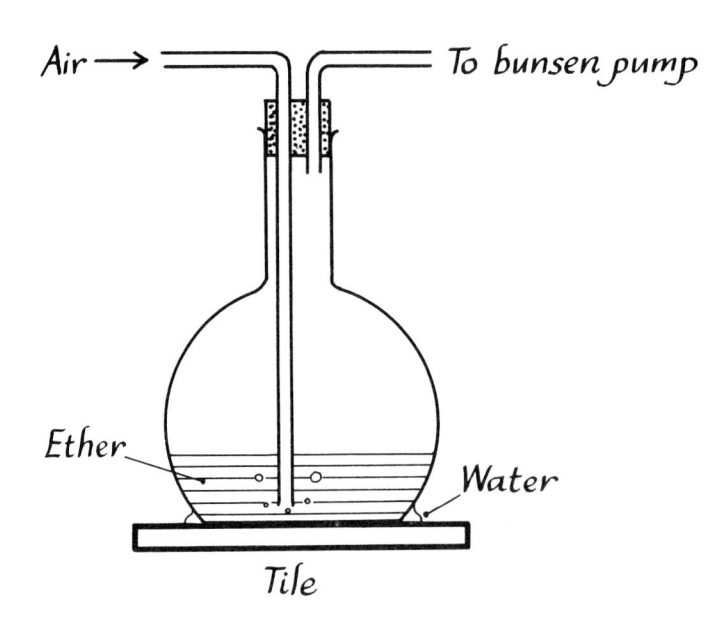

Figure 7.8 Apparatus for Investigation 7f

Investigation 7g. Comparison of evaporation rate and pressure

Half fill a spherical flask with warm water (about 50 °C). Fit the flask with a double-holed bung with a thermometer through one hole so that the bulb of the thermometer is in the water and a piece of glass tubing through the other hole, as shown in Figure 7.9. Connect the piece of glass tubing to a vacuum pump with a length of thick-walled rubber tubing and note the temperature of the water. Watch the thermometer, switch on the vacuum pump and note any change in temperature. If a vacuum pump is not available, a bunsen pump may be used, but the results are not so spectacular. Can you explain why the reduction in pressure causes this temperature change? Why would thin-walled rubber tubing be unsuitable for this investigation?

The cooling effect of evaporation is constantly being used by the human body. During any form of strenuous exercise, the body sweats. As the sweat evaporates, it cools the skin which, in turn, absorbs heat from the blood in the capillary blood vessels just beneath the skin, thus reducing the blood temperature. If this did not happen, any form of exercise would be immediately followed by a rapid rise in blood temperature.

When concrete is laid in hot weather, it is often covered with wet sacking or a layer of wet sand. This covering layer is kept wet to ensure that the concrete sets slowly. If setting concrete is allowed to

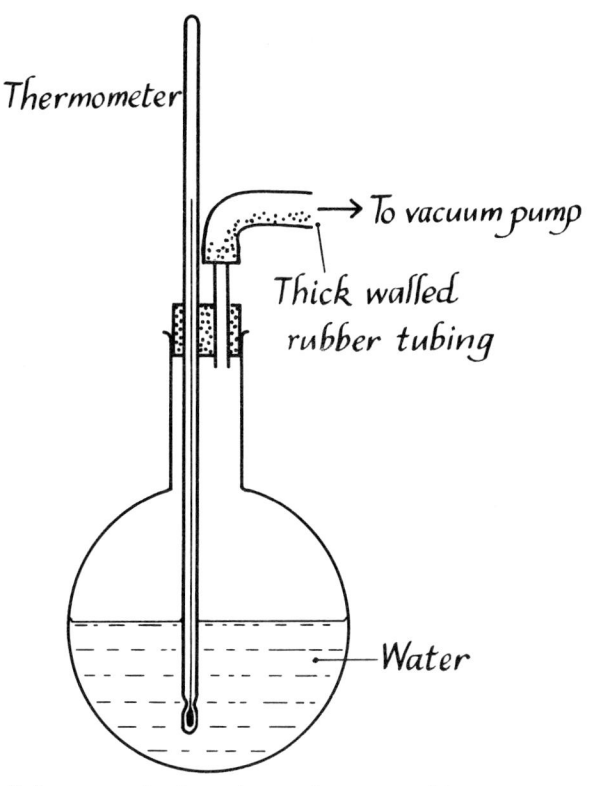

Figure 7.9 Apparatus for Investigation 7g

get too hot, some of the water in the mixture is removed by evaporation, and the concrete sets too quickly, producing a weaker finished product.

When the face is dabbed with Eau de Cologne or after-shave lotion, it feels cold; this has a refreshing effect. What have these two liquids in common that produces this effect?

Before the domestic refrigerator came into common use, many people kept their perishable foods in a box supported over a tray of water. The box was covered with a piece of sacking or other absorbent material, the ends of which dipped into the water, as shown in Figure 7.10. So long as there was water in the tray, it was soaked up by the absorbent material and evaporated. This evaporation kept the inside of the box cool. The same principle can be used to keep milk cool: a bottle of milk is placed in a small tray of water and covered with an upturned piece of unglazed pottery of the same shape as, but a little larger than, the bottle.

The rate of cooling depends on the rate of evaporation; this depends on four factors:

a. The nature of the liquid. Volatile liquids evaporate faster and produce a greater cooling effect.

b. The ratio between the volume and the surface area. For a fixed volume, the greater the surface area, the greater will be the temperature drop (e.g. a cup of hot soup will cool faster if it is poured into a soup plate).

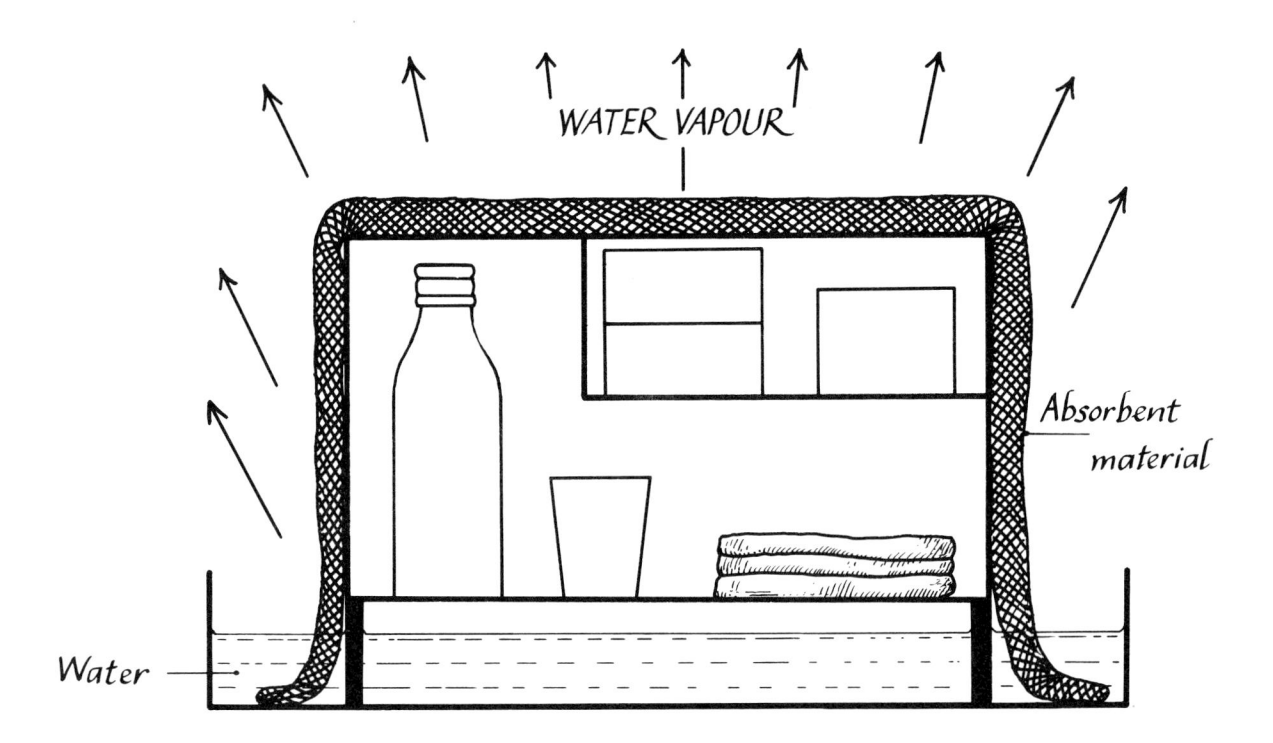

Figure 7.10 A simple food cooler

c. The air pressure on the surface of the evaporating liquid. An increase in the pressure reduces the rate of evaporation. In order to escape from the surface, only the fastest moving particles are able to overcome the increased pressure. Particles which are moving just fast enough to escape under normal pressure are unable to do so under increased pressure; so they remain in the liquid, thus reducing the rate of evaporation and also the rate of cooling. Under conditions of reduced pressure, the particle speed required to escape from the surface is less; consequently, more particles are able to escape. This increases the rate of evaporation and the rate of cooling.

d. The amount of vapour in the air in contact with the evaporating surface. As the amount of vapour increases, the rate of evaporation decreases until the air is saturated. When this occurs, evaporation will cease.

7.4. The refrigerator

The forerunner of the modern domestic refrigerator was the ice-box. This consisted of a double-walled container with lagging (an insulating material such as asbestos wool or kapok) between the inner and outer walls. Food was stored in the lower part of the ice-box and blocks of ice were placed on a sloping shelf at the top. The whole cabinet was covered by a lagged lid to reduce heat transfer from the surroundings to the inside of the ice-box (see Figure 7.11). Heat from the air in the cabinet was absorbed by the ice,

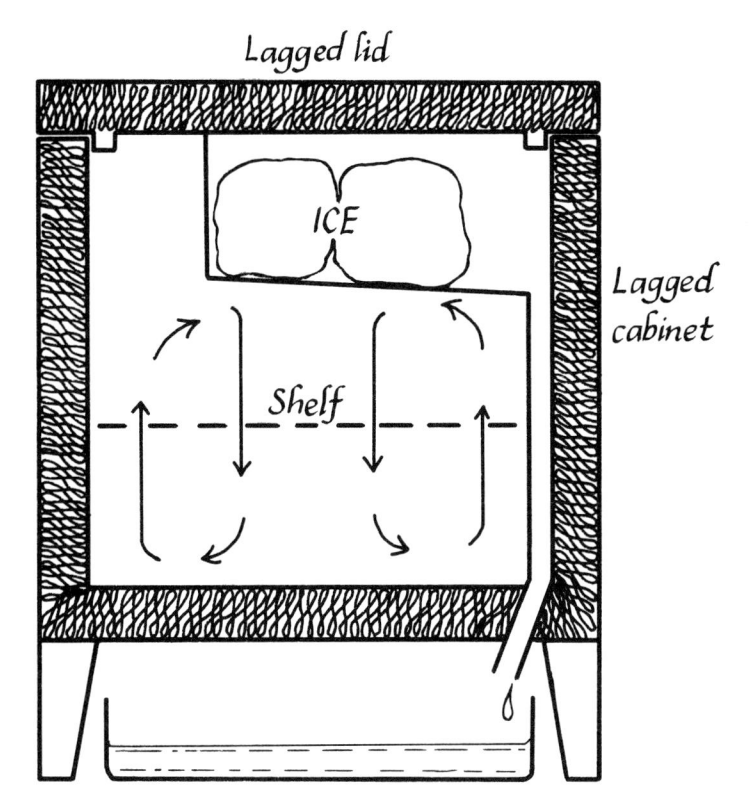

Figure 7.11 The ice-box

which slowly melted. The cold air circulated round the food in the cabinet by convection. Water from the melting ice ran down grooves in the sloping shelf and passed down a pipe into a drip-tray underneath.

Portable ice-cream containers use a similar system, using blocks of solid carbon dioxide (often called 'dry ice', sold under various trade names such as 'Drikold') instead of blocks of ice. There are several advantages in this method:

a. Since solid carbon dioxide changes directly from a solid to a gas without passing through a liquid stage, there is no possibility of spillage.

b. This change of state from solid to gas occurs at $-79\ °C$. Solid carbon dioxide, therefore, keeps the contents of the cabinet at a lower temperature than would be possible with ice.

c. Solid carbon dioxide is less dense than ice, thus reducing the overall weight of the container; this weight reduces still further as the carbon dioxide escapes as a gas.

The function of a refrigerator is to reduce the temperature inside the cabinet. This is achieved by transferring heat energy from the inside of the cabinet to the outside. The substance used as a transport medium for this heat transfer is called the **refrigerant**. Typical refrigerants are ammonia (NH_3), sulphur dioxide (SO_2) and freon (CCl_2F_2). These substances are gases at normal atmospheric pressure and temperature, but can be liquefied by an increase in pressure.

The commonest type of domestic refrigerator uses the compression system. Inside the cabinet is the cooling unit (the evaporator) and outside the cabinet is the condenser. These two units are connected by narrow-bore tubing to a compressor and a valve, to form a continuous circuit. A simple compression refrigerator system using the 'flooded' type of evaporator is shown in Figure 7.12.

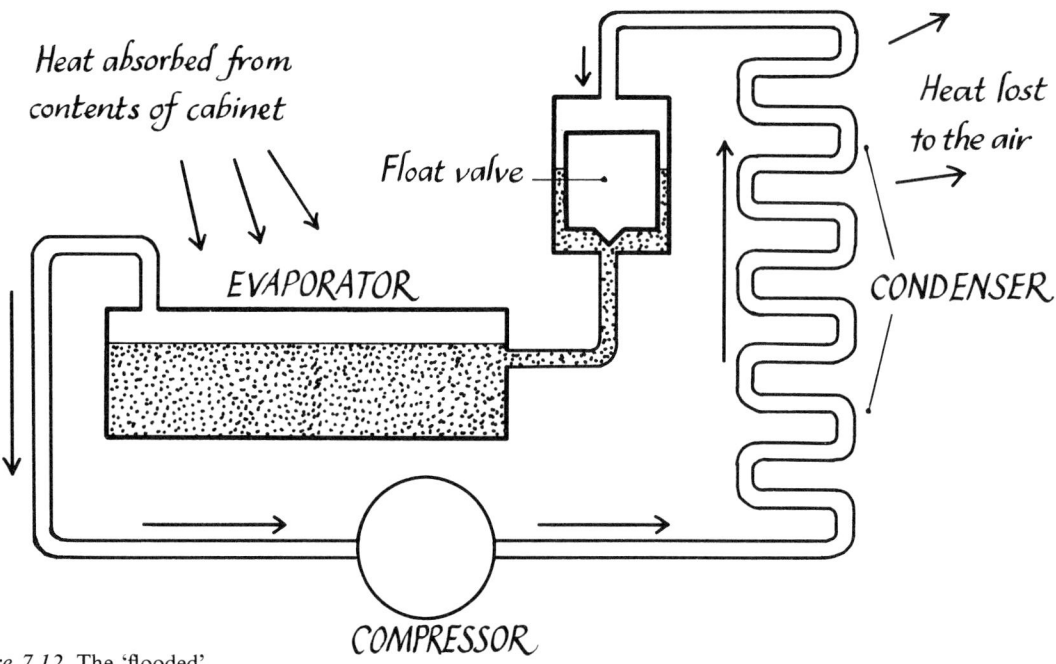

Figure 7.12 The 'flooded' evaporator compression refrigerator

The liquid refrigerant in the evaporator absorbs heat from the contents of the cabinet, causing the refrigerant to evaporate. The compressor, which is driven by an electric motor controlled by a thermostat inside the cabinet, pumps the heat-laden vapour from the evaporator to the condenser. This compression heats the vapour to a temperature higher than the temperature of the air surrounding the condenser. As this hot vapour loses heat to the air, its temperature drops and then, by losing its latent heat, it liquefies. The passage of the liquid refrigerant from the condenser to the evaporator is controlled by a float valve. When the level of the liquid in this valve drops, the valve closes, thus enabling the compressor to build up the necessary vapour pressure. When the level of the liquefied refrigerant is high enough in the valve, it opens, thus allowing the refrigerant to return to the evaporator.

A more complicated compression refrigerator system using the 'dry' type of evaporator is shown in Figure 7.13.

In this type of refrigerator, the liquid refrigerant is stored in a tank in the base of the refrigerator. Between this tank (called the receiver) and the evaporator is an expansion valve. The function of

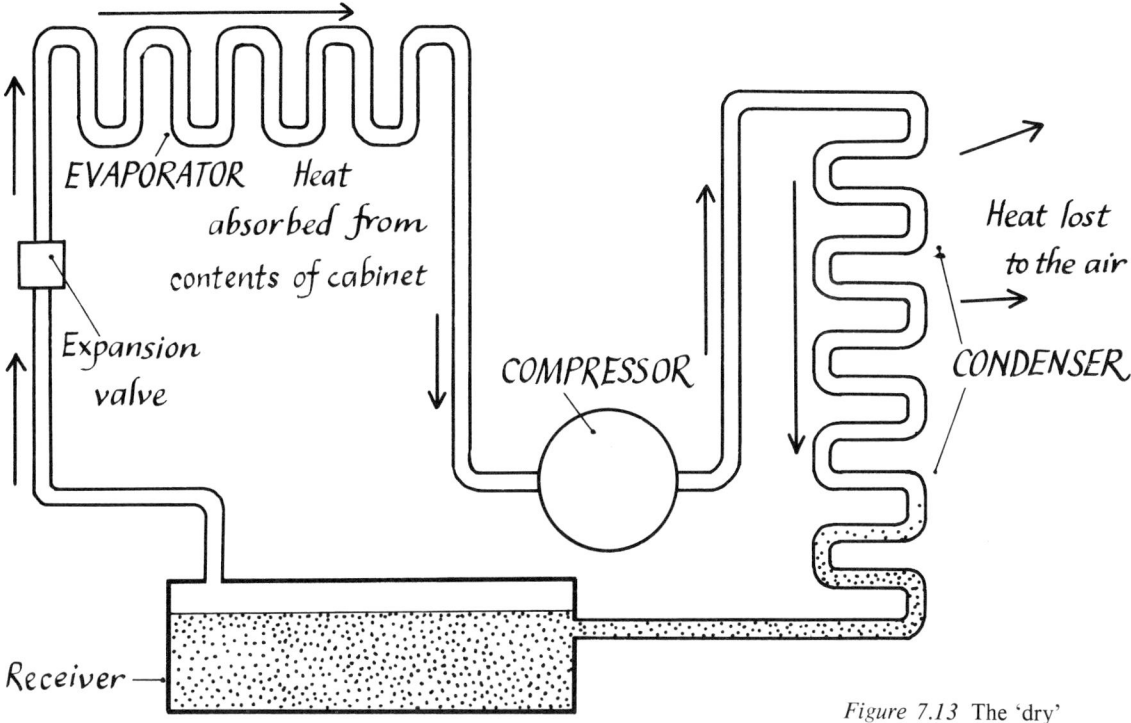

Figure 7.13 The 'dry' evaporator compression refrigerator

this valve is to restrict the flow of refrigerant, thus ensuring that the refrigerant in the evaporator is at a low pressure. Because of this low pressure, the refrigerant is at a low temperature and is able to absorb heat from the contents of the cabinet. As the refrigerant is pumped to the condenser by the compressor, its temperature rises so that it can lose heat in the condenser and pass back to the receiver, thus completing the cycle.

Water in the air inside the cabinet condenses on the evaporator in the form of frost. From time to time, this frost is removed by switching the refrigerator off and allowing the frost to melt. The water thus formed is collected in a drip-tray.

7.5. Heat balance

When a mixture of petrol vapour and air is exploded in the cylinder of an internal combustion engine, a great deal of heat is produced. Some of this heat is used to expand the gases inside the cylinder, forcing the piston down and, consequently, turning the

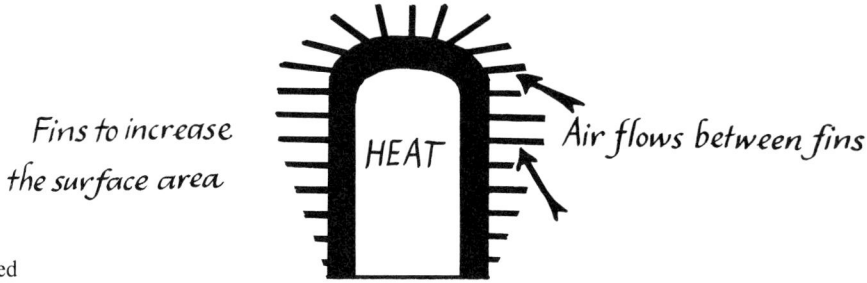

Figure 7.14 An air-cooled cylinder

Fins to increase the surface area — HEAT — Air flows between fins

crankshaft. The surplus heat increases the temperature of the cylinder block. In order to prevent an excessive increase in temperature, which would cause the pistons and cylinders to expand and eventually seize (fit so tightly that the pistons could not move up and down inside the cylinders), some form of cooling system must be used. In an air-cooled engine, the outsides of the cylinders are shaped in the form of fins, which increase the surface area (see Figure 7.14). Air passing between these fins absorbs heat, thus reducing the temperature of the cylinders.

In a water-cooled engine, heat is absorbed from the cylinder walls by water, which flows round them in a water jacket. By convection, the heated water rises through the top hose to the tank at the top of the radiator. As the water passes downwards through the radiator core, it is cooled by the passage of air through the core. The cooled water then passes through the bottom hose and back to the cylinder block, thus completing the cycle (see Figure 7.15).

Figure 7.15 The cooling system of a water-cooled engine

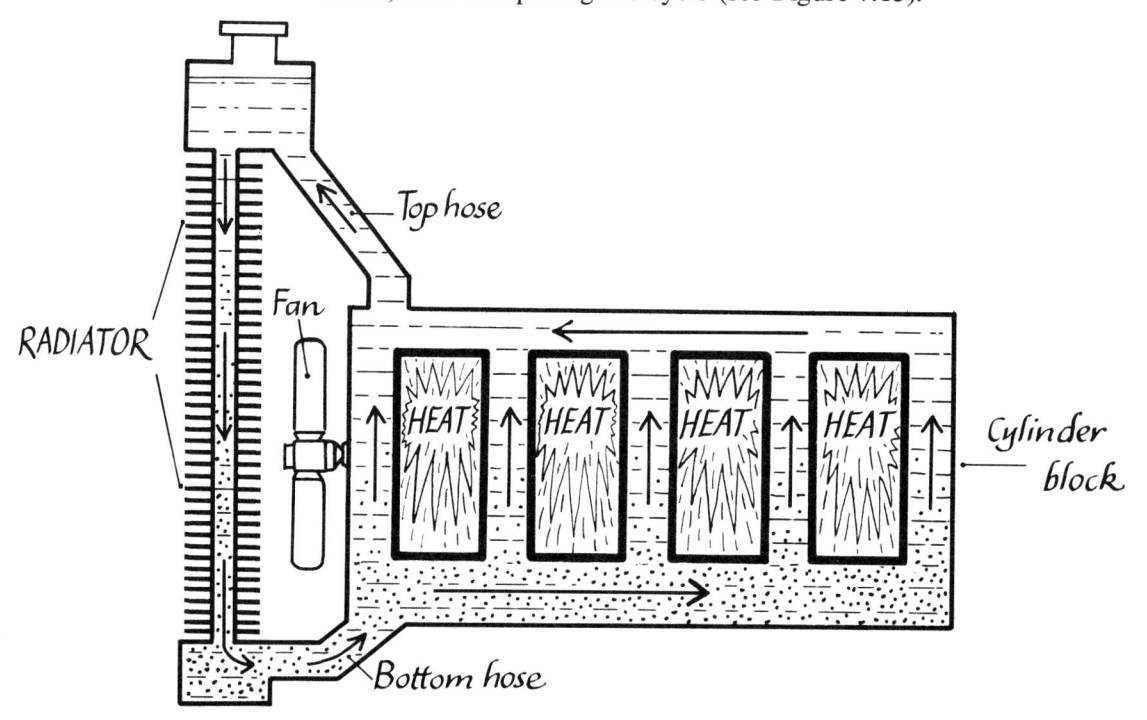

The flow of water in a water-cooled engine is controlled by a thermostat fitted between the cylinder block and the top hose. When the water in the cylinder block is cold, the thermostat is closed, thus preventing the flow of water to the radiator. When the water in the cylinder block reaches a predetermined temperature (usually about 80 °C), the thermostat opens, allowing the water to pass to the radiator. This ensures a comparatively constant temperature within the cylinder block. The function of the fan (which is driven by the engine) is to ensure a flow of air through the radiator when the vehicle is stationary with the engine running.

In a water-cooled engine, the water is used as a medium to transfer heat from the cylinder block to the air. In doing this, it performs the same function as the refrigerant in a refrigerator.

All birds and mammals are warm-blooded. This means that the blood temperature of a species of bird or mammal remains almost constant, no matter what the temperature of its surroundings may be. One of the most reliable symptoms of ill-health in a warm-blooded animal is an abnormal temperature (either higher or lower than it should be). The blood temperature of cold-blooded creatures, however, is approximately the same as the temperature of their surroundings.

Investigation 7h. Daily blood temperature variation

When you are in a normal state of health, take your temperature with a clinical thermometer as soon as you get up in the morning and then every two hours throughout the day until you go to bed. Continue taking your temperature every two hours during the day for several days, keeping a record of times and temperatures. Then plot a graph of your readings. At what time during the day is your temperature lowest? At what time of the day is your temperature highest? By how many degrees does your temperature vary?

Our bodies are heated in two ways:
a. By absorbing radiant heat from the sun or a radiant source, such as an electric fire or a gas fire.
b. By oxidizing the carbon in the fuel foods that we eat (see Chapter 10).

Most of the heat that we require is obtained by the oxidation of carbon. This oxidation takes place in the body cells, particularly in the muscle cells, and is called **respiration**. The oxygen needed for this process is extracted from the air by the lungs and is transported to the cells by the bloodstream.

More heat is needed to maintain the blood temperature in cold weather than in warm weather because the body loses heat at a faster rate. This extra heat is obtained by eating a larger proportion of the fuel foods (see Chapter 10) and by exercise.

When the muscles are exercised, the cells oxidize carbon faster, thus releasing energy faster. In order to do this, additional oxygen must reach the muscle cells. The blood can only carry a certain amount of oxygen, so the blood supply to the muscles must be increased. This is achieved by the heart, which beats faster, thus causing an increase in the breathing rate to maintain the oxygen supply to the blood.

Heat is given out by the body in a number of ways. We lose heat to our surroundings by conduction, convection and radiation in the same way as any object which is at a higher temperature than its surroundings loses heat. These heat losses can be limited by wearing clothes which insulate the body. Because air is a poor conductor of heat, clothes which trap large amounts of air are efficient insulators. Not only do they reduce heat loss by conduction but, because the flow of air is restricted by the fabric, heat loss by convection is also reduced. It is for these reasons that a string vest is so effective in keeping us warm in cold weather.

Birds increase the thickness of their insulating layer of feathers by fluffing up their feathers in cold weather. When the weather becomes warmer, they moult and the feathers lie flatter against the skin. Furry animals also increase the thickness of their insulating layer by making each hair project from the skin at a steeper angle. This is achieved by the contraction of muscles connected to the **follicle** (the tube through which the hair grows) of each hair. You can see the effect of these muscles when you have 'goose pimples'. Fur coats, fur boots and fur gloves may look attractive, but they would keep us warmer if they were made to be worn inside out, as the following parody on Longfellow's 'Song of Hiawatha' shows:

'The Modern Hiawatha'

When he killed the Mudjokiwis,
Of the skin he made him mittens,
Made them with the fur side inside,
Made them with the skin side outside,
He, to get the warm side inside,
Put the inside skin side outside,
He, to get the cold side outside,
Put the warm side fur side inside.
That's why he put the fur side inside,
Why he put the skin side outside,
Why he turned them inside outside.

Anon.

Heat losses from animals without fur are reduced by a layer of fat beneath the skin. This is apparent in animals such as pigs, whales and seals.

A great deal of heat is lost through the evaporation of perspiration (sweat). In warm weather or during vigorous exercise, the skin perspires freely. As this perspiration evaporates, heat is taken from the blood in the capillaries. These dilate (become larger), supplying more heat to the skin. The dilation of the capillaries has the effect of making the skin redder. During strenuous exercise, it is important to wear lightweight, loose-fitting clothing to enable the perspiration to evaporate freely. Because it takes time for the capillaries to return to their normal size at the end of strenuous exercise, it is advisable to reduce heat loss by putting on some warm clothing. This prevents over-cooling and a rapid drop in blood temperature.

Heat is also lost when we breathe, since exhaled air is warmer than inhaled air. This is particularly noticeable during frosty weather, when the warm water vapour in the exhaled air can be seen to condense into droplets of water on contact with the cold air. Dogs and other furry animals lose a greater proportion of heat in this way than we do, because they have fewer sweat glands. This is the reason why dogs pant. Can you suggest why a healthy dog usually has a cold, wet nose?

Although heat is lost by the excretion of urine and faeces, this has little or no effect on body temperature because an equal mass is also removed from the body at the same time (in the same way that a cup-full of water removed from a bath-full has no effect on the temperature of the water in the bath).

The maintenance of body temperature depends on the balance of heat input and heat output: if heat input exceeds heat output, body temperature will rise; if heat output exceeds heat input, body temperature will drop (see Table 7.3).

TABLE 7.3. HEAT BALANCE OF THE BODY

Heat Input	Heat Output
Radiant heat	Conduction, convection and radiation
Oxidation of food	Evaporation of perspiration, exhaled air

Test your understanding

1. Why does a plate of soup cool faster than the same amount of soup at the same temperature in a cup?
2. Explain why snow remains on the roofs of some houses longer than on others.
3. Explain why:
    (a) Buildings in the tropics are usually white.
    (b) In hot weather we usually wear light-coloured clothing.
    (c) Gardeners often put soot on their gardens.

(d) In hot weather the glass in a greenhouse is often splashed with whitewash.
(e) Oil pipelines are painted white or silver.
(f) On the outside of a house, cream paintwork lasts longer than black paintwork.
(g) A silver teapot keeps tea hotter for a longer time than a black one.

4. Why does a summer shower cool the air?
5. Why does a refrigerator warm the air in the room in which it is being used?
6. How does frost form in a refrigerator?
7. In what way are the functions of the refrigerant in a refrigerator, the water in a water-cooled engine and the blood in the body similar to each other?
8. Why is it necessary to empty the bladder more frequently in cold weather than in hot weather?
9. Why is it necessary to eat more food in winter than in summer?
10. What is wrong with the following statement: 'Clothes make you warm'?

Chapter 8

Mechanics

8.1. Speed and velocity

If you walk a distance of eight kilometres and the journey takes you two hours, your average speed for the journey is four kilometres per hour (4 km h^{-1}). To find the average speed of a moving object, we divide the total distance travelled by the total time taken to cover that distance.

$$\text{Average speed} = \frac{\text{Distance travelled}}{\text{Time taken}}$$

When we talk of the **velocity** of a moving object, we mean the speed *in a particular direction.*

Suppose that B is a town due north of another town A and that the two towns are connected by two roads. The straight road is 10 km long while the winding road is 30 km long (Figure 8.1). If a car takes the straight road and completes the journey in ten minutes, its *average speed* for the journey will be 60 km h^{-1}. Since the straight

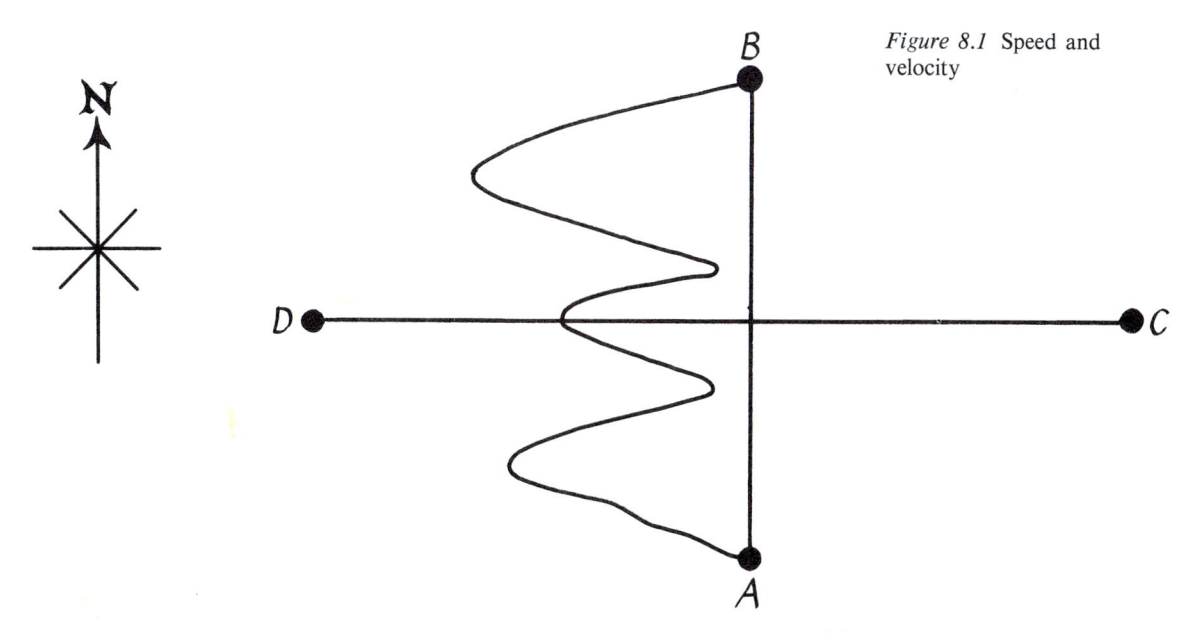

Figure 8.1 Speed and velocity

road runs in a north–south direction, the *velocity* of the car will be 60 km h^{-1} *in a northerly direction.*

If a second car took the winding road and took thirty minutes to reach B, its average speed would also be 60 km h^{-1} but, since it would only have travelled 10 km in a northerly direction, its average velocity in that direction would be 20 km h^{-1}. If another car travelled in a westerly direction from C to D, it would have no velocity in the northerly direction, no matter what its speed was.

8.2. Mass and inertia

Sir Isaac Newton defined the **mass** of a body as being the amount of matter in the body. This is not a very satisfactory definition, because it does not state what is meant by 'matter', nor say how it may be measured. However, it is easy to see that one metre of copper wire

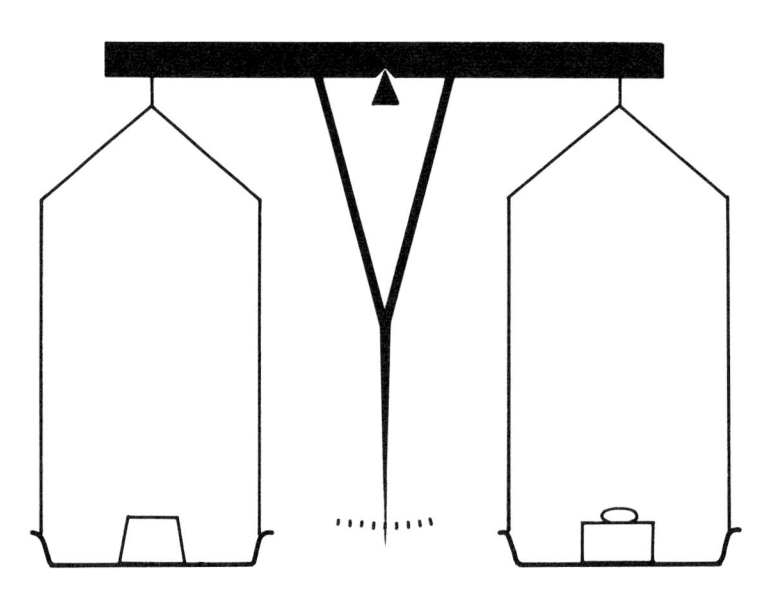

Figure 8.2 Comparing masses

would have one hundred times the mass of one centimetre of the same wire.

The unit of mass is the **kilogramme**. A piece of platinum, which is kept at Sèvres, near Paris, is the standard against which any kilogramme mass is compared. In order to compare masses, a beam balance may be used. When balance is obtained, the mass in the right-hand scale pan is equal to that in the left-hand scale pan (see Figure 8.2).

Investigation 8a. Inertia

Suspend a brick and a piece of expanded polystyrene, which has similar dimensions to the brick, by separate cords to a beam. Give each a gentle push to start it moving. Which required the bigger push?

Start each of the objects swinging and then stop each in turn by 'catching' it. Which was the more difficult to stop?

All bodies possess the property of resisting any attempt to start them moving or, if they are already moving, of resisting any attempt to change their motion. This property is called **inertia**. The greater the mass of a body the greater is its inertia.

8.3. Weight

The earth exerts a pull on all objects on, or near, its surface. This pull on an object, which acts towards the centre of the earth, is called the **weight** of the object. The weight of an object is proportional to its mass.

In Chapter 1 we said that pushes and pulls are called forces. When we measure weight, we are measuring a force and must use units of force to do so. The unit of force is the **newton (N)**.

Investigation 8b. Falling bodies

Stand on a bench and hold a large marble in one hand and a small ball bearing in the other. Hold both hands at the same height above the floor and drop the objects together. Repeat this several times. Does the mass of an object appear to have any effect on how fast it falls?

Now, drop a marble and a feather at the same time. Which strikes the floor first? Can you suggest why this is so?

Investigation 8c. The falling feather

Place a feather in a tube, as shown in Figure 8.3, and invert the tube. Notice how the feather falls.

Draw the air out of the tube by means of a vacuum pump and close the clip. Invert the tube once more. What do you notice? Can you explain this?

8.4. Forces and acceleration

Investigation 8d. The ticker tape timer

This instrument has a striker which can be made to tap a piece of carbon paper fifty times a second. A length of paper tape is drawn under the carbon paper so that the timer marks a series of dots on the tape (see Figure 8.4).

Pull the tape through the timer at a constant speed. Notice that the

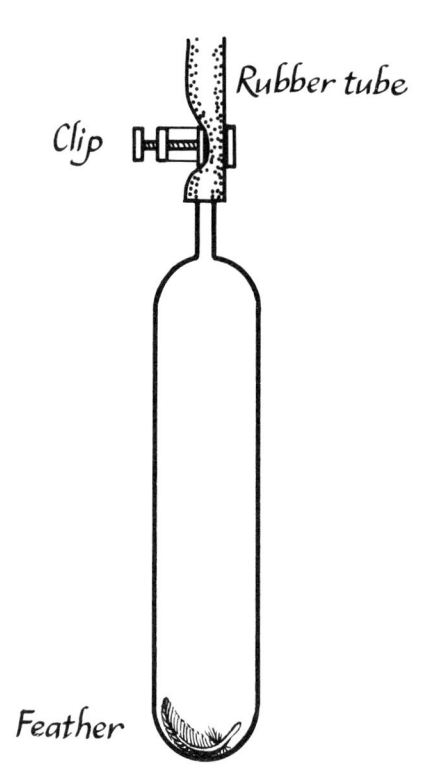

Figure 8.3 The falling feather

dots are evenly spaced. Put a new length of tape in the timer and attach it to a trolley (or a roller skate) by means of 'Sellotape'. Fasten a dynamometer (an instrument using the same principle as a spring balance but calibrated in newtons) to the other end of the trolley, as shown in Figure 8.5, and pull the trolley along the bench by means of the dynamometer. Try to keep the dynamometer reading constant during the pull. Stop the timer, remove the tape and examine it. What does this tell you about the motion of the trolley?

Repeat the investigation, but this time keep a higher reading on the balance than before. Compare the two tapes.

When a body increases its velocity, it is said to **accelerate**. Acceleration is defined as the rate of change of velocity. If, at one

Figure 8.4 The ticker tape timer

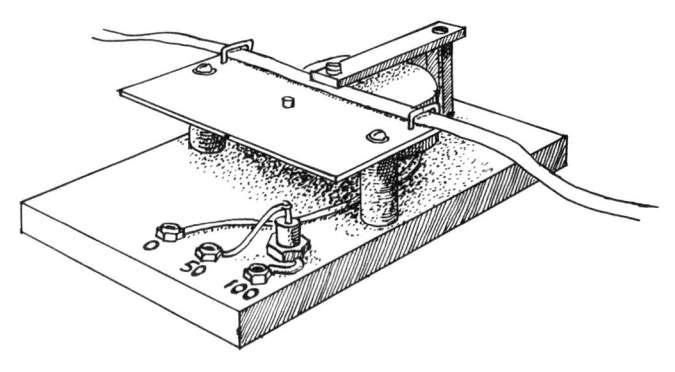

Figure 8.5 Using the trolley

instant, a body was moving at a velocity of 5 m s⁻¹ and, one second later, its velocity had increased to 7 m s⁻¹, it would have increased its velocity by two metres per second *in one second*. Its velocity change would be two metres per second, but the acceleration would be stated as two metres per second per second (written 2 m s⁻²).

From Investigation 8d you should have observed that when a constant force is applied to a body, producing motion, the body accelerates. For a particular body (of constant mass) the acceleration is proportional to the force applied.

Investigation 8e. Gravitational force

Set up a ticker tape timer near the edge of a bench, as shown in Figure 8.6. Attach a 10 g mass to the end of the tape. Start the timer

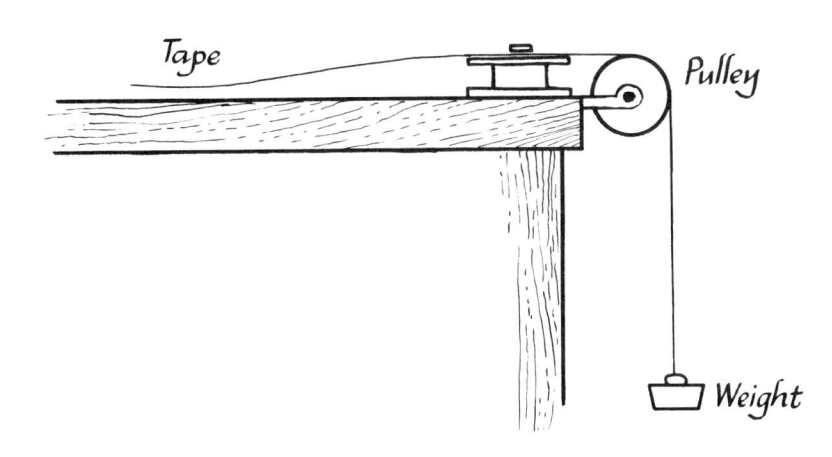

Figure 8.6 Gravitational force

and allow the mass to fall to the floor. Stop the timer, remove the tape and examine it.

Repeat the experiment, using first a mass of 50 g and then a mass of 100 g.

When bodies fall under the earth's gravitational force, they accelerate at approximately 10 m s⁻². Table 8.1 shows how a falling body accelerates.

TABLE 8.1. ACCELERATION DUE TO GRAVITY

Time (s)	Velocity (m s^{-1})	Acceleration (m s^{-2})
0	0	—
1	10	10
2	20	10

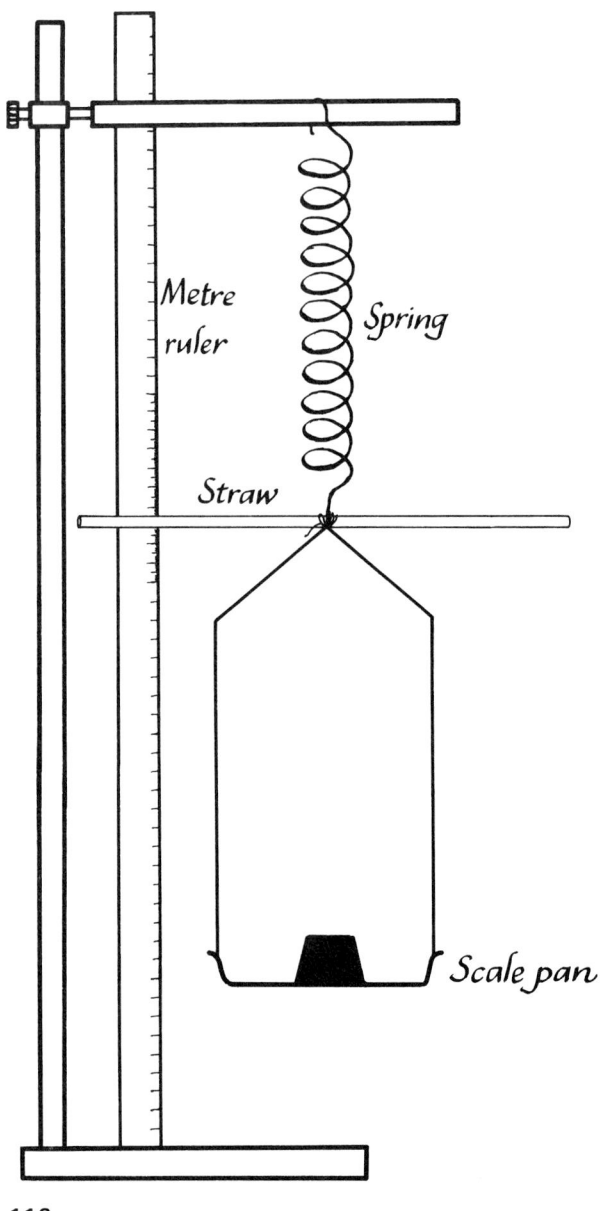

Figure 8.7 The spring balance

8.5. Measuring forces

A force of one newton is defined as that force which will produce an acceleration of one metre per second per second when it is applied to a mass of one kilogramme. Thus, when a mass of 1 kg falls under gravity, the force involved is 10 N.

Investigation 8f. The spring balance

Support a helical spring in a retort stand and support a metre ruler behind it. Attach a scale pan to the lower end of the spring and fasten a drinking straw to the top of the scale pan by means of 'Sellotape'. The general arrangement is shown in Figure 8.7. Note the reading on the ruler indicated by the straw. Put a mass of 5 g on the pan and note the new position of the straw. The difference between this reading and the last will show how far the spring has stretched.

Increase the mass in the scale pan to 10 g and note the extension which this produces. Continue to increase the mass in 5 g steps, until the mass on the pan is 100 g.

The force of gravity acting on 100 g is 1 N, so we can write down the force acting on the spring in newtons. Draw up a table of the results (see Table 8.2).

TABLE 8.2. RESULTS OF INVESTIGATION 8f

Force (N)	Scale Reading (mm)	Extension (mm)
0	400	—
0·05	470	70
0·1		
0·15		

Draw a graph, plotting mass against extension, as shown in Figure 8.8.

The extension of a spring is directly proportional to the force producing the extension. However, if too great a force is applied, the spring may lose its elastic properties. The spring balance makes use of the elastic properties of a spring, and is a device for measuring force. Remember that the force acting on a mass due to gravity is the weight of the body. Thus, while a beam balance is used to measure *mass*, a spring balance is used to measure *weight*.

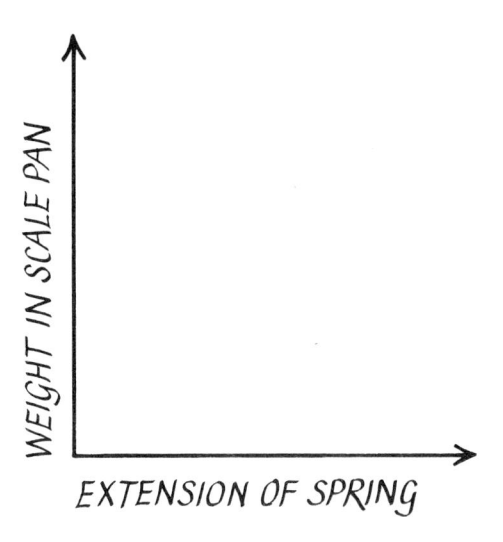

Figure 8.8 A spring balance graph

[Graph with vertical axis labeled "WEIGHT IN SCALE PAN" and horizontal axis labeled "EXTENSION OF SPRING"]

8.6. Work

In order to move an object, mechanical work has to be done on it; it is necessary to apply a force in order to produce motion. We measure mechanical work by multiplying together the force applied and the distance through which the object (to which the force is applied) has moved.

Work done = Distance moved × Force applied

If a force is applied to an object but no motion occurs, no mechanical work is done.

The unit of work is the **joule (J)**. If a force of one newton acts through a distance of one metre, one joule of work is done.

Suppose that a crane lifts a mass of 1 000 kg to a height of 10 m. The work done by the crane has been acting in opposition to the force of gravity acting on this mass.

Force required to lift mass of 1 000 kg = 1 000 × 10 N
Distance through which force has acted = 10 m
Work done = 10 × 10 000 J
= 100 000 J

If a boy having a mass of 50 kg runs up a flight of stairs having a vertical height of 3 m, the work done against gravity will be:

50 × 10 N × 3 m = 1 500 J
FORCE × DISTANCE = WORK DONE

8.7. Energy

Energy is the capacity for doing work. The units used in the measurement of energy are, therefore, the same as the units used to measure work.

If a crane is supporting a mass of 1 000 kg, 10 m above the ground, the load is said to have an energy of 100 000 J with respect to the ground. If the cable were to break, the mass would fall to the ground, doing this amount of work.

While the mass is supported above the ground it is said to possess **potential energy**. If it falls, the energy due to its motion is called **kinetic energy**. While it is falling its potential energy is decreasing while its kinetic energy is increasing. At all times during the fall, the

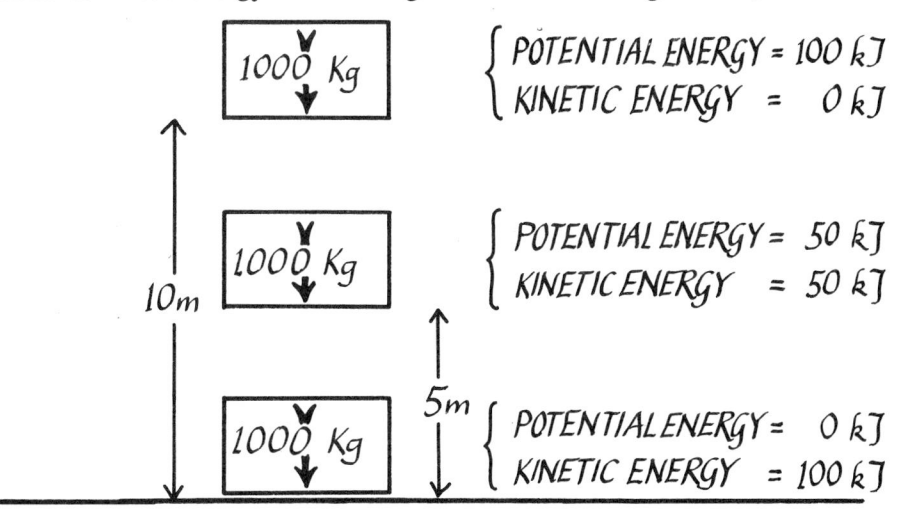

```
           1000 Kg          { POTENTIAL ENERGY = 100 kJ
                            { KINETIC ENERGY   =   0 kJ


           1000 Kg          { POTENTIAL ENERGY = 50 kJ
   10m                      { KINETIC ENERGY   = 50 kJ


           1000 Kg    5m    { POTENTIAL ENERGY =   0 kJ
                            { KINETIC ENERGY   = 100 kJ
```

sum of the potential and kinetic energies will be equal to the original amount of potential energy before the fall began (see Figure 8.9).

Figure 8.9 Kinetic energy and potential energy

Figure 8.10 illustrates more examples of kinetic energy and potential energy.

To summarize:
a. **Potential energy is due to the position of the body.**
b. **Kinetic energy is due to the motion of the body.**

8.8. Power

Power is the rate of doing work. The unit of power is the **watt**. When one joule of work is done in one second, the power is one watt. Large amounts of power are measured in kilowatts.

$$1 \text{ kilowatt} = 1\,000 \text{ watts}$$

If a crane lifts a mass of 1 000 kg to a height of 10 m in 5 s, then the power will be as follows:

Force required $= 1\,000 \times 10 = 10\,000$ N
Distance moved $= 10$ m
Work done $= 10\,000 \times 10 = 100\,000$ J
Time taken $= 5$ s
Power generated $= \dfrac{100\,000 \text{ J}}{5 \text{ s}} = 20\,000$ W (20 kW)

Figure 8.10 Examples of kinetic energy and potential energy

8.9. The moment of a force

Investigation 8g. Moments

For this investigation you will need a metre ruler with a hole bored through the middle. Pass a pin through this hole and into a cork. Support the cork in a retort stand. If the ruler tilts, a small piece of 'Plasticine' may be attached to the 'high' side in order to balance it.

Fasten loops of thread on to masses of 50 g and 100 g. Use these to hang the masses from the ruler, as shown in Figure 8.11. Adjust

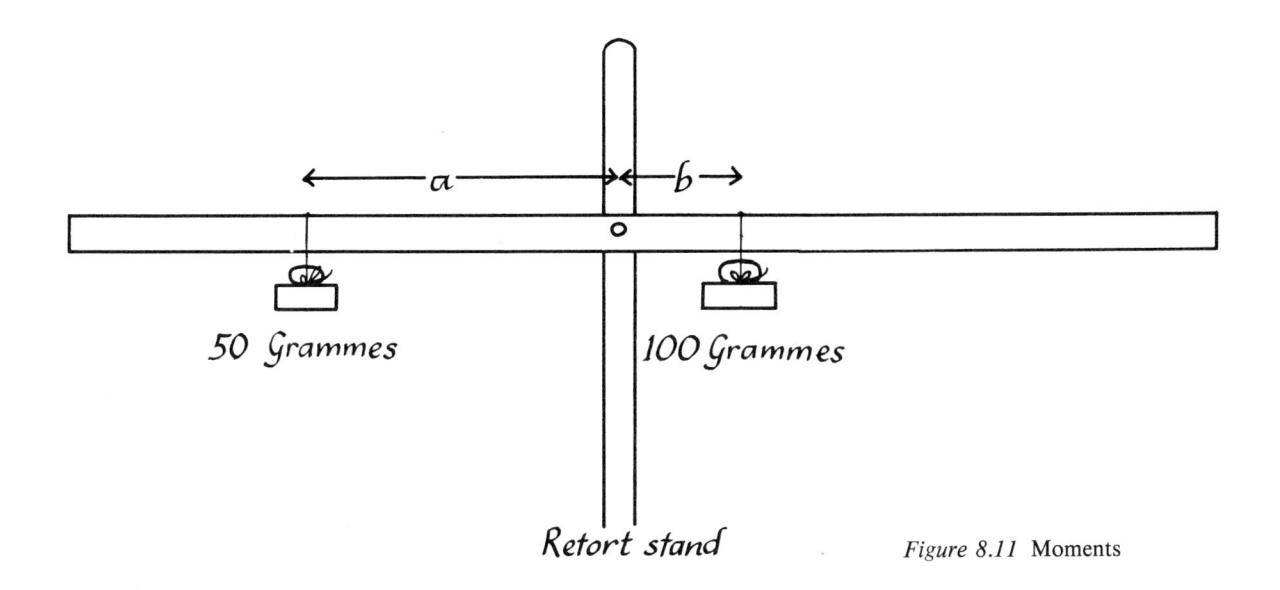

Figure 8.11 Moments

the positions until the ruler balances, and note the distance of each mass from the balance point.

Move the 50 g mass to a new position. Re-position the 100 g mass to balance the ruler again. Once more, note the distance of each mass from the balance point.

Repeat this for several other positions of the masses and draw up a table of results (see Table 8.3).

TABLE 8.3. RESULTS OF INVESTIGATION 8g

Left-hand Side			Right-hand Side		
Force on 50 g Mass	Distance (mm)	Force × Distance	Force on 100 g Mass	Distance (mm)	Force × Distance
0.5 N	a	$0.5 \times a$	1.0 N	b	$1.0 \times b$

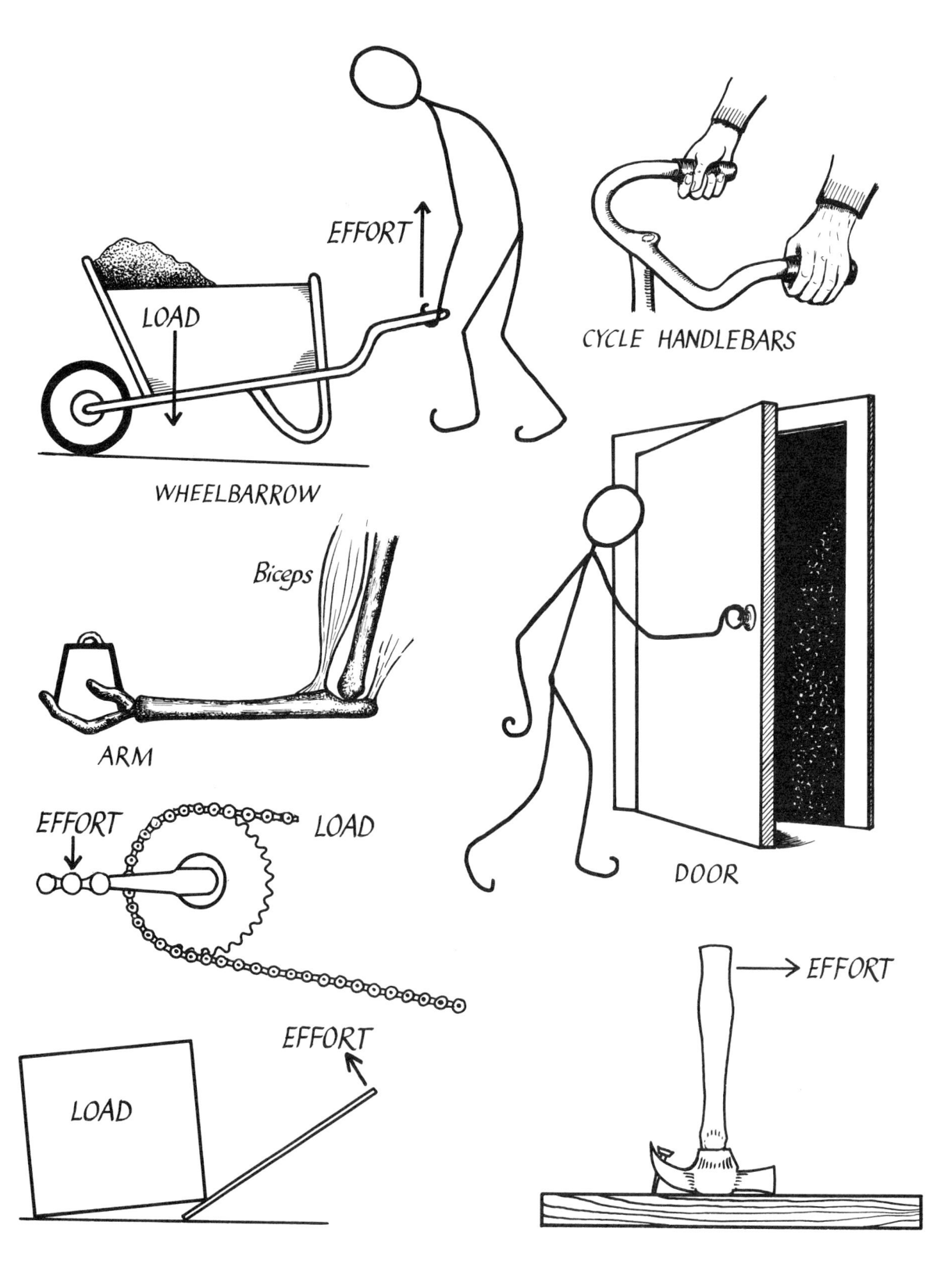

Figure 8.12 Levers

The **moment of a force** is the product of the force and the perpendicular distance through which it acts. A moment produces a turning effect, either in a clockwise direction or an anti-clockwise direction. When the ruler used in Investigation 8g balanced, the clockwise moment was equal to the anti-clockwise moment. This principle is used in the beam balance, where equal masses balance each other by being placed at equal distances from the balance point. It should be remembered that equal masses have the same weight *when they are under the same gravitational influence.*

In Chapter 1 we noted that, when dealing with work, newtons × metres = joules. When dealing with the moment of a force, we must leave the answers in newton-metres.

8.10. Machines

A machine may be regarded as a device which enables work to be done more conveniently.

One of the simplest and most commonplace machines is the lever (see Figure 8.12). This makes use of the principle of moments, usually by causing a small force acting through a large distance to balance a large force acting through a small distance. The balance point of a lever is called the *fulcrum.* Find the fulcrum in each of the diagrams in Figure 8.12.

Another simple machine is the pulley. Figure 8.13 shows how a single pulley simply changes the direction of a force.

If two or more pulleys are suitably arranged, they may be used to

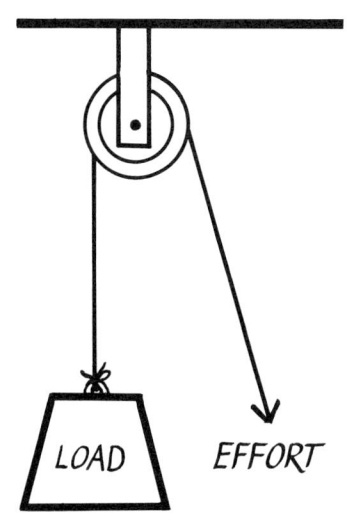

Figure 8.13 A single pulley

raise a large load by the application of a small force. Figure 8.14 shows a suitable arrangement. It can be seen that when the applied force (or effort) pulls out a metre of string, the load is lifted through a distance of only a quarter of a metre. This is because each of the four lengths of string supporting the load has been shortened by an equal amount (25 cm).

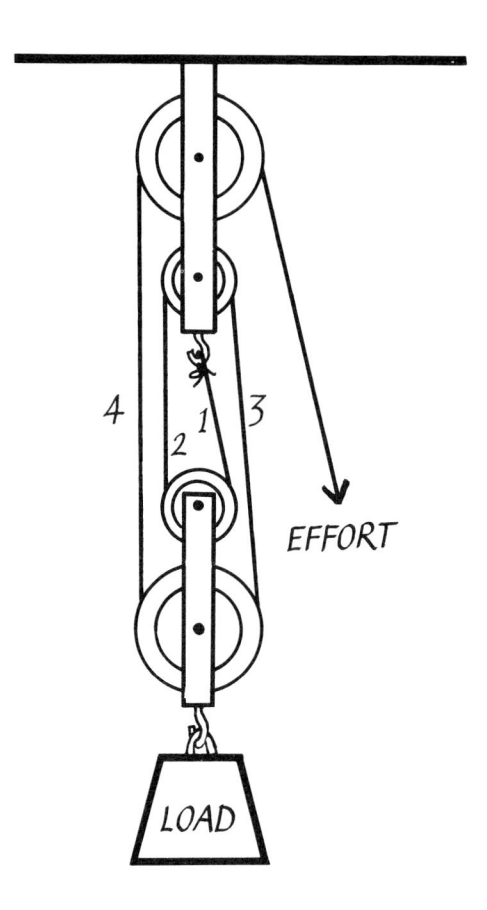

Figure 8.14 A pulley system

The ratio between the distance moved by the effort and the distance moved by the load is called the **velocity ratio** of the machine. The velocity ratio of the pulley system shown in Figure 8.14 is 4:1.

The ratio between the load and the effort is called the **mechanical advantage** of the machine.

The ratio between the useful work obtained from a machine and the work put into the machine is called the **efficiency** of the machine. This is often expressed as a percentage. No machine has an efficiency of 100 per cent. This is because some of the work put into the machine is converted, by friction, into heat. In the case of the pulley system, the lower pulley block as well as the load has to be raised, which leads to a further reduction of efficiency.

8.11. Some other simple machines

a. The wheel and axle

Figure 8.15 illustrates an example of this type of machine being used to raise a bucket of water from a well. The effort moves through a large circle in order to wind one small turn of rope on the axle, which is of small diameter. Although the effort moves through a greater distance than the load, the effort is smaller than the load.

Another example of this type of machine is the carpenter's hand brace. Make a list of any other examples you can think of.

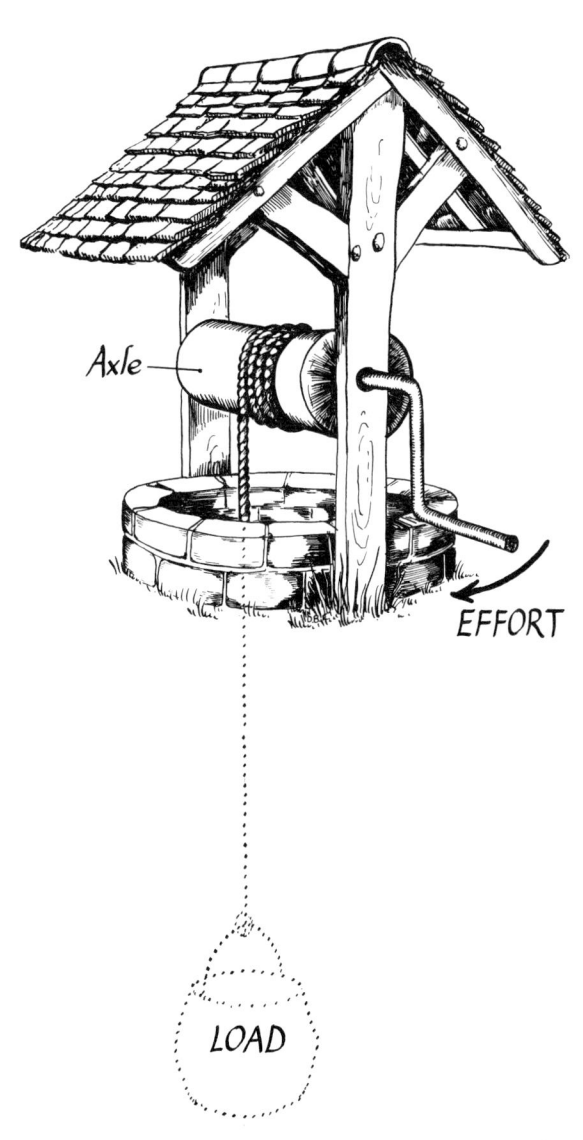

Figure 8.15 The wheel and axle

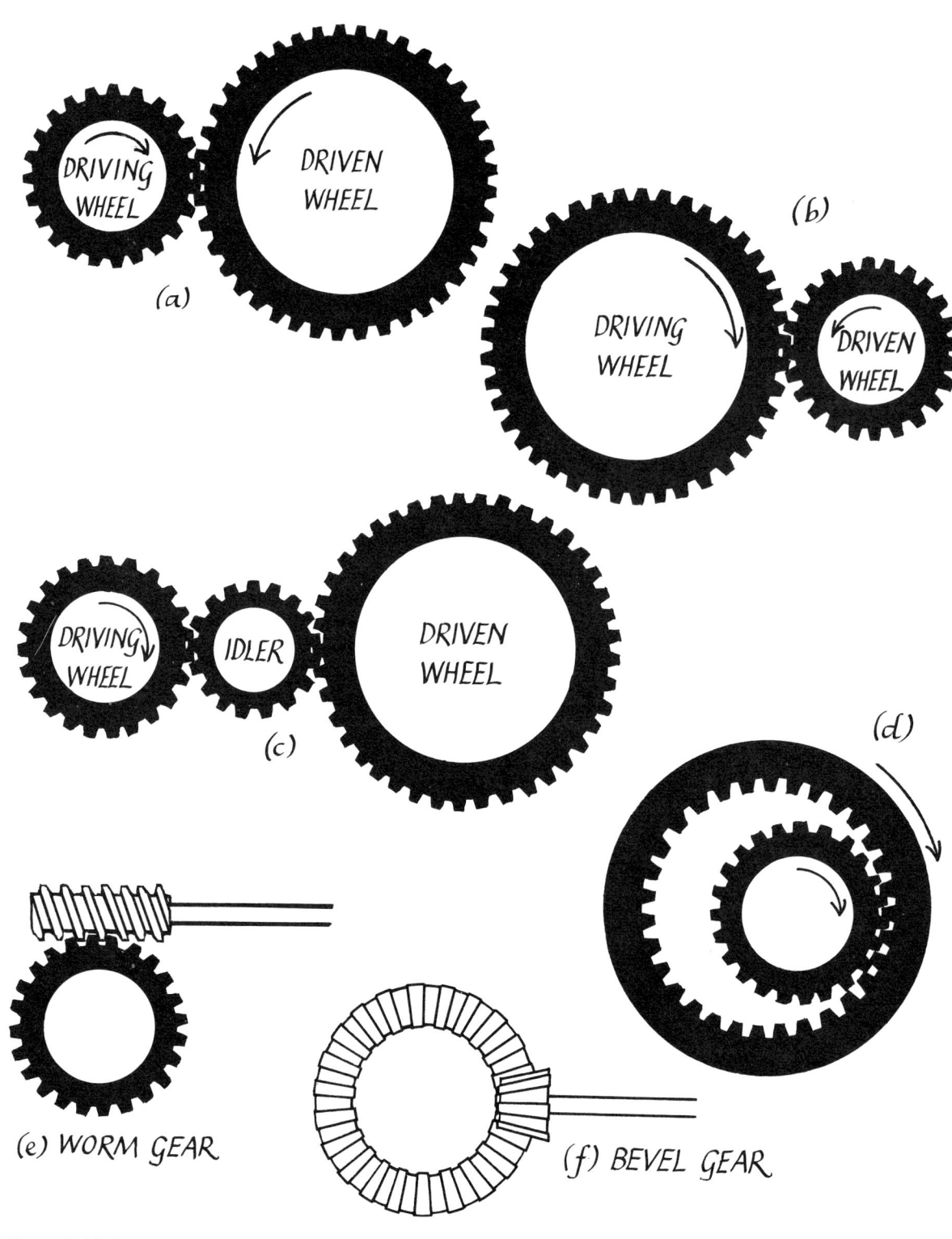

Figure 8.16 Gear systems

b. Gear systems

A small effort may be used to turn a small gear wheel rapidly, which in turn may be used to drive a large gear wheel slowly. In this case, the large wheel will be able to exert a force greater than the effort applied to the small gear wheel.

On the other hand, if a large gear wheel is used to drive a smaller wheel, the smaller wheel will turn faster than the larger one but will exert a smaller force.

Figure 8.16 shows various arrangements of gears. What is the purpose of the 'idler' wheel in Figure 8.16 (c)?

c. The inclined plane

Investigation 8h. The inclined plane

For this investigation you will need a trolley (or a roller skate), a board about 1 m long, a dynamometer and a few books.

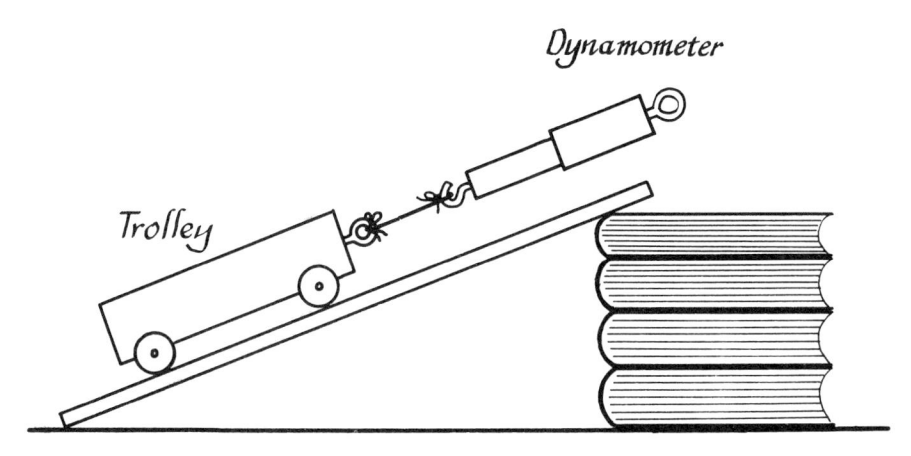

Figure 8.17 The inclined plane

Put the pile of books under one end of the board so that it is inclined to the bench (see Figure 8.17). Attach a dynamometer to the trolley and use this to draw the trolley up the board at a steady rate. Note the balance reading.

Remove some of the books so that the board is less steeply inclined and repeat the investigation.

Weigh the trolley with a dynamometer and compare this with the force required to draw the trolley up the plane.

The inclined plane is a kind of machine. Do your results confirm this?

Cut a triangle of paper, as shown in Figure 8.18, mark a pencil line along the long edge of the paper and then wrap the paper round

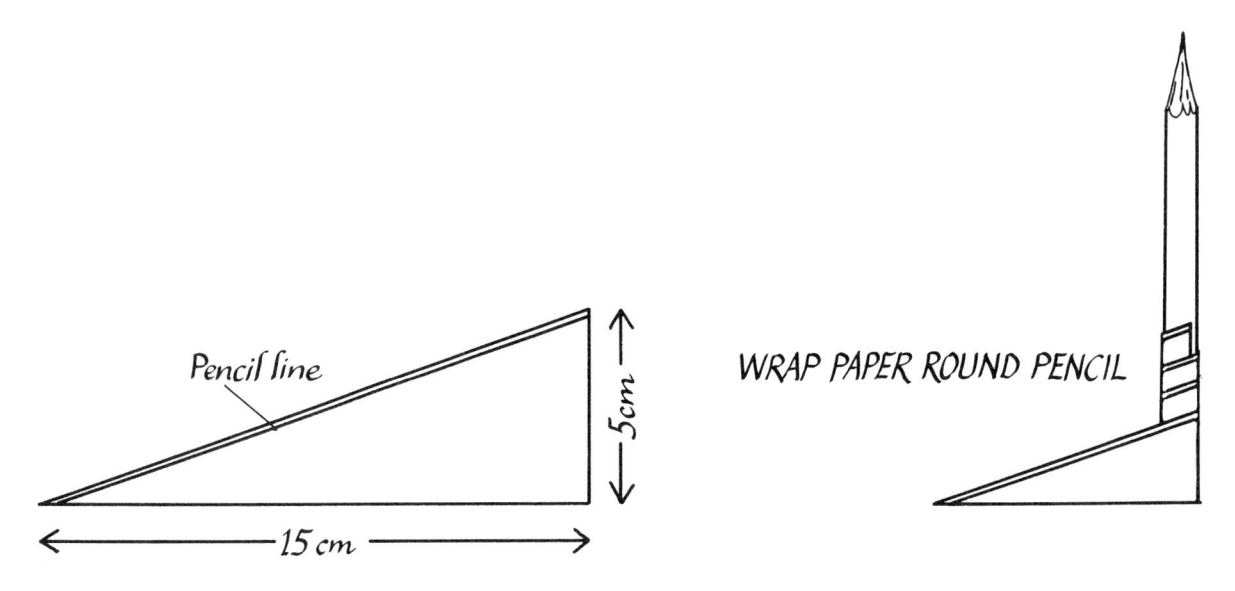

Figure 8.18 Wrapping the triangle round the pencil

the pencil. What common mechanical device does this suggest to you?

Figure 8.19 shows some devices which make use of the inclined plane. Can you think of any more?

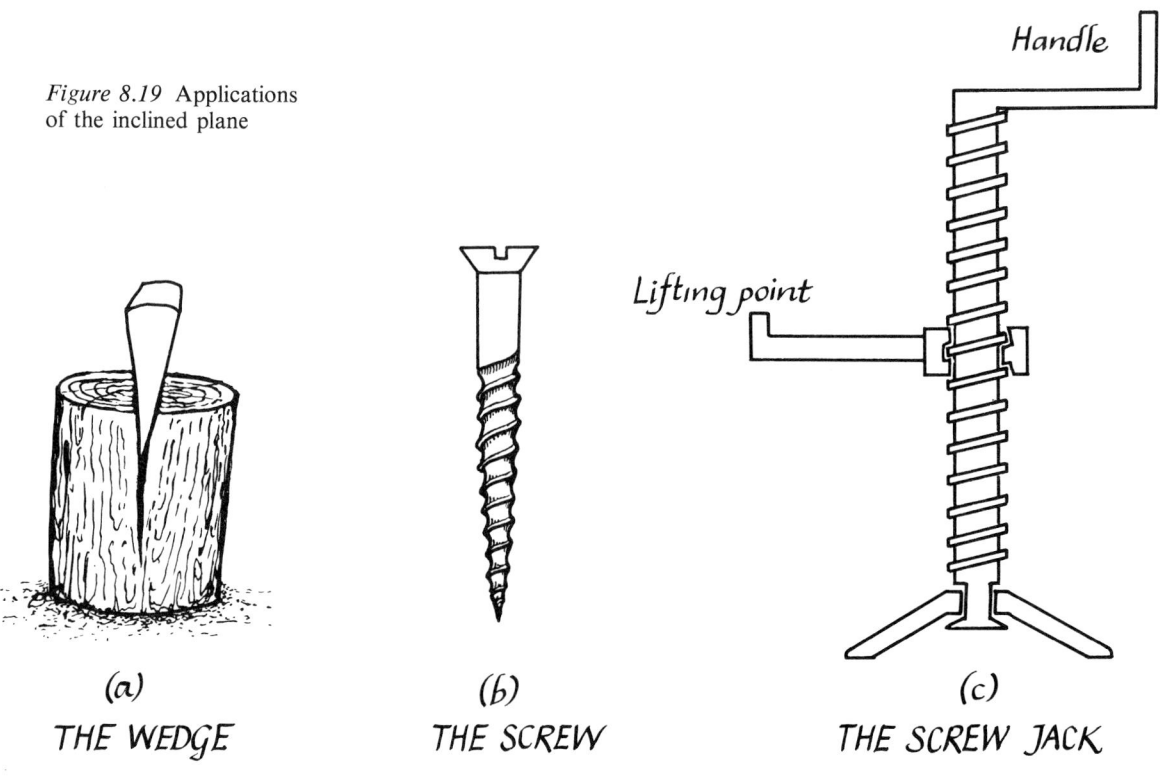

Figure 8.19 Applications of the inclined plane

(a) THE WEDGE

(b) THE SCREW

(c) THE SCREW JACK

d. The hydraulic jack

This device makes use of the fact that liquids are incompressible. Figure 8.20 shows two cylinders of different diameters, each fitted with a piston. The cylinders are connected by a tube. The cylinders and connecting tube are filled with a liquid.

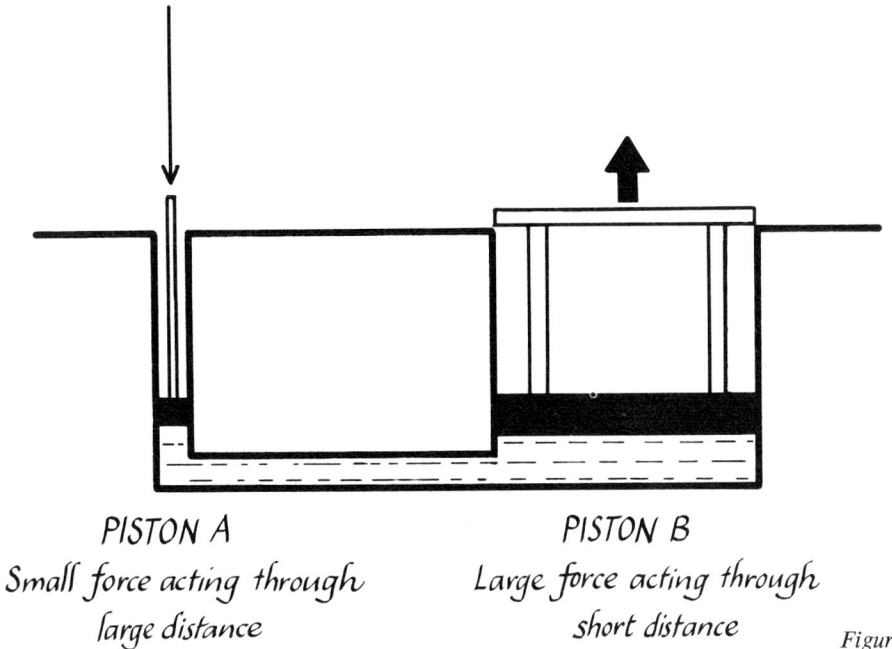

PISTON A
Small force acting through large distance

PISTON B
Large force acting through short distance

Figure 8.20 Transmission of energy through a liquid

The **pressure** on a surface is the force acting on unit area of that surface. Thus, if a force of 200 N acts on a surface having an area of 5 m², the pressure on the surface will be 40 N per square metre (40 N m^{-2}).

$$\text{Pressure} = \frac{\text{Force}}{\text{Area}}$$

If a force acts on piston A (Figure 8.20), a pressure is produced which is transmitted through the liquid. This pressure will then be exerted on piston B, which has a larger surface area than piston A. The total force acting on piston B will be greater than the force acting on piston A, but A will have to move through a greater distance than B.

Figure 8.21 represents a practical hydraulic jack. A lever is used to apply a force to a small piston which is fitted in a short cylinder. A reservoir and valves are used to produce a series of short strokes which raise the larger piston.

On the down stroke, valve P opens but Q is closed; on the up stroke, P is forced shut but Q opens and more liquid is drawn into

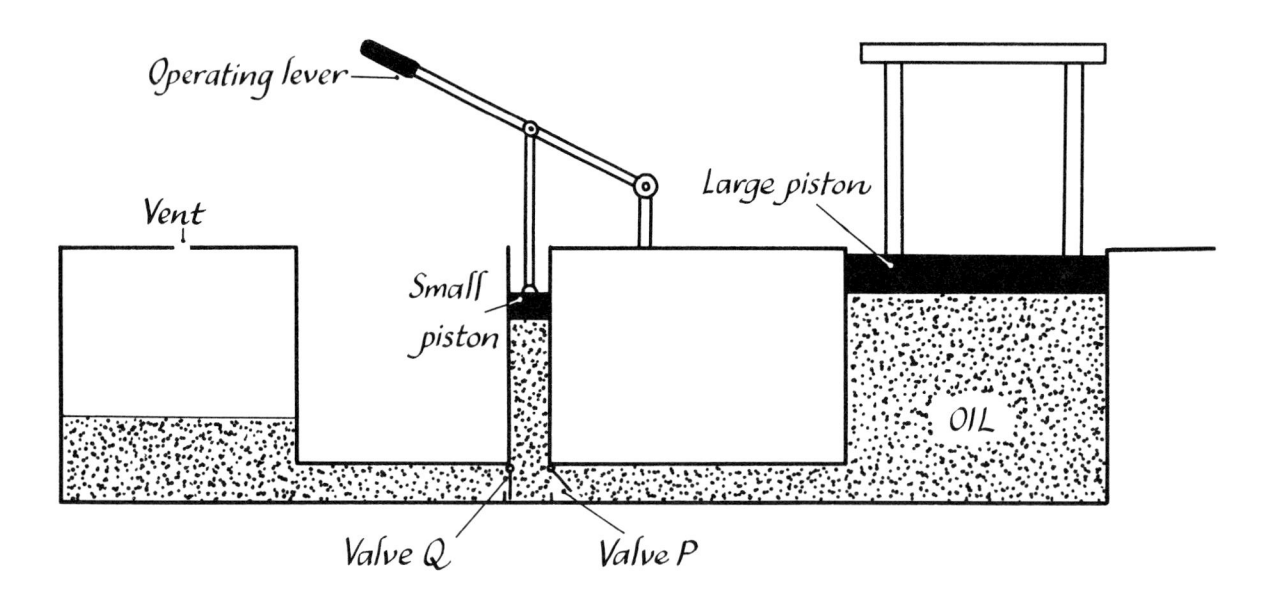

Figure 8.21 The hydraulic jack

the small cylinder. The liquid used is oil, and precautions are taken to ensure that this does not leak through the pistons.

8.12. Friction

The purpose of a machine is to convert mechanical energy into a more convenient form. Some energy may also be converted from one form into another less useful form. For example, when one surface rubs against another, mechanical energy is converted into heat energy. Surfaces in contact with one another tend to resist any attempt to make them slide over one another. This resistance is called **friction**. Friction causes mechanical energy to be converted into heat energy. There is friction between the moving parts of any machine and it is this which absorbs some of the mechanical energy put into the machine, rendering the machine less than 100 per cent efficient.

Investigation 8i. Friction

For this investigation you will need two blocks of wood (approximate dimensions: 5 cm × 10 cm × 15 cm), some string, a pulley, some weights and a sheet of glass.

Set up the apparatus, as shown in Figure 8.22, with the largest surface of one of the blocks in contact with the bench. Hang weights on the end of the string until the block just begins to slide along the bench. Calculate the total force used.

Repeat the investigation with the block resting on one of its smaller surfaces.

Repeat both investigations with the block resting on a sheet of glass placed on the bench.

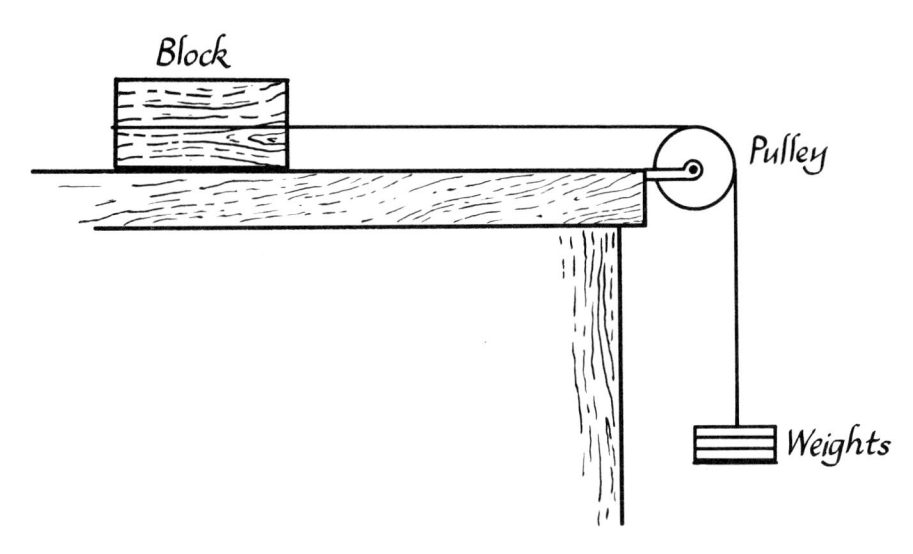

Figure 8.22 Friction

Then repeat the whole series of investigations, this time placing the second wooden block on top of the first.

The friction between two surfaces depends on the force acting between the surfaces and on the nature of the surfaces (whether they are rough or smooth), but it does not depend on the area of surface in contact. Does your investigation confirm this?

Friction is essential between the wheels of a car and the road. If the road becomes icy, the friction is reduced and the car is more liable to skid. Friction between the moving parts of the car engine wastes mechanical energy by converting it into heat and also tends to wear away the moving parts. Lubricating oil is used to reduce friction.

8.13. Momentum

Moving bodies are said to possess **momentum**. The momentum of a body is the product of its mass and its velocity.

Momentum = Mass × Velocity

If a rolling marble strikes a stationary marble, the total momentum possessed by the two marbles after the collision is very nearly equal to their total momentum before the collision. Figure 8.23 illustrates this.

If no mechanical energy was wasted during the collision, the total momentum after the collision would be *exactly* equal to the total momentum before the collision. This is the law of **conservation of momentum**.

If you drop a tennis ball to the ground, it will bounce, but it will

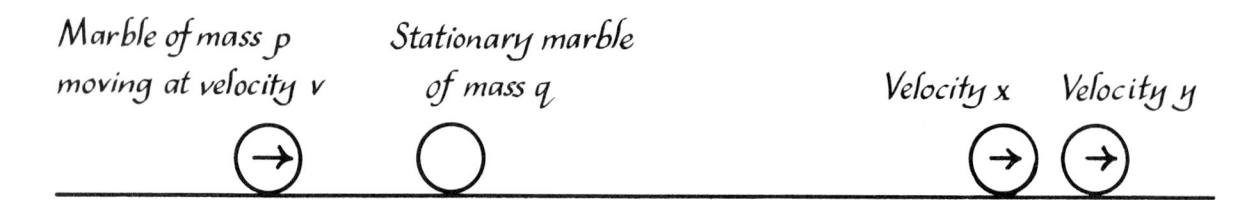

Marble of mass p moving at velocity v Stationary marble of mass q Velocity x Velocity y

Total momentum before impact $= p \times v$ Total momentum after impact $= (p \times x) + (q \times y)$

$$p \times v = (p \times x) + (q \times y)$$

Figure 8.23 Conservation of momentum

not reach the height from which it was dropped. Some energy is used up during the impact with the ground, so that the momentum after the impact is, in this case, less than the momentum before impact (see Figure 8.24). Where do you think the 'missing' energy has gone?

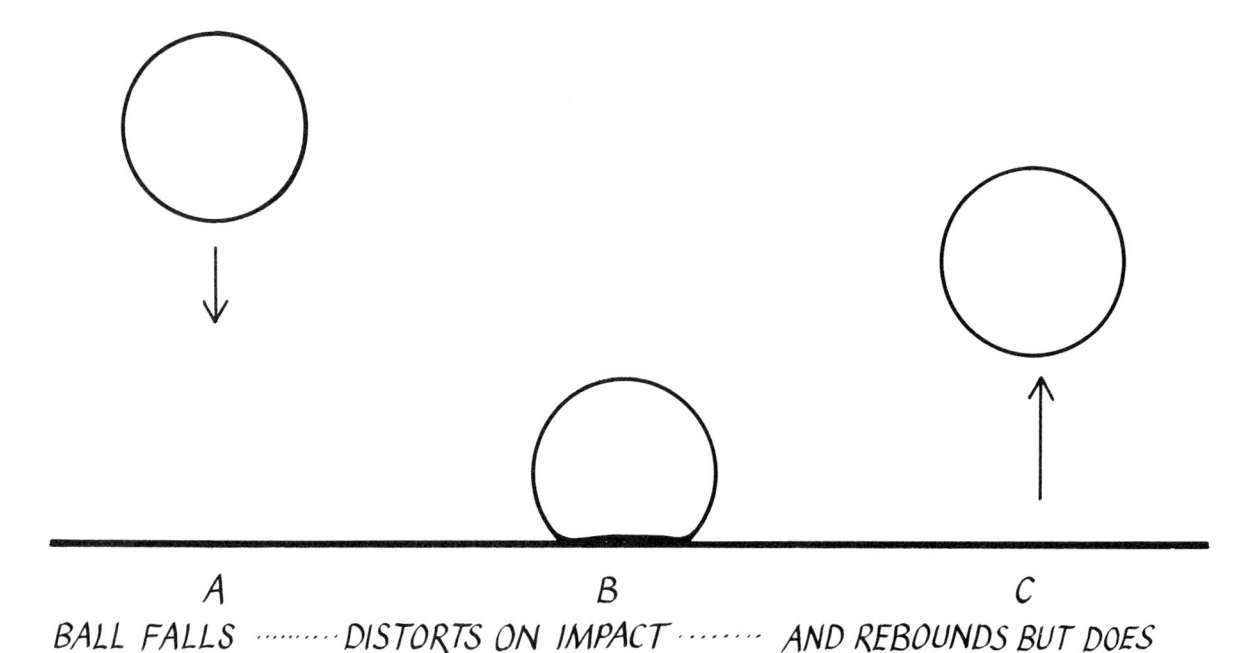

A — BALL FALLS ········ B — DISTORTS ON IMPACT ········ C — AND REBOUNDS BUT DOES NOT REACH FORMER HEIGHT

Figure 8.24 The bouncing ball

8.14. Newton's laws of motion

During the seventeenth century Sir Isaac Newton stated three laws of motion. These explain the behaviour of moving bodies, including the motion of the planets moving in their orbits around the sun.

Law 1. A body will remain stationary, or will continue moving at constant velocity, unless some force is applied to it.

Law 2. The size of the applied force is directly proportional to the rate of change of momentum which it produces.

Law 3. To every action there is an equal and opposite reaction.

The first law deals with the idea of inertia, which we considered in Section 8.2. As stated there, the inertia possessed by a body is directly proportional to its mass.

If a body changes its velocity, the amount of momentum possessed by the body also changes. When we talk of the '*rate of change of momentum*' of such a body, we mean *the mass of the body × the rate of change of velocity*. The rate of change of velocity is acceleration (see Section 8.4), so Newton's second law may be stated as:

Force ∝ Mass × Acceleration

We made use of this relationship in defining the newton.

Figure 8.25 The rocket (an example of Newton's third law)

The third law deals with equal and opposite actions. Let us look at a few examples. If a 1 kg mass of iron rests on a bench, it is exerting a downward force of 10 N. Since the iron does not sink through the bench, it follows that the bench must be exerting an upward force of 10 N on the iron. If you stretch an elastic band by exerting a force of 2 N, the elastic band will exert a force of 2 N in the opposite direction.

Forces always act in pairs, one opposing the other. When a rocket is launched, the chemical energy of the fuel produces mechanical energy in the exhaust gases. The force of the escaping exhaust gases is accompanied by an equal and opposite force, which acts on the rocket. It is this reation which lifts the rocket (see Figure 8.25).

8.15. Gravity

All bodies exert a force of attraction on all other bodies. The size of this attractive force depends on the masses of the bodies concerned and on the square of the distance separating them. Such attractions are called **gravitational attractions**. The gravitational force between the bodies is directly proportional to their masses and inversely proportional to the square of the distance separating the bodies.

$$\textbf{Gravitational force} \propto \frac{M_1 \times M_2}{d^2}$$

where M_1 and M_2 are the masses of the bodies and d is the distance separating the bodies.

An object situated near the earth's surface is attracted by all other objects but, because of the very great mass of the earth, the attraction between the earth and the object is very much larger than other attractions. If the object is free to move, it will fall to the earth's surface as a result of gravitational attraction. The object moves towards the earth, rather than the earth moving towards the object, because the earth, having a very large mass, possesses a very great inertia.

As stated in Section 8.3, the gravitational attraction between any object and the earth exerts a force on the object which we call the *weight* of the object. An object near the surface of the moon would be attracted to the surface of the moon by gravitational attraction but, because the mass of the moon is smaller than the mass of the earth, the force of gravitational attraction would be less. Therefore, the weight of the object on the moon would be only about one-sixth of its weight when situated near the earth's surface.

8.16. Projectiles

Investigation 8j. Falling marbles

Hold a marble in each hand. At the same time, try to drop one while you throw the other horizontally (see Figure 8.26). Listen as they strike the floor. Repeat this several times. What conclusion do you reach about the times taken for the marbles to reach the ground?

When any object is thrown in any direction, other than vertically, the path which it takes is affected by the earth's gravitational attraction. The path taken by a **projectile** (any object thrown in this way)

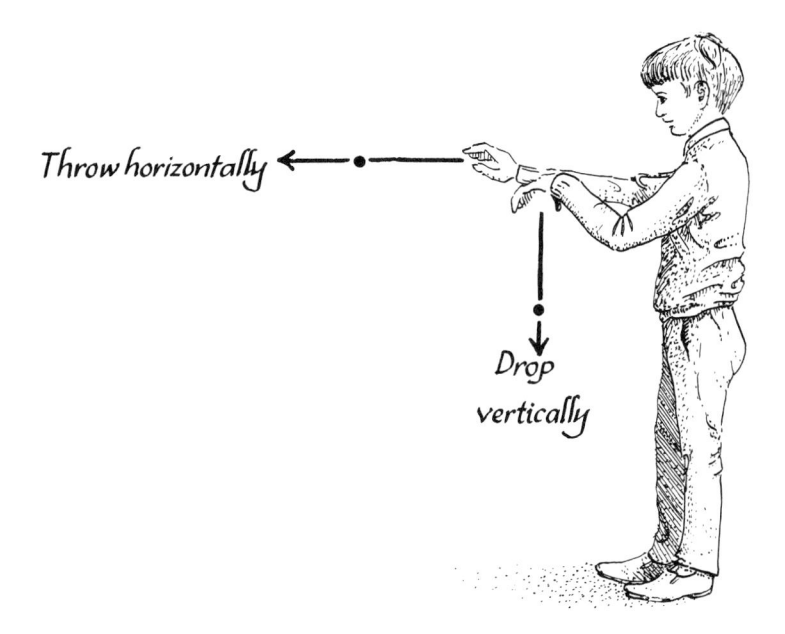

Figure 8.26 Projectiles

is called its **trajectory**. The geometrical name for the shape of a trajectory is a **parabola**. Figure 8.27 shows the trajectories of various projectiles.

If we examine the trajectory of a shell fired horizontally from a gun, we find that the greater the muzzle velocity, the shallower the trajectory will be, and the shell will travel further before striking the ground. Newton suggested that if the forward motion was great enough, the projectile would 'keep falling over the edge of the world' and would never strike the ground. Such a projectile would go into a circular orbit around the earth. The speed required to do this is about

Figure 8.27 Trajectories

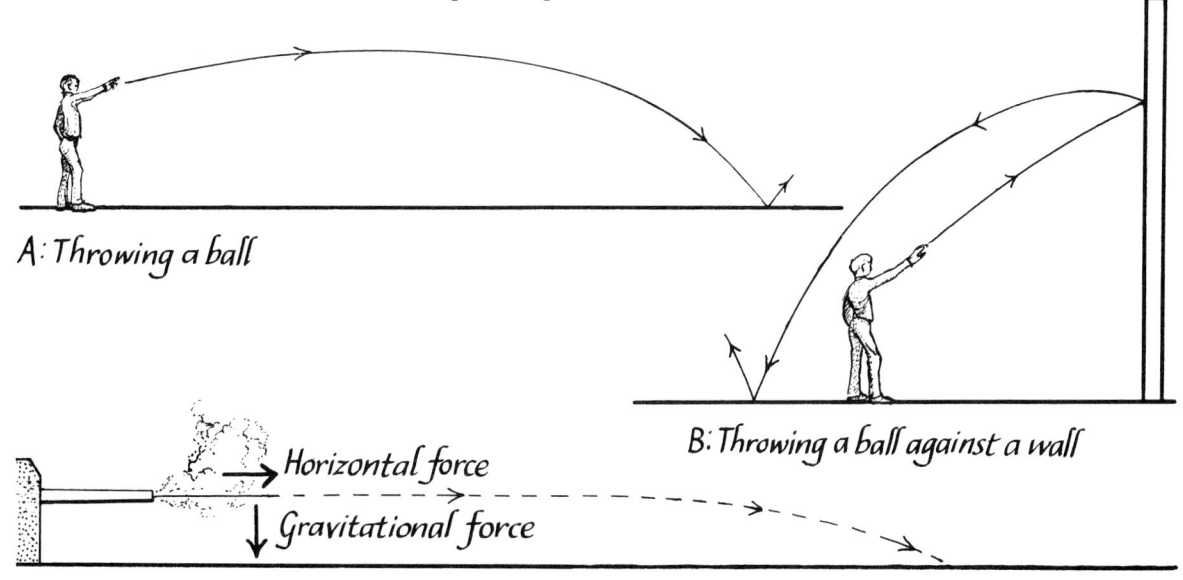

A: Throwing a ball

B: Throwing a ball against a wall

C: Gun firing a shell horizontally

29 000 km h^{-1}. A speed of 40 000 km h^{-1}, or more, would lead to the projectile following a very elliptical orbit. This principle is used in space flight.

Test your understanding

1. A motorist travels a distance of 100 km. If the journey takes two hours, what is his average speed? If, on the return journey, a traffic hold-up delays him, so that his average speed for the first 50 km is 25 km h^{-1}, at what speed must he complete his journey in order to maintain the same average speed as before? (*There is a catch. Explain your answer.*)
2. Complete the statement:
   The force acting on a body due to gravitational attraction is called the ..... of the body.
3. What is the difference between potential energy and kinetic energy? Give three examples of each.
4. How much work is done when a mass of 10 kg is lifted to a height of 10 m? If this is to be done in a time of 5 s, what power must be developed?
5. What is meant by 'the moment of a force'? Explain how you could use a broom handle and a 1 kg bag of sugar to find the mass of a brick. (You may assume that string and a ruler are available.)
6. Make a list of as many different levers as you can see in the room in which you are sitting.
7. Sketch a gear system having a velocity ratio of 4:1. Indicate which is the driving wheel and which is the driven wheel, and state the number of teeth on each. What would be the effect of placing an 'idler' between these two wheels?
8. Complete the statement:
   Friction converts ..... energy into ..... energy. How may the friction between two surfaces be reduced? Give two examples where friction is an advantage and two examples where it is a disadvantage.
9. Rocket power is the only method of travelling in space. Why is this so?

Chapter 9

Engines

An engine is a device for converting one form of energy into another. Mechanical energy is usually required from the engine, and some other form of energy is usually supplied to the engine.

9.1. Hero's engine

One of the earliest engines was invented by Hero, who lived in Egypt during the second century A.D. Hero made use of the fact that when water is converted into steam the steam occupies a much greater volume than the water from which it was made. The steam was produced in a container, from which it was allowed to escape through small tubes positioned so that the escaping steam caused the container to rotate. This engine, invented nearly two thousand years ago, was jet propelled. Hero called it the 'Ball of the Winds'.

Investigation 9a. Jet propulsion

Blow up a balloon and release it. The motion of the balloon is caused by the escaping air, but the balloon moves in the opposite direction to the moving air. This is an example of Newton's third

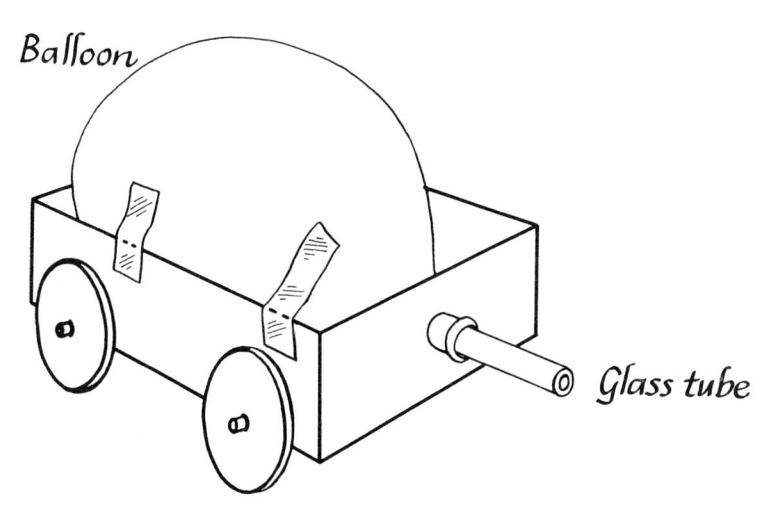

Figure 9.1 A jet-propelled car model

law of motion (see Section 8.14), in which the force exerted by the escaping air causes a reaction in the opposite direction. It is this reaction which moves the balloon.

A model jet-propelled car may be made by fitting a small cardboard box with wheels and making the propulsion unit from a balloon fitted with a short length of tubing. Pieces of 'Sellotape' should be used to keep the balloon in place. Figure 9.1 shows the general arrangement.

Investigation 9b. A model of Hero's engine

Obtain a tin can and make holes on opposite sides of it, just large enough to take single-holed rubber bungs. Fit short lengths of glass tubing into the bungs. The tubes should have right-angled bends in

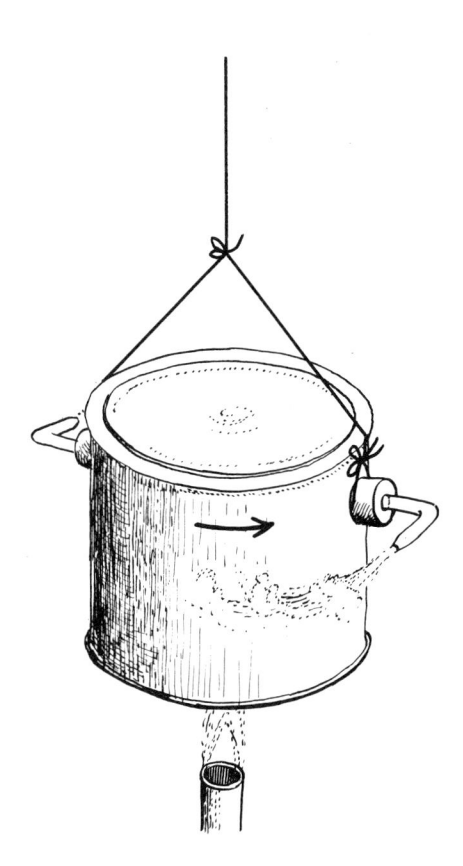

Figure 9.2 A model of Hero's engine

them and should be drawn to a jet at one end. Fit the bungs so that the jets point in opposite directions. Suspend the tin from a retort stand by means of a cord. Pour water into the tin until it is about a quarter full, and push the tin lid into place. Heat the bottom of the tin, using a small bunsen flame (see Figure 9.2).

9.2. Windmills and water-wheels

Both windmills and water-wheels are simple contrivances for harnessing available mechanical energy to drive machines. Both were extensively used in the past for grinding grain to produce flour. Very few are still in use, but there are still many disused windmills and water-wheels to be seen in some parts of the country.

Investigation 9c. A model windmill

Cut a piece of paper to form a square with sides 20 cm long. Mark the diagonals in pencil and cut down from each corner for a distance of 9 cm. Bend each alternate free corner so that they overlap at the centre of the paper. Pass a pin through the centre, as shown in Figure 9.3. Hold the pin between the fingers while you walk. Notice the direction of rotation of the 'windmill'.

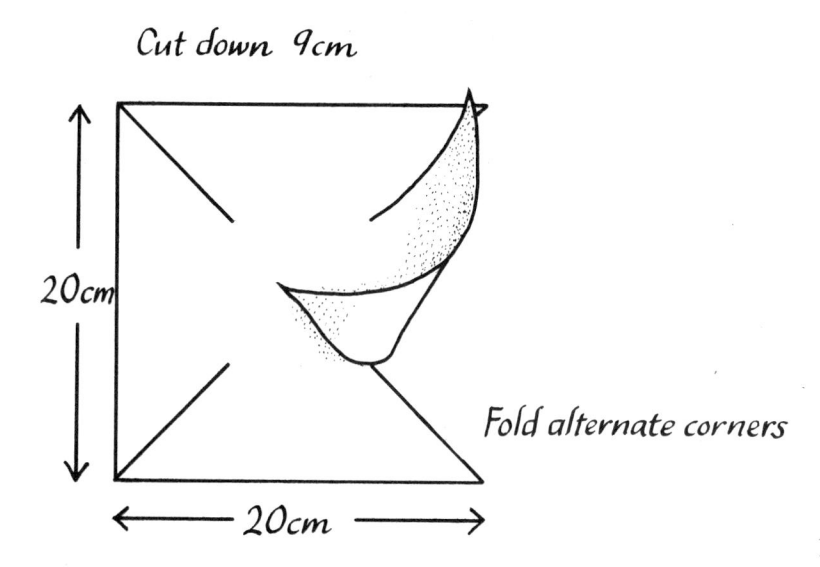

Figure 9.3 Making a model windmill

Remove the pin and gum the overlapping corners to the centre of the paper. When the gum is dry pass a long pin through the windmill and into the centre of a piece of light wood (balsa wood) of about 3 cm square section and about 20 cm long (see Figure 9.4). Lift the wood as high as you can and allow it to fall to the ground.

The helicopter makes use of the windmill principle.

Investigation 9d. A model water-wheel

Obtain a cotton reel and cut eight slots in it, as shown in Figure 9.5. Fit strips of metal sheeting into the slots, and pass a short length of dowel rod through the hole in the reel to act as an axle. Arrange

Figure 9.4 Controlled descent

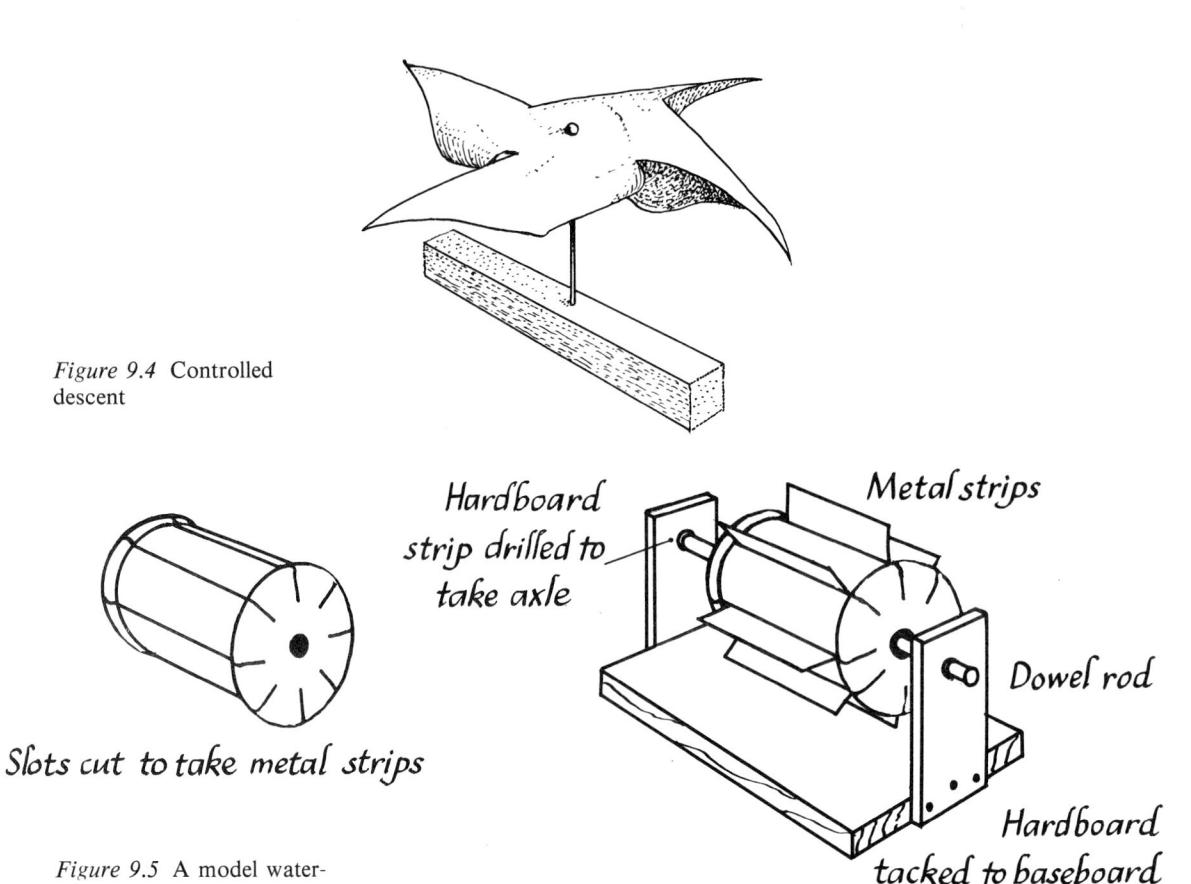

Figure 9.5 A model water-wheel

Slots cut to take metal strips

Hardboard strip drilled to take axle

Metal strips

Dowel rod

Hardboard tacked to baseboard

this on a baseboard, place the model in a sink and pass a jet of water on to one side of the wheel.

9.3. The steam engine

When water is heated until it becomes steam, the volume of the steam is much greater than that of the original water.

Investigation 9e. Steam pressure

★ WARNING. *This should be done as a demonstration by the teacher.*

A boiling can is about a third filled with water, and a cork is placed in the can. The can is then placed on a tripod and supported by a retort stand. The can is then heated (see Figure 9.6).

During this investigation you should keep well away from the apparatus and there should be no obstruction near the cork.

The property demonstrated in Investigation 9e is used in the steam engine. One common type of steam engine uses steam pressure to drive a piston back and forth in a cylinder. Steam is passed through a valve into the cylinder to force the piston down. The valve then closes and another opens to admit steam to the other side of the piston, thus driving it back (see Figure 9.7).

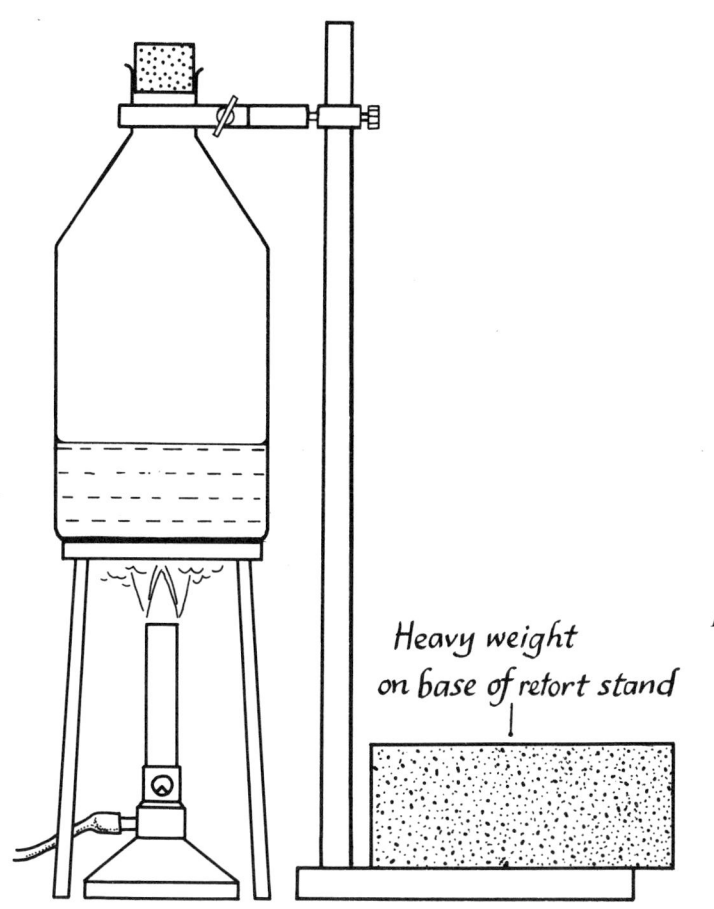

Figure 9.6 Steam pressure

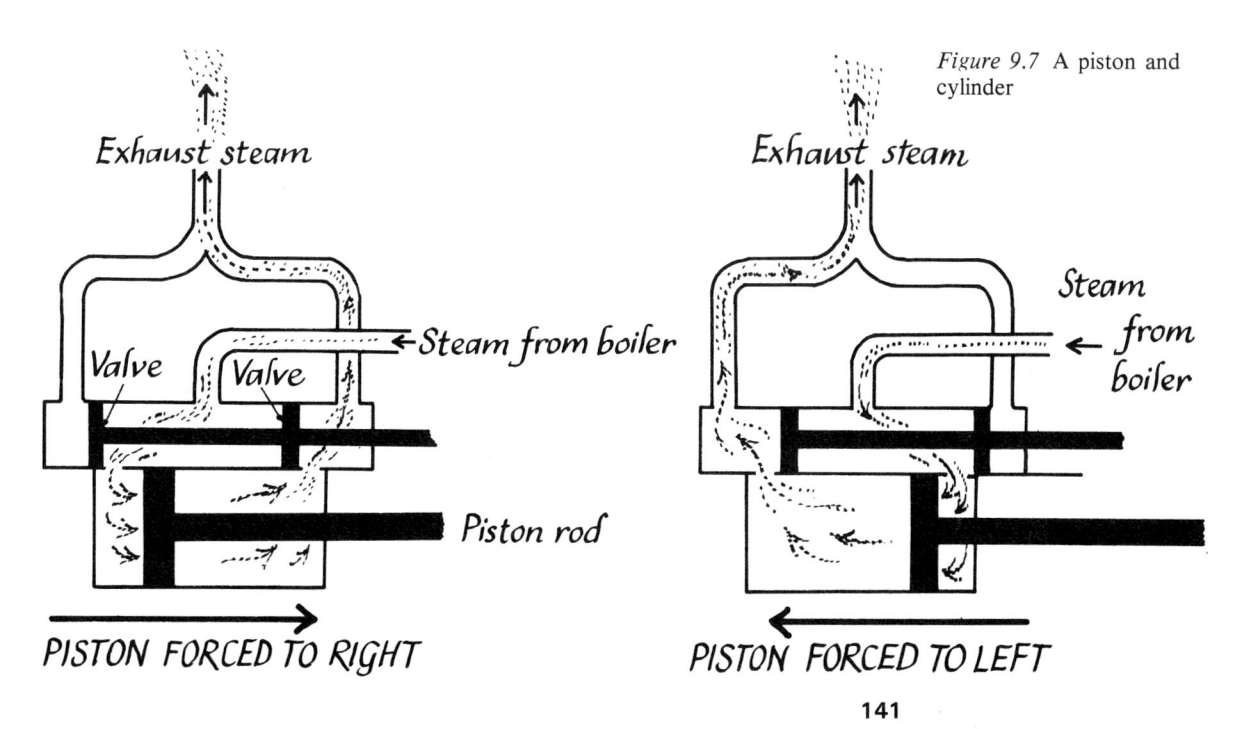

Heavy weight on base of retort stand

Figure 9.7 A piston and cylinder

Exhaust steam

Valve Valve ← Steam from boiler

Piston rod

PISTON FORCED TO RIGHT

Exhaust steam

← Steam from boiler

PISTON FORCED TO LEFT

The to-and-fro motion so produced is called **reciprocating** motion, and may be converted into rotary motion by means of an **eccentric** (see Figure 9.8).

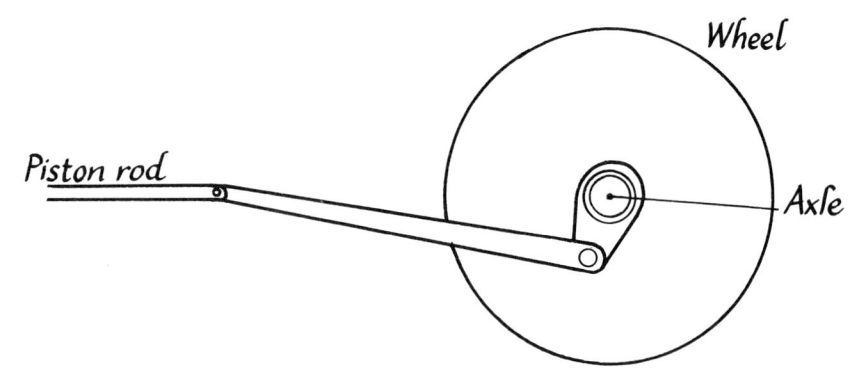

Figure 9.8 The eccentric

9.4. The steam turbine

In the steam turbine (Figure 9.9) steam is made to turn wheels directly, in much the same way as water is used to drive a waterwheel. The motion obtained from a turbine is rotary motion.

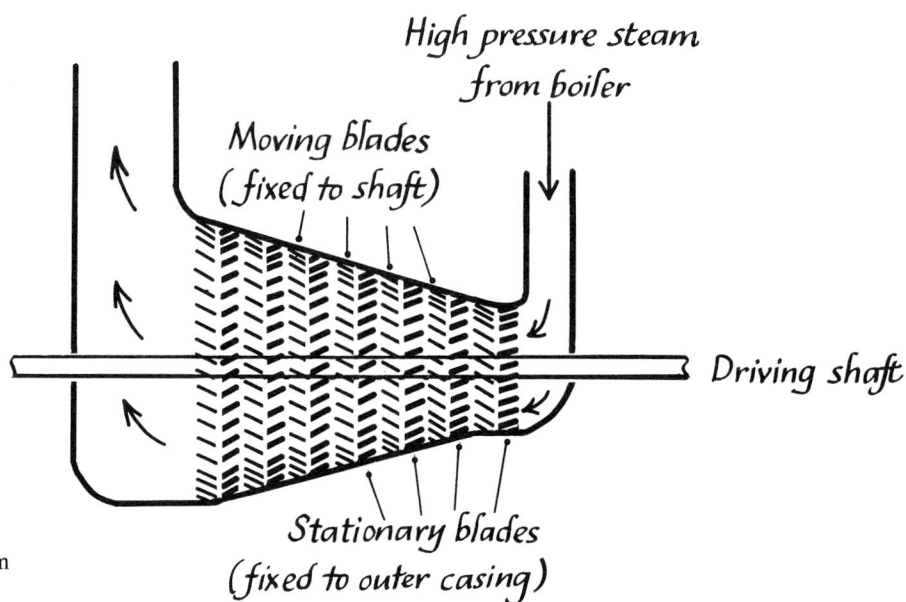

Figure 9.9 The steam turbine

9.5. The internal combustion engine

In the steam engine heat is provided from outside the engine in order to provide steam to drive a piston down a cylinder. In the internal combustion engine fuel is burned inside the cylinder to produce a large volume of gas which drives the piston down the cylinder.

One of the commonest forms of internal combustion engine is the

four-stroke petrol engine. We shall deal with this at some length in Section 9.7.

A mixture of petrol vapour and air is fed into the cylinder, where it is exploded by means of an electric spark. This produces a quantity of hot gas, which drives the piston down the cylinder, thus providing mechanical energy. Most of the energy produced is in the form of heat and much of this is wasted. After the expanding gases have been used to drive the piston down they are removed as exhaust gases.

Petrol is a hydrocarbon, a compound consisting of the elements hydrogen and carbon. When a hydrocarbon is burned, oxides of carbon and hydrogen are produced. Thus, the gases present in the exhaust are carbon dioxide, carbon monoxide and water vapour. It should be noted that carbon monoxide is extremely poisonous and, for this reason, it is most dangerous to run an internal combustion engine in an enclosed space, unless special arrangements are made to carry out the exhaust gases. (Compare this with Section 6.1.)

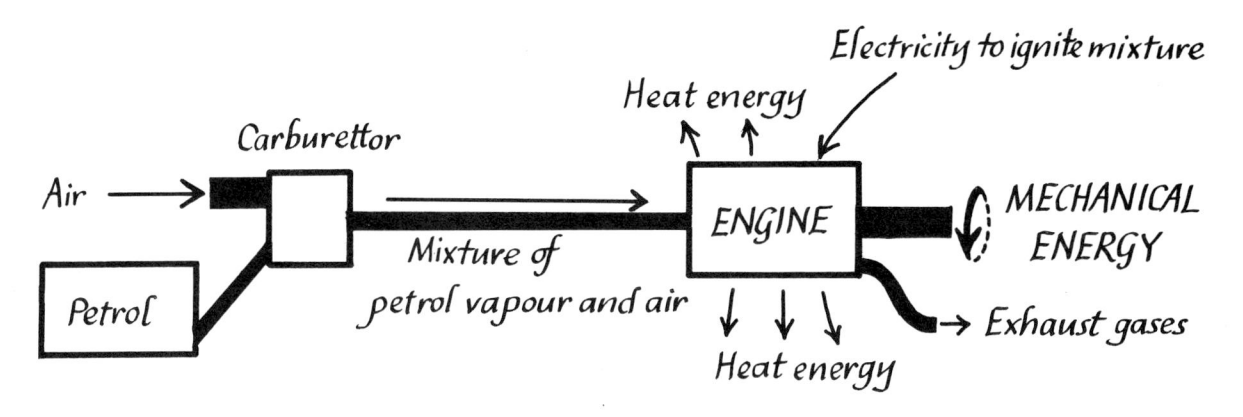

Figure 9.10 represents the energy conversions which take place.

Figure 9.10 Energy conversions in the petrol engine

9.6. The carburettor

The fuel which is supplied to the cylinders is petrol vapour, which must be mixed with air. The carburettor is a device which produces this mixture. Figure 9.11 shows a simple form of carburettor.

Petrol is pumped into the float chamber from the petrol tank. A valve, controlled by the float, ensures that a constant level is maintained in the float chamber. Petrol is fed from the float chamber through a jet. When the engine is started, the piston moves down the cylinder, causing a partial vacuum. Air is forced, by atmospheric pressure, past the jet and the petrol is vaporized. The throttle is a shutter controlling the supply of fuel to the engine, thus affecting the speed. Most modern carburettors are more complex than the one illustrated, but they work on the same principle.

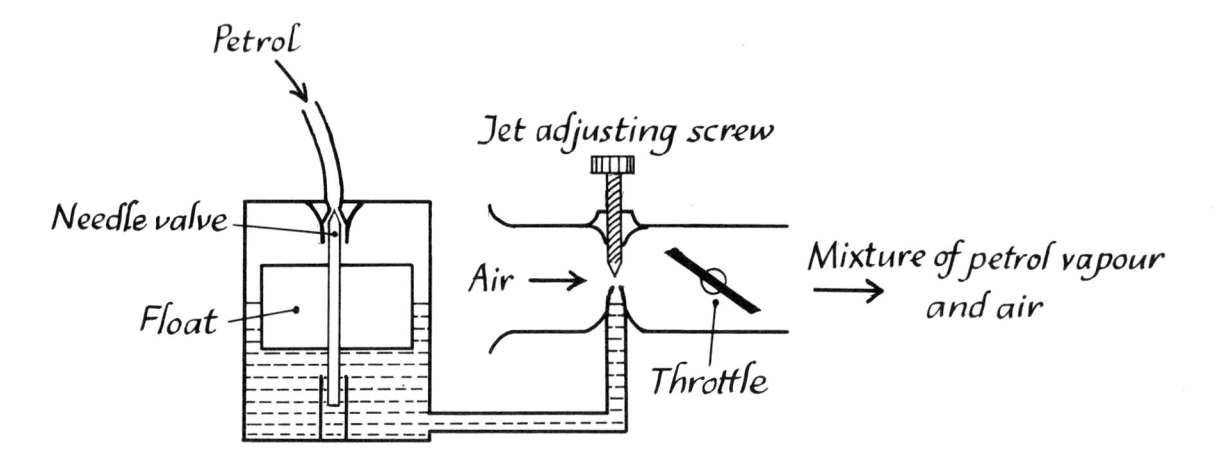

Figure 9.11 A simple carburettor

9.7. The four-stroke cycle of a petrol engine

Most car engines have four or six cylinders. Each of these is fitted with two valves, one to allow the petrol vapour and air to enter the cylinder (the inlet valve), and the other to allow the exhaust gases, which are produced when this mixture is burned, to escape (the exhaust valve).

A piston is fitted into the cylinder, springy metal rings ensuring a close fit. A connecting rod joins the piston to the crankshaft, so that as the piston moves up and down, the crankshaft is made to rotate (see Figure 9.12).

The four-stroke cycle is as follows:

a. When the engine is started the piston moves downwards, the

Figure 9.12 A cylinder and crankshaft

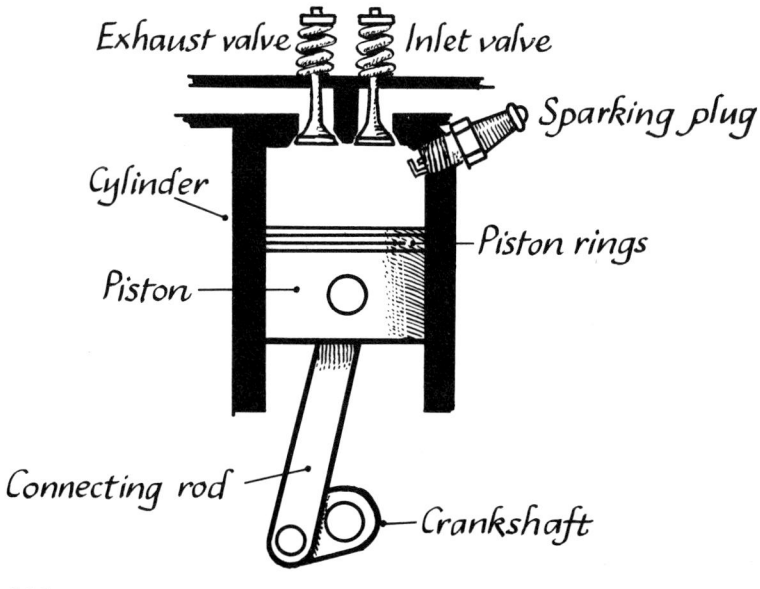

Figure 9.13 The four-stroke cycle

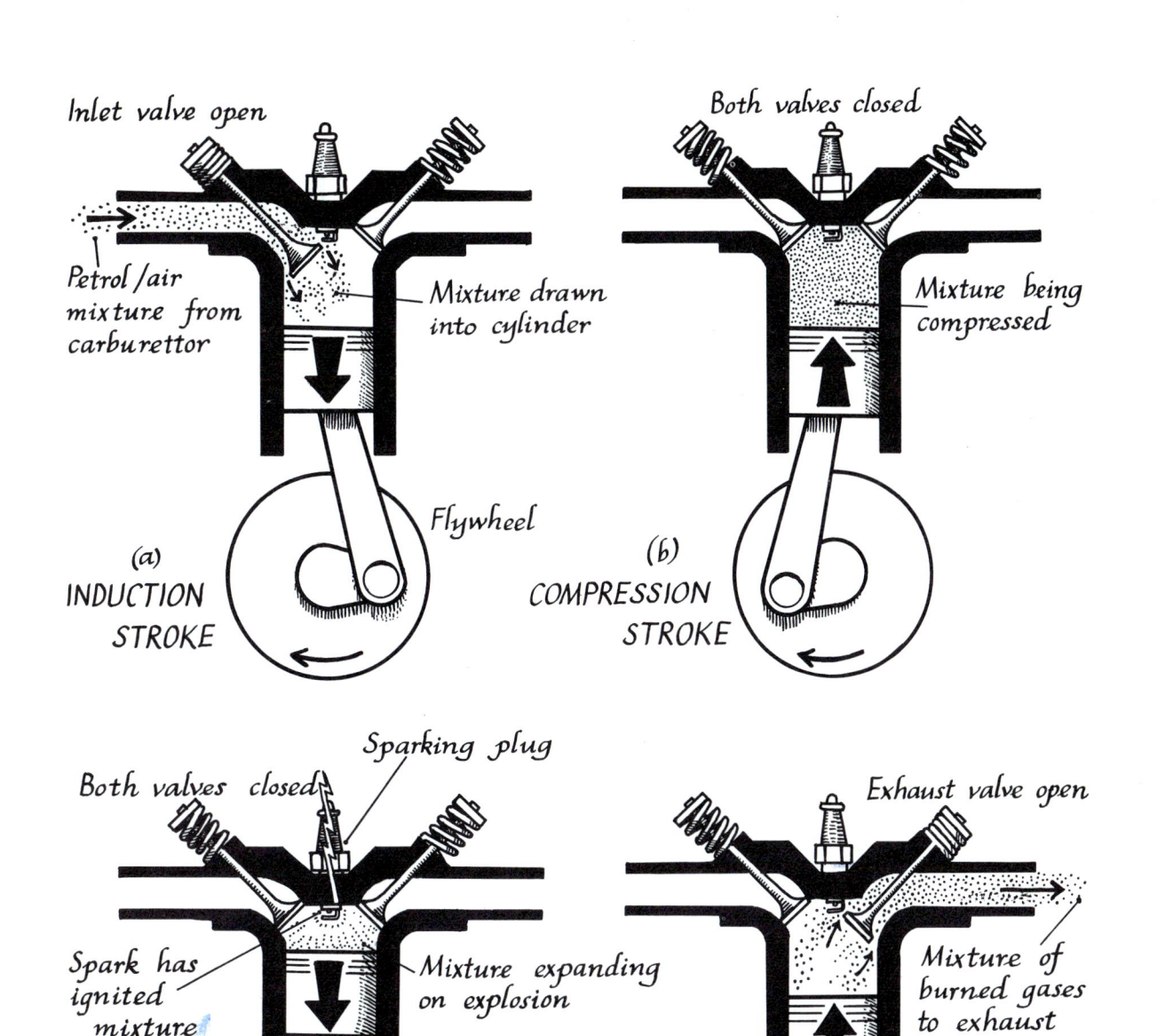

inlet valve opens and fuel and air are drawn into the cylinder. This is the **induction** stroke.

b. The piston then moves upwards while both valves are closed, so that the fuel and air mixture is compressed. This is the **compression** stroke.

c. The gases are then ignited by means of an electric spark and an explosion results, which drives the piston down the cylinder. This is the **power** stroke.

d. The piston then returns up the cylinder while the exhaust valve is open, driving the exhaust gases out of the cylinder. This is the **exhaust** stroke.

Figure 9.13 shows the four strokes.

The whole sequence is then repeated.

The power stroke is the one which drives the engine. The smooth running of the engine is assisted by the attachment of a heavy flywheel on the crankshaft to keep it turning smoothly between power strokes. In engines having more than one cylinder the power strokes of the cylinders follow one another; this also greatly assists the smoothness of running.

9.8. Cooling the engine (See also Section 7.5)

Over 70 per cent of the energy obtained when the fuel is exploded is in the form of heat. This is unwanted heat and must be removed if the engine is not to be damaged by 'seizing up'. Some engines are air-cooled. These have metal fins attached to the cylinder block, thus increasing the surface area considerably. As air passes over these fins it becomes warm, and so heat is conducted away from the cylinder block. This system is commonly used in motor-cycle engines.

In a water-cooled engine, water is circulated in a jacket surrounding the cylinder block by means of a pump. The water conducts heat away from the cylinder block and is then made to pass through a radiator consisting of a large number of fine tubes between which air can pass. As air passes between the radiator tubes the air becomes warmer and conducts heat away from the water. A fan is fitted behind the radiator to help draw air between the tubes.

Figure 9.14 shows typical cooling arrangements.

9.9. Producing the spark

In order to produce a spark to ignite the fuel, a high voltage supply of electricity is needed. The car has a battery and a dynamo (or alternator) to charge the battery; but these are capable of supplying only six or twelve volts, and a much higher voltage is needed to produce the spark.

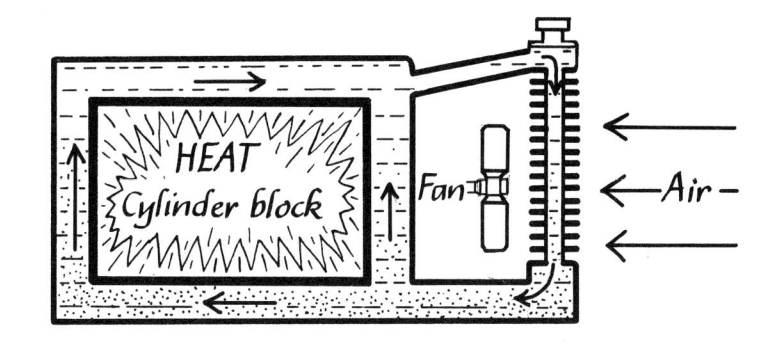

AIR COOLING   WATER COOLING

The necessary voltage is obtained by using an **ignition coil**. This is really a kind of induction coil (see Chapter 17), in which a low voltage is passed through a coil of wire having a fairly small number of turns. This produces a strong magnetic field, which cuts across a coil having a very large number of turns, and momentarily produces a very high voltage in that coil. The **distributor** is a mechanically operated device which connects each sparking plug in turn to the ignition coil as the piston approaches the top of the cylinder on a compression stroke.

Figure 9.15 shows the ignition coil and distributor head.

Figure 9.14 Air cooling and water cooling

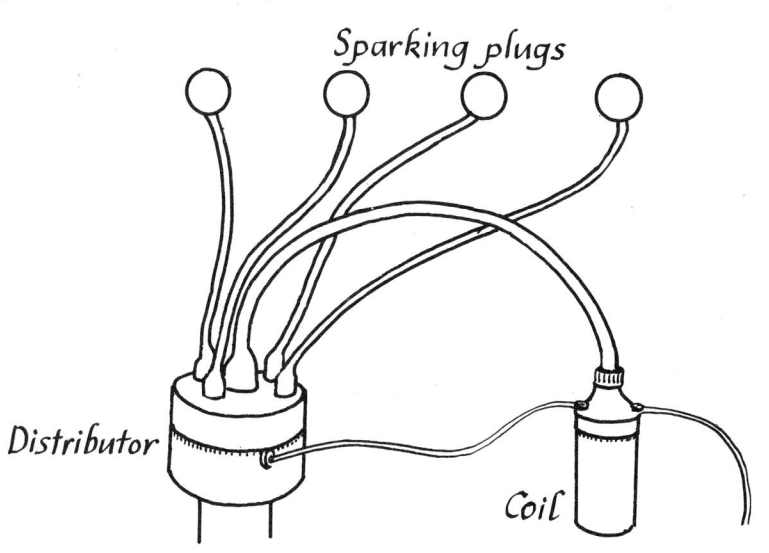

Figure 9.15 The ignition coil and distributor head

9.10. Operating the valves

The inlet and exhaust valves are operated mechanically by means of a system of camshafts and cams (see Figure 9.16).

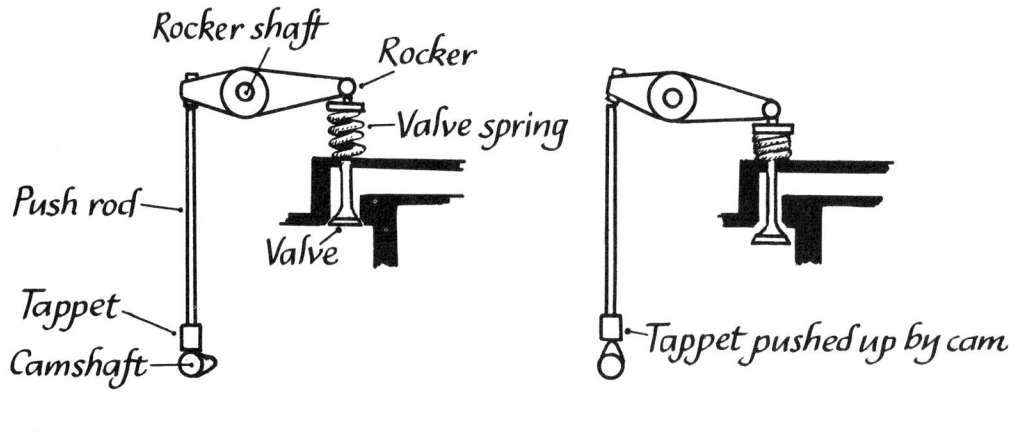

Figure 9.16 Operating the valves

VALVE SHUT **VALVE OPEN**

9.11. Starting the engine

The internal combustion engine is not self-starting. Before it can start, the engine must be turned so that the induction stroke can draw the petrol vapour and air mixture into the cylinders. This initial turning may be done with a starting handle, 'kick-start' or by means of an electric motor (the so-called self-starter).

9.12. Other types of petrol engine

a. The two-stroke engine

Many small engines are of this type. Usually, valves are not employed, the flow of gases to and from the cylinder being through **ports** which are opened and closed by the movement of the piston.

Figure 9.17 (a) shows a cylinder with a mixture of petrol vapour and air compressed at the top of the cylinder. At this stage, the inlet port is open and a fresh mixture enters the lower part of the cylinder.

When the piston is forced down the cylinder, the exhaust gases escape through the exhaust port, while the mixture enters the upper part of the cylinder through the transfer port (Figure 9.17 b and c).

The complete cycle takes only two strokes.

b. The Wankel engine

This engine employs a specially shaped piston which moves with a rotary motion, controlling ports for the inlet of fuel and the removal of exhaust as it does so. Figure 9.18 shows the general arrangement.

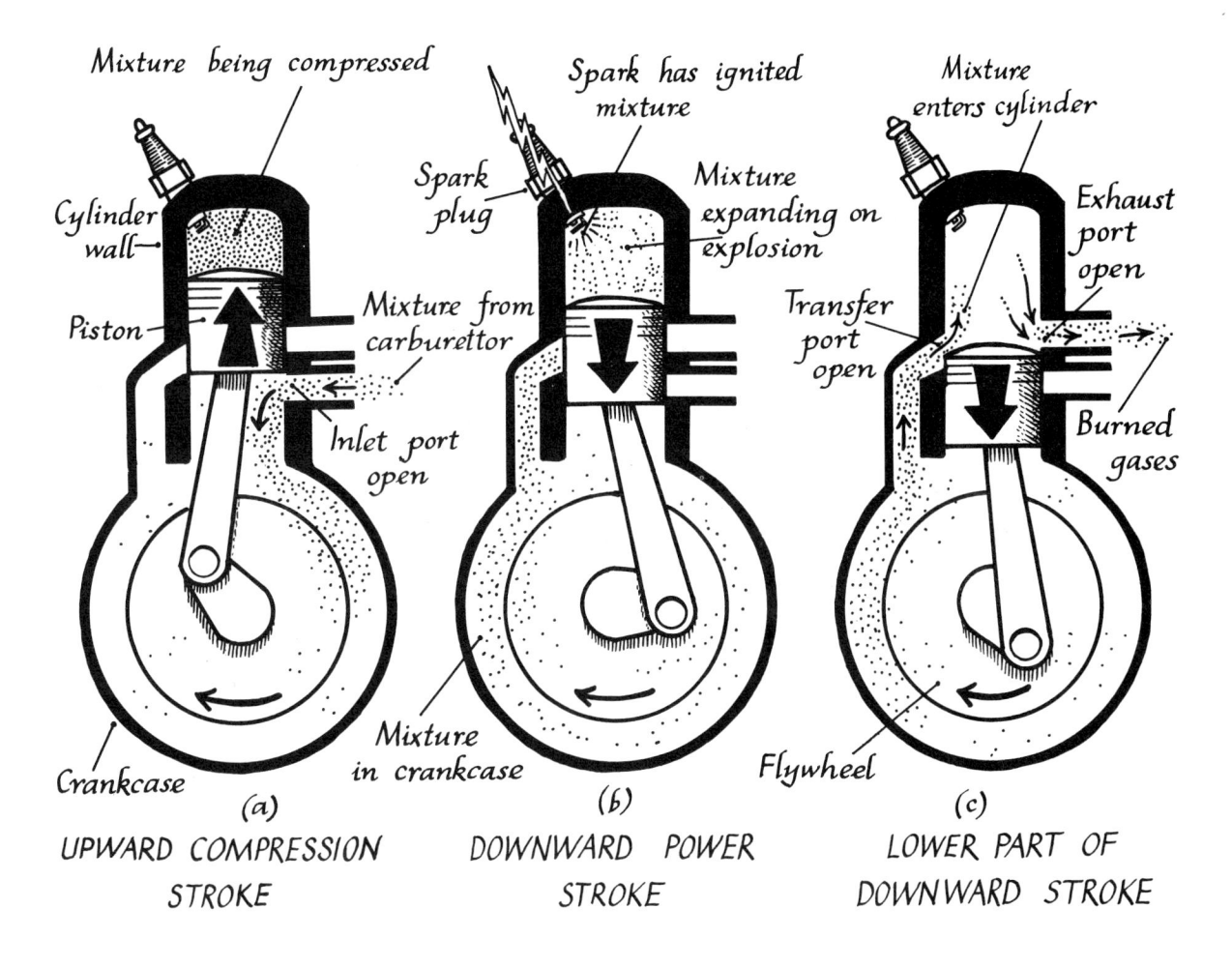

Figure 9.17 The two-stroke cycle

(a) UPWARD COMPRESSION STROKE

(b) DOWNWARD POWER STROKE

(c) LOWER PART OF DOWNWARD STROKE

9.13. The diesel engine

This type of engine uses a heavier oil as fuel, and has neither a carburettor nor an electrical ignition system. Air is drawn into the cylinder and is greatly compressed, so that it reaches a very high temperature. A fine spray of oil is then injected into the cylinder. When the oil meets the hot, compressed air it ignites violently, driving the piston down the cylinder (Figure 9.19).

9.14. Gas turbine and jet propulsion engines

These are internal combustion engines which do not require pistons or other reciprocating parts. Air is pumped into the engine and compressed. A spray of fuel oil is then injected into the air and the mixture is ignited electrically.

In the case of the jet engine (see Figure 9.20), the escaping exhaust gases provide the jet, but in the gas turbine engine (see Figure 9.21), these gases are made to drive a turbine before escaping.

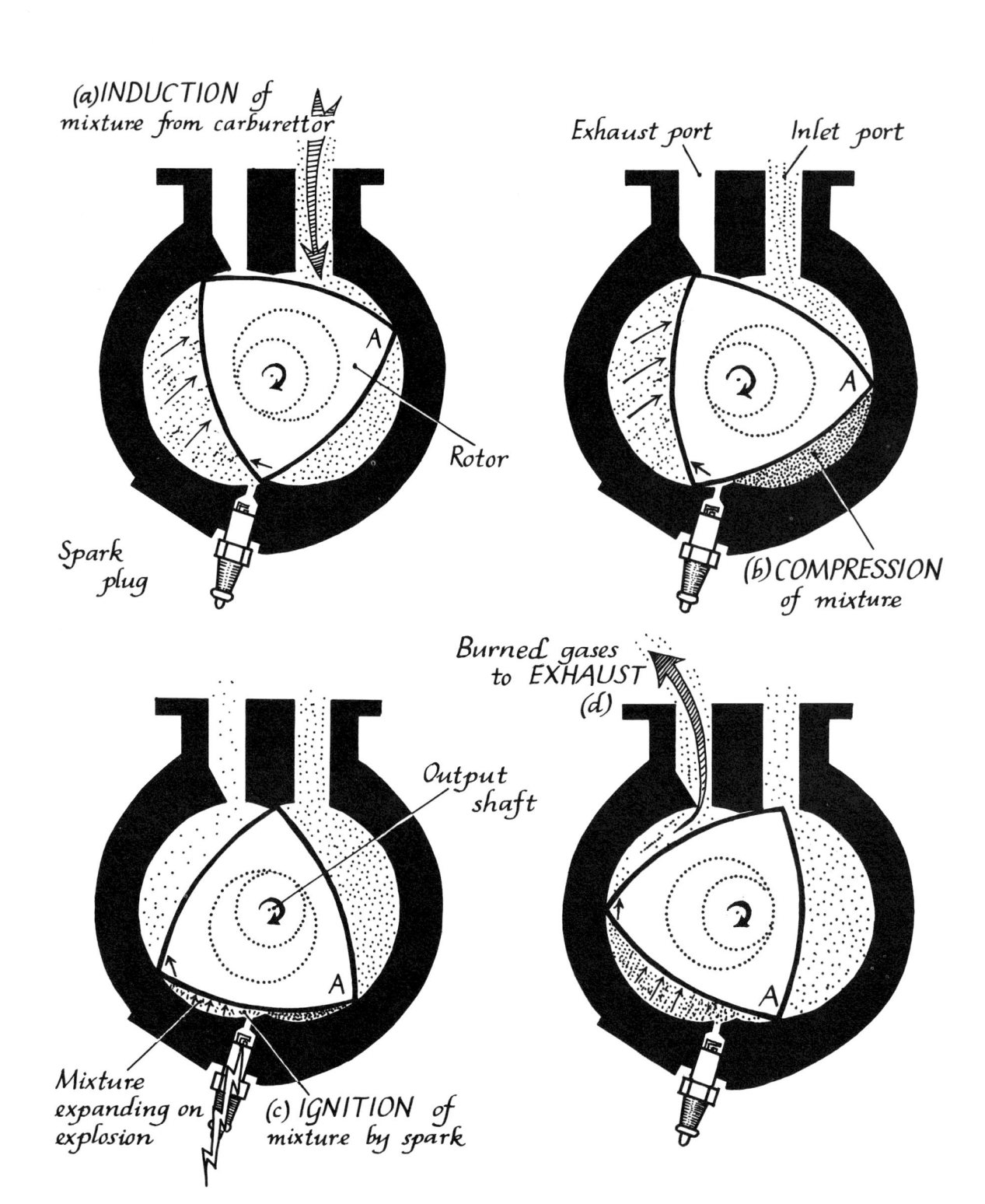

Figure 9.18 The Wankel engine

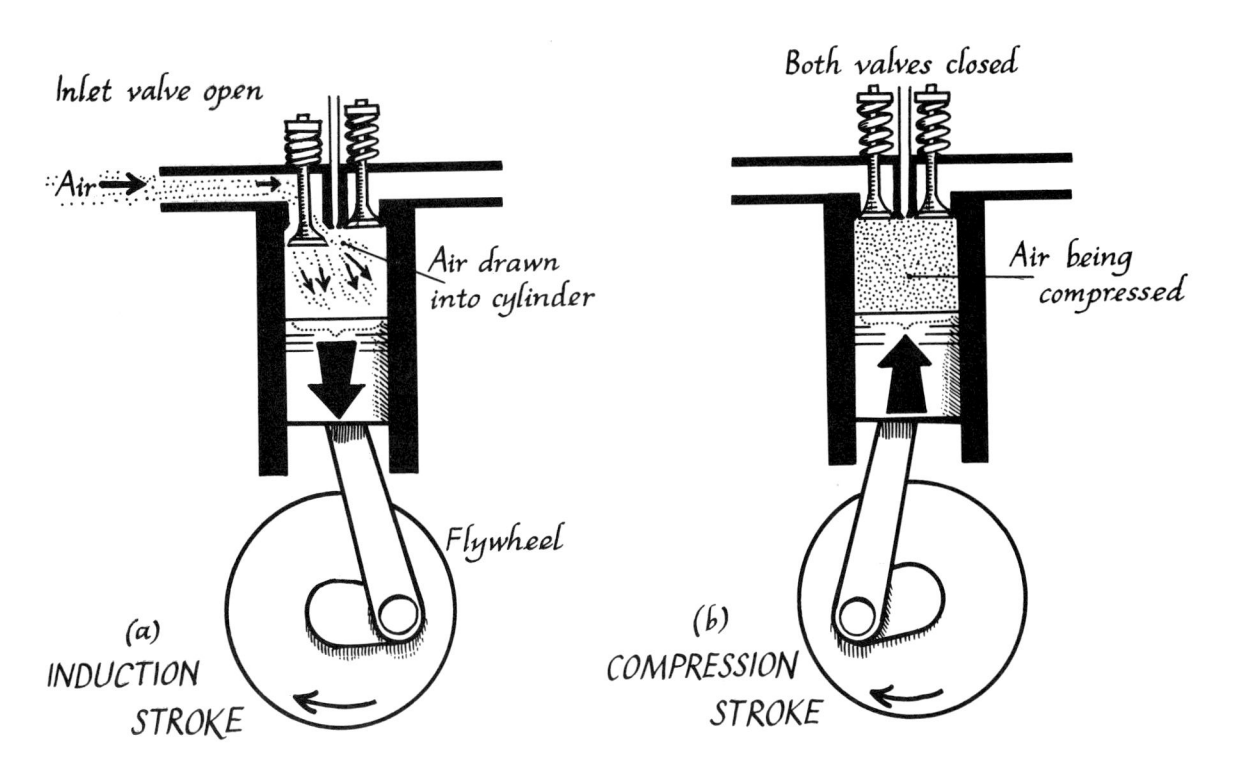

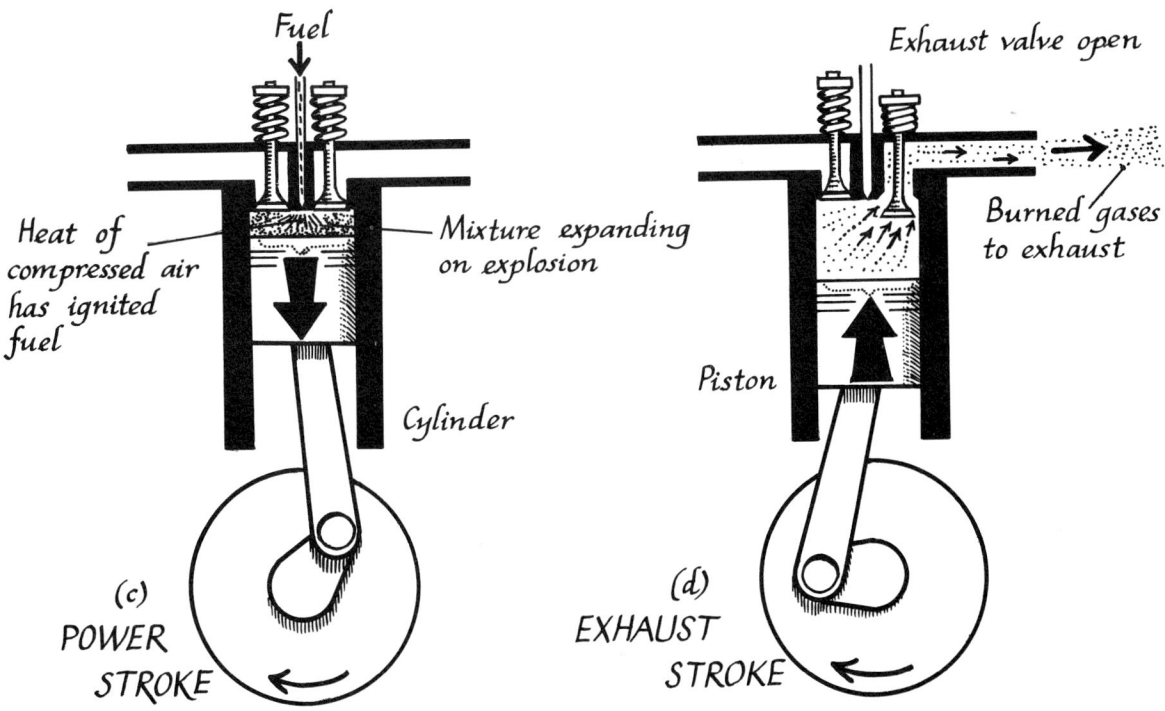

Figure 9.19 The four-stroke cycle of a diesel engine

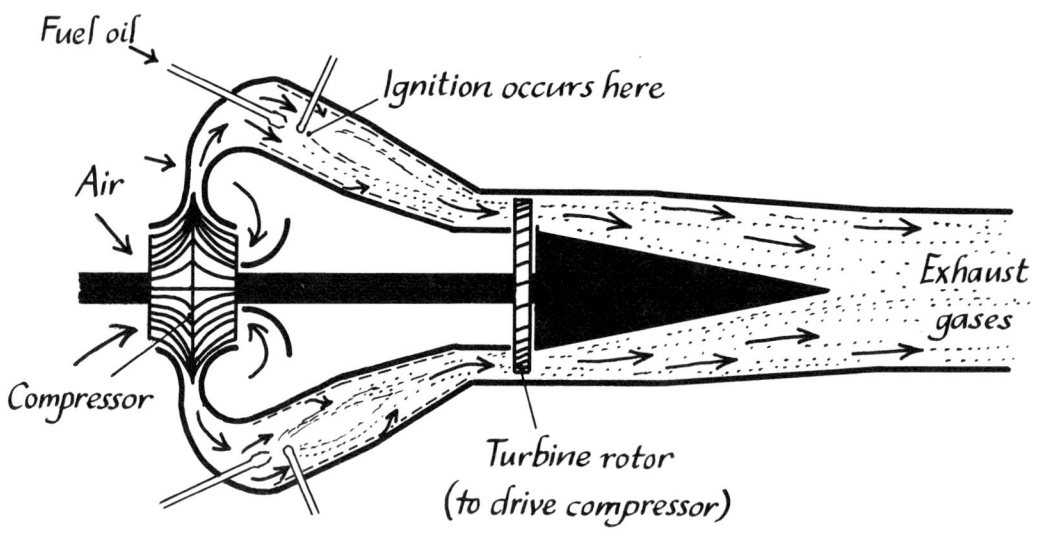

Figure 9.20 A jet engine

In both cases, the compressor, which draws in the air and compresses it, is kept in operation by the expanding gases.

9.15. Rocket propulsion

The rockets used to put artificial satellites into orbit around the earth, and to travel to the moon and the planets, carry their own fuel, and the oxygen which is necessary to burn the fuel. The fuel and the oxygen are mixed and ignited in a combustion chamber, and the

Figure 9.21 The gas turbine engine

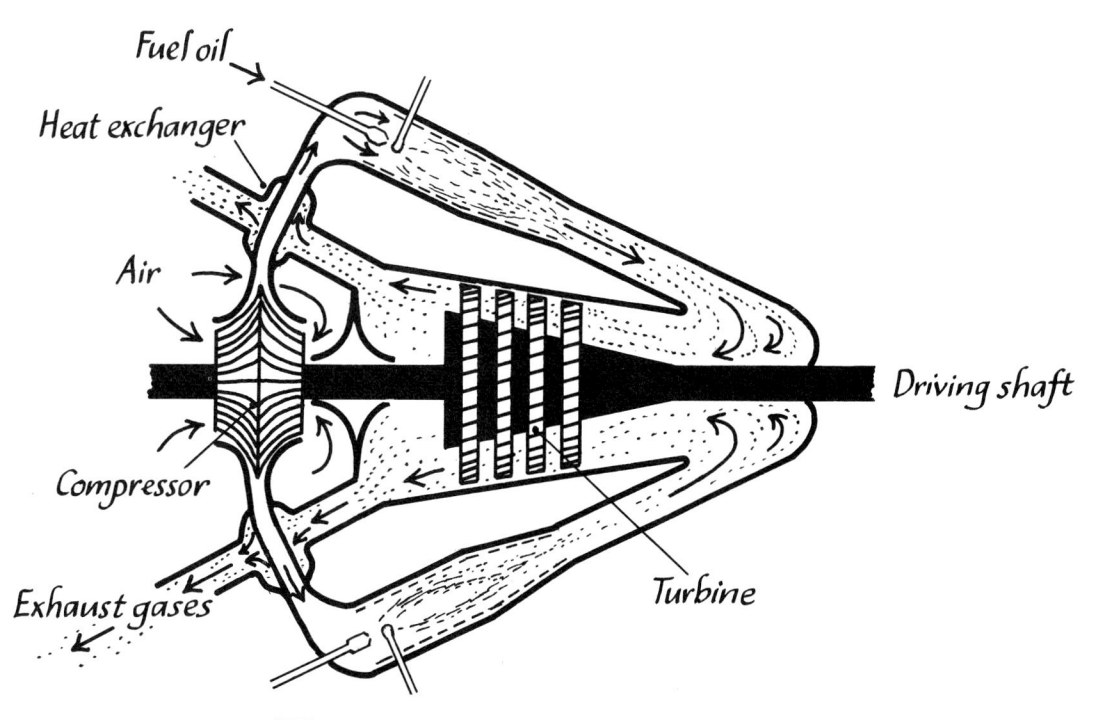

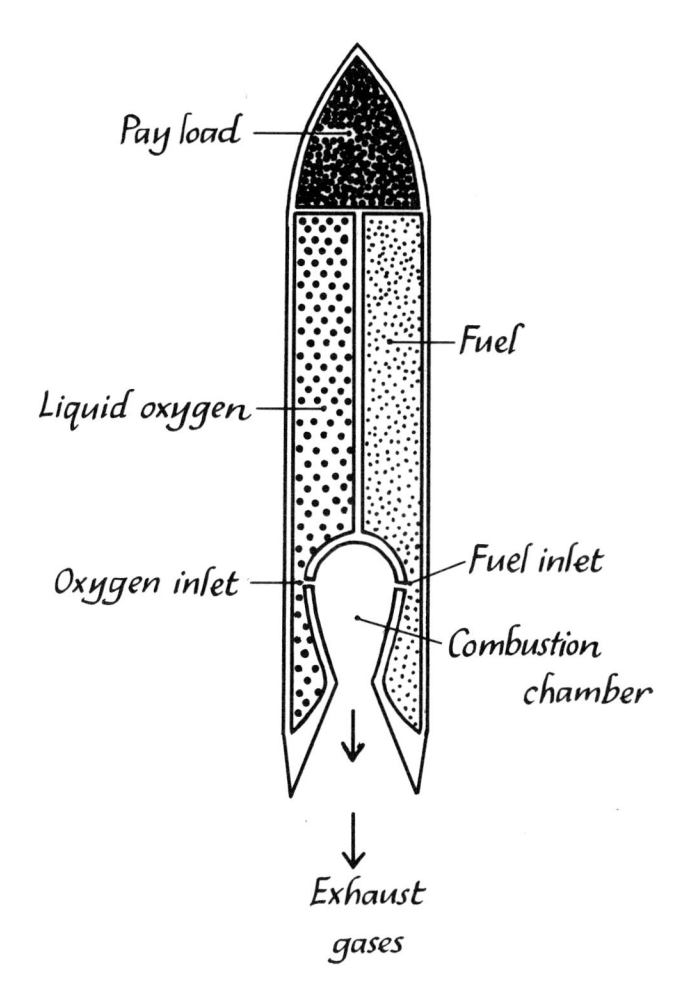

Figure 9.22 A single-stage rocket

exhaust gases provide the jet to move the rocket. Rockets depend on Newton's law of action and reaction (see Section 8.14).

Most of the space in the rocket is taken up by the fuel, the space available for carrying men and materials (the pay load) being relatively small (see Figure 9.22).

Rockets are often built in stages, each stage being a separate rocket engine. When the fuel in one stage has been used, that stage is released and the next is fired. By this means, unwanted mass is discarded.

Test your understanding

1. A steam engine makes use of the fact that steam occupies a much greater ..... than the water from which it was made.
2. The to-and-fro motion of a piston is called ..... motion.
3. While a piston engine produces ..... motion, a turbine produces ..... motion.

4. In a four-stroke engine the strokes are called:
   (a) ....., (b) ....., (c) ....., (d) ......
   Describe briefly what occurs at each stroke.
5. The device used to mix petrol vapour and air is called a ......
6. The gases produced when petrol is burned are:
   (a) ....., (b) ....., (c) ......
   The gas ..... is very poisonous, and for this reason it is extremely dangerous to run a car engine in a closed garage.
7. Describe briefly, with the aid of diagrams, (a) a gas turbine engine, (b) a jet engine, (c) a modern rocket.
8. Describe a piston and cylinder suitable for use in a steam engine.
9. With the aid of diagrams, explain the difference between a four-stroke engine and a two-stroke engine.
10. Why must a rocket be used in order to travel to the moon?

Chapter 10

The Basic Foodstuffs

There are six basic foodstuffs, three of which provide the body with fuel and body-building materials. These three are called the **nutrients**. The other three provide no nourishment, but are essential for good health. (See Table 10.1.)

TABLE 10.1. THE BASIC FOODSTUFFS

	Basic Foodstuff		Purpose
Nutrients	Carbohydrates (starch and sugar) Fats	Fuel foods	Warmth and energy
	Proteins		Body-building and body-repair
Non-nutrients	Water Vitamins Mineral salts		Health

10.1. The fuel foods

The fuel foods, **carbohydrates** and **fats**, are the foods which perform the same function for animals (*and we are animals*) that coal and coke perform for a boiler (i.e. they are sources of energy). This energy is released when they combine with oxygen.

Investigation 10a. What does bread contain?

Put a small piece of bread into a test-tube and hang a strip of dry cobalt chloride paper from the rim of the test-tube so that it reaches half-way down into the tube, as shown in Figure 10.1. (Cobalt chloride is a compound which is blue when dry and turns pink in the presence of water.)

Now heat the test-tube over a bunsen burner until no further change takes place. Do you notice any substance appearing on the

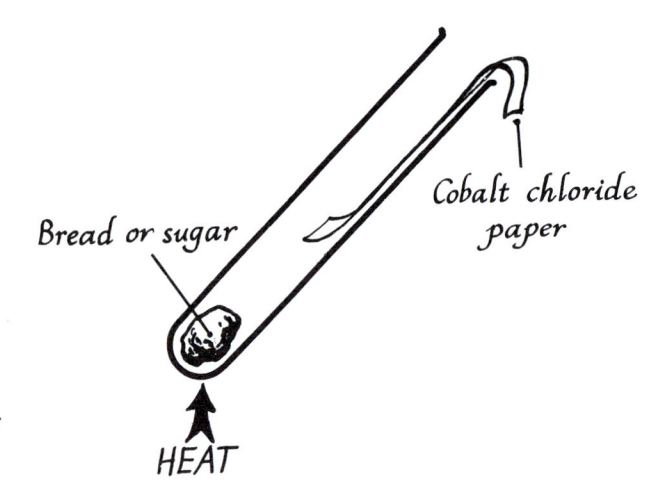

Figure 10.1 Apparatus for Investigations 10a and 10b

cold part of the inside of the test-tube? What colour is the cobalt chloride paper? What does this indicate? What colour is the substance left at the bottom of the test-tube? What do you think it is?

Investigation 10b. What does sugar contain?

Repeat Investigation 10a, using sugar instead of bread. You should obtain similar results because bread (which contains starch) and sugar are both carbohydrates and therefore contain the same three chemical elements. Can you name these three elements?

Foods which provide us with carbohydrates are:
a. **Starch**. Root vegetables, potatoes, cereals (grain) and all foods made from cereals, e.g. bread, biscuits, cake, macaroni and spaghetti.
b. **Sugar**. All fruits (even lemons), most root vegetables, milk and all foods made with sugar, e.g. sweets, cakes, jam, chocolate, syrup and treacle.

The non-carbohydrate fuel foods are fats. There are two main types:
a. **Animal fats**. All meat (even the leanest meat contains fat), fish, milk, butter, lard, eggs, poultry, etc.
b. **Vegetable fats**. All cereals, nuts, seeds, olive oil, coconut oil, margarine, etc.

10.2. The body-building foods

The body-building and body-repair foodstuff is **protein**.

Investigation 10c. What does protein contain?

Repeat Investigation 10a, using a small piece of cheese instead of bread (cheese is rich in protein). Do you get the same result? While you are heating the cheese, notice the nasty smell. This is ammonia, a nitrogen compound. There are four main elements present in protein. Can you name them?

Burn some small pieces of wool, hair, feather, leather and fingernail. Do you notice the smell of ammonia? These substances contain protein, although they are not very appetizing!

There are two types of protein:
  a. **Animal protein** (sometimes called first-class protein). All meat, fish, eggs, poultry, milk, butter, cheese, lard, etc.
  b. **Vegetable protein** (sometimes called second-class protein). All cereals, nuts and seeds, particularly the seeds of leguminous plants such as peas and beans.

As you will see, there is a great similarity between the sources of proteins and the sources of fats.

10.3. The non-nutrient foodstuffs

Although most people take one of the basic foodstuffs for granted, it is most important for survival. It is possible to keep alive without carbohydrates, fats and proteins for much longer than without—can you guess?—**water**. The body is able to store reserves of the nutrients, but it loses water all the time by excretion, in the form of urine from the bladder, faeces from the bowel, perspiration from the skin, water vapour in expired air from the lungs and even tears from the tear glands.

Of course, you could not live for very long on a diet of water and nothing else but, without water, your body would be unable to make use of any of the other foodstuffs. Fortunately, most foods contain water—even 'dried' foods. The water contained in our normal food is not sufficient by itself, and the average person needs to drink about a litre of water every day to maintain normal health. The water can be in any form—milk, tea, coffee, cocoa, lemonade, ice-cream, jelly, etc.

The last two of the basic foodstuffs, although essential to health, are only required in relatively small quantities. They do not provide warmth and energy like the carbohydrates and the fats, nor are they materials for body-building and body-repair like the proteins.

In addition to common salt (sodium chloride), which is deliberately added to many foods during cooking, the body needs other **mineral salts**, the most important being the salts of calcium, iron and phosphorus.

Calcium and phosphorus salts are necessary for the healthy growth of bones, teeth, skin, nails and hair. Calcium salts are obtained from hard water, milk, all dairy produce, green vegetables and fruit. Phosphorus salts are obtained from milk, eggs, fish, peas, green vegetables and fruit.

Iron salts are important for the formation of the red cells of the blood. Iron salts are obtained from meat (particularly liver), eggs, water cress, spinach, etc.

The remaining basic foodstuff is a group called the **vitamins**. The different vitamins are identified by letters, A, B, C, D, etc., and, although these are the main divisions, there are some which have different types within the division, e.g. B_1, B_2, B_3, etc. The function of vitamins cannot easily be described by what they do for the body but can be described by the illnesses which they prevent.

Vitamin A prevents rickets and certain eye diseases. It is present in milk, all dairy produce, fish oils, liver, green vegetables, etc.

Vitamin B. Different types prevent beri-beri, loss of weight, loss of appetite and constipation. Vitamin B is found in cereals, brown bread (much more than in white bread), peas, beans, yeast and yeast extracts, milk, eggs, liver, etc.

Vitamin C prevents scurvy. It is found in milk, fresh vegetables and fresh fruit (particularly the citrus fruits, such as oranges, lemons, grapefruit and limes).

Vitamin D prevents rickets and is necessary for the healthy growth of bones and teeth. It is found in milk, butter, cheese, fish, liver, eggs, etc. Vitamin D is also produced in the body by the action of ultraviolet rays (from the sun) on a substance called **ergosterol**, which is present in the fatty layer just under the skin.

10.4. Food-testing

Investigation 10d. To show the presence of starch

Place a small quantity of food which contains starch (e.g. bread) in a test-tube, cover the food with water and allow it to soak for about ten minutes. Now, add one or two drops of a solution of iodine in potassium iodide and note the change in colour. If you find that the new colour is too dark for you to be certain of what colour it is, dilute it with water. To ensure that it is the starch that is responsible for the colour change and not the water, add a few drops of iodine solution to some water in a test-tube. If this produces no effect, the colour change must have been caused by the presence of starch. (This latter procedure is known as a *control experiment.)*

Investigation 10e. To find out if sugar contains starch

Put some sugar and water into a test-tube. Add a few drops of iodine solution. Is there any colour change? Does sugar contain starch?

Investigation 10f. To show the presence of sugar

Place a small quantity of food which contains sugar (e.g. syrup) in a test-tube and add sufficient water to cover it. Put the test-tube aside. Into another test-tube, pour a depth of about 1 cm of Fehling's solution No. 1, followed by the same amount of Fehling's solution No. 2, and shake the test-tube until the liquid is clear. Now, pour the contents of this test-tube into the test-tube containing the food and water. Gently heat the test-tube and note any colour changes. Can you think of a suitable control experiment?

(*Note*. The Fehling's test will only indicate the presence of certain types of sugars (the reducing sugars). It will not indicate the presence of non-reducing sugars, such as sucrose (found in cane sugar). To convert a non-reducing sugar into a reducing sugar which will produce a colour change with Fehling's solution, add a few drops of dilute hydrochloric acid, heat and then allow to cool.)

Investigation 10g. Testing for sugar

Repeat Investigation 10f, using each of the following: glucose, saccharin and glycerine (all of which taste sweet—try them). Which of these substances contain sugar?

(*Note*. Test-tubes used for Investigations 10f and 10g should be washed out as soon as possible, otherwise the insides will be stained. This stain can be removed by filling the test-tubes with dilute hydrochloric acid and then rinsing them with water.)

Investigation 10h. To show the presence of fat

Press a piece of food which contains fat (e.g. nuts) between two sheets of filter paper. Remove the food and gently heat one of the sheets over a radiator or a low flame. Hold the paper up to the light. Can you see where the food was? Is the paper which was in contact with the food lighter or darker than the surrounding paper? As a control, put a drop of water on a filter paper and heat the paper.

Investigation 10i. To show the presence of protein

Place a piece of food containing protein in a test-tube, cover the food with water and allow it to soak. Carefully add a few drops of concentrated nitric acid and gently heat the test-tube. Note any colour change. Bring the contents of the test-tube to the boil and allow to cool. Now, add a depth of about 1 cm of ammonia solution and note any colour change. As a control, try this test without the protein.

★ WARNING. *Concentrated acid must only be used under strict supervision.*

This method of testing for protein is called the **Xanthoproteic** test and is effective with most proteins. Unfortunately, there are some proteins which show no reaction with this test. If you find that this is so, add **Millon's reagent** to the soaked food, and heat.

Investigation 10j. Simple analysis of some foods

Now that you know the tests for starch, sugar, fats and proteins, make a copy of Table 10.2 and indicate which of these are present in each of the samples of food.

TABLE 10.2. RESULTS TABLE FOR INVESTIGATION 10j

Food	Carbohydrate		Fat	Protein
	Starch	Sugar		
Milk				
Bread				
Rice				
Bacon				
Cheese				
Fish				
Egg				
Peas				
Potato				
Apple				

10.5. A balanced diet

For the average person to keep healthy, it is not only important to eat the correct amount of food, but also the correct proportions of each of the nutrients together with an adequate supply of water and traces of the necessary mineral salts and vitamins. This balance of food is difficult to maintain in any single meal, but should be aimed at in a day's intake of food.

The proportions of the nutrients which should be taken in daily are:

        Carbohydrates   60%
        Fats   20%
        Proteins   20%

Figure 10.2 shows the proportions in a balanced diet.

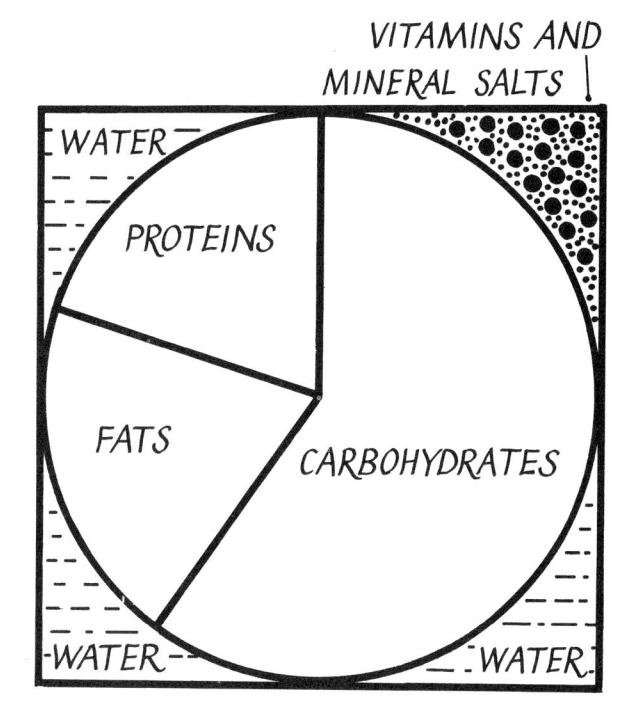

Figure 10.2 A 'square' meal

10.6. The importance of protein

Which is the correct way to describe each of the following dishes:
a. Roast beef and two veg. or two veg. and roast beef?
b. Sausage and mash   or mash and sausage?
c. Chips and fish   or fish and chips?
d. Salad and ham   or ham and salad?
e. Curry and rice   or rice and curry?

You will notice that all of these dishes sound correct when you start with the protein food. Ask other people to name some main-course dishes and you will find that the first food they mention will be the protein one.

Investigation 10k. To find the relative costs of the nutrients

Make a copy of Table 10.3 and fill in the price per kilogramme of each of the foods. You will probably find that the prices of many of these foods will vary from place to place, from season to season and according to quality, so try to obtain an average price for each food.

10.7. Burning food

In many ways the human body is like an engine. We use fuel foods (carbohydrates and fats) to produce warmth and energy; the steam engine uses coal or coke and the motor-car engine uses petrol as a

TABLE 10.3. THE COST OF NUTRIENTS

	Food	Price per Kilogramme	Average Price per Kilogramme
CARBOHYDRATES	Bread		
	Sugar		
	Rice		....
	Potato		
	Bananas		
FATS	Butter		
	Margarine		....
	Lard		
PROTEINS	Beef		
	Pork		
	Eggs		....
	Cheese		
	Fish		

fuel. We need water, mineral salts and vitamins to keep us in good health; both the steam engine and the motor-car engine need water, oil and grease to keep them in working order. However, there is a big difference: we use proteins for body-building and body-repair, but there is nothing that will do this for the steam engine or the motor-car. When parts of these wear out, they must be removed and replaced. A mini-car will never grow into a Rolls Royce!

When a fuel is burned, it combines with oxygen, producing an oxide, water and energy (usually in the form of heat). This process is called oxidation.

$$\text{Fuel} + \text{Oxygen} \rightarrow \text{Oxide} + \text{Water} + \text{Energy}$$

Similarly, when the body converts a fuel food into energy, it does so by combining it with oxygen, producing carbon dioxide and water as waste products.

Since burning a fuel and the conversion of food into energy are both the same process, we can compare the energy value of foods in the same way as we compare the heating value of fuels. The energy value of a food can be measured in **joules** (J) or **kilojoules** (kJ). To make the figures more convenient, we shall use kilojoules in this chapter.

10.8. Food energy values

When the body oxidizes 1 gramme of pure carbohydrate, about 17 kJ of energy are released. Although its main function is for body-building and body-repair, pure protein also releases about 17 kJ of energy when 1 g is oxidized. Pure fat, on the other hand, releases about 38 kJ when 1 g is oxidized.

Of course, these figures are only correct for the oxidation of *pure* carbohydrates, proteins and fats. The foods in our normal diet also contain water, mineral salts and vitamins, none of which has any energy value.

TABLE 10.4. SOME FOODS AND THEIR ENERGY VALUES

Food	Carbohydrates (%)	Fats (%)	Proteins (%)	Energy Value (kJ g^{-1})
Butter	0	82	2	31·4
Cheese	2	34	26	17·7
Porridge	70	9	12	17·4
Sugar	100	0	0	17·0
Honey	81	0	0	13·8
Figs	75	0	4	13·4
Fish	0	16	36	12·2
White bread	54	1	9	11·1
Steak	0	18	19	11·0
Cream	4	19	2	9·2
Eggs	0	10	15	6·4
Bananas	22	1	1	4·3
Milk	5	4	3	2·9
Apples	1	0	14	2·5
Potatoes	12	0	2	2·4
Onions	10	0	1	1·9
Runner beans	8	0	3	1·9
Strawberries	8	0	1	1·5

Table 10.4 shows the approximate percentages of carbohydrates, fats and proteins in some common foods, arranged in energy-value order with the foods having the highest energy values at the top. Anything less than 1 per cent is shown as 0 per cent. The mineral salt and vitamin content of most foods is normally less than 1 per cent, so these have been omitted from the table. The remaining percentage in each food is, of course, water.

10.9. How much food do we need?

It is impossible to give a simple straightforward answer to this question, just as it would be impossible to answer the question: 'How much petrol does a car need?'. You would need to know a great deal more about the car: the size of the engine, the weight of the car, the weight of the passengers, the speed it is to be driven, the distance it is to be driven, whether the route is level or hilly, whether the speed is constant or with frequent stops and starts, how well the driver can drive, the condition (efficiency) of the engine, etc.

The amount of food needed by human beings also depends on a number of factors: age, sex, state of health, occupation, climate, etc. The daily requirement of a man in normal health and doing moderate work is about 165 kJ for each kilogramme of his body weight. A woman under the same conditions needs about 132 kJ for each kilogramme of her body weight. The variation in the amount of food needed is shown in Table 10.5.

TABLE 10.5. AVERAGE DAILY FOOD REQUIREMENTS

	Kilojoules per Day	
	Male	Female
Heavy manual worker	25 000	—
Manual worker	19 000	—
Average worker	12 500	10 000
Age 14 to 21	12 500	10 000
Age 10 to 14	8 400	8 400
Age 5 to 10	6 300	6 300
Age 2 to 5	5 000	5 000
Age under 2	3 800	3 800

From the point of view of energy value alone, the average daily requirements of a boy of fifteen years could be obtained from any of the following:

>   400 g of butter
>   740 g of sugar
>   1·13 kg of bread
>   2·9 kg of bananas
>   4·3 kg of milk (about 4 litres)
>   5·2 kg of potatoes
>   8·3 kg of strawberries

However, none of these would provide a balanced diet with the correct proportions of carbohydrates (60 per cent), fats and proteins (20 per cent each).

The body is able to convert carbohydrates (particularly the sugars) into muscle fuel fairly quickly; this is why athletes often eat glucose before an athletic event. Unfortunately, if this energy is not used, the body can only store limited amounts, as glycogen, in the liver.

Although fats produce over twice as much energy for a given weight as carbohydrates, this energy takes some time before it becomes available. When a food containing fat is eaten, the fat is stored in a layer just under the skin and round the kidneys. This storage fat must then be converted into an energy source.

The rates at which energy is released from carbohydrates and fats can be compared with the rates at which heat is produced by burning paper and burning coal.

Because man is a warm-blooded animal whose blood temperature remains fairly constant (at about 37 °C) no matter what the temperature of the surroundings may be, the amount of fuel needed to maintain this temperature varies—more fuel in cold weather and less fuel in hot weather.

Test your understanding

1. What three chemical elements are contained in all carbohydrates?
2. Why is water such an important part of our diet?
3. How could you prove that a sample of 'dried' fruit contained water?
4. What is meant by a 'balanced diet'?
5. How could you test a sample of food for the presence of (a) starch, (b) a reducing sugar, (c) a non-reducing sugar, (d) fat, (e) protein?
6. What is oxidation? Give two common examples.
7. Explain the differences you would expect to find between the diets of a lumberjack in Canada and a clerk in India.
8. In a weight-reducing diet, fried food should be avoided. Explain why this is so.
9. Why is porridge more common in Scotland and the north of England than it is in the southern counties?
10. Approximately how many kilojoules per day does a man weighing 65 kg need if he is in normal health and doing a moderate amount of work? How many does a woman of the same weight and doing the same job need?
11. You will notice that in Table 10.4 the foods have been arranged in energy-value order. Make a copy of Table 10.6 and complete the columns.

TABLE 10.6

Carbohydrate Order	Fats Order	Protein Order	Water Order
Sugar	Butter	Fish	Strawberries

Chapter 11

Food Sources

On this planet, we rely for our food resources upon the carbon, nitrogen, water and oxygen cycles, and the radiant energy of the sun. The sun's radiant energy is essential for the functioning of all these cycles.

You will know, from Investigations 10a and 10b, that carbohydrates are compounds of carbon, hydrogen and oxygen and, from Table 10.4, that the foods which are rich in carbohydrates are of vegetable origin. How does a plant obtain its raw materials?

11.1. Water in plants

All plants need a constant supply of water which they absorb through their roots.

Figure 11.1 Water absorption by plants

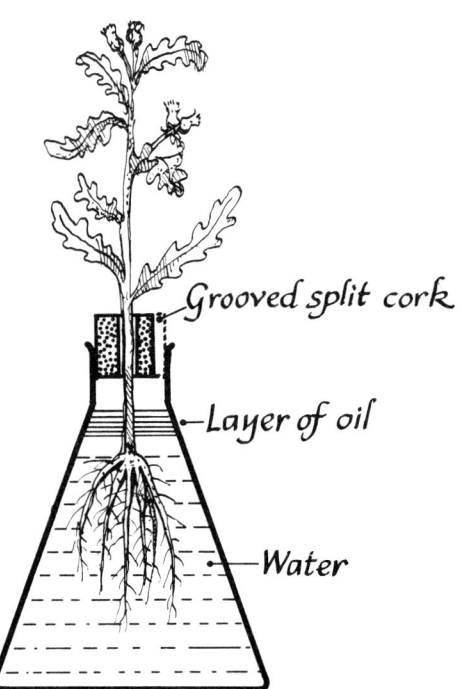

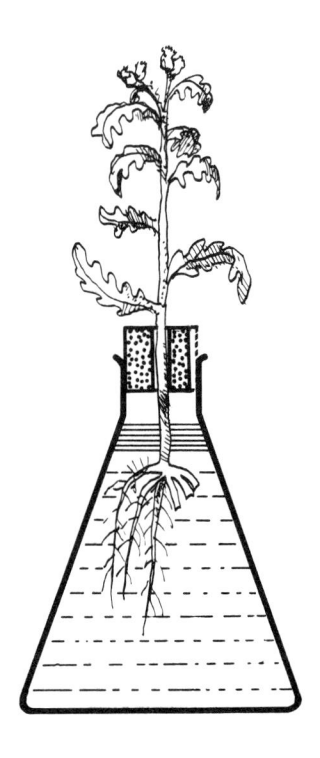

Investigation 11a. Water absorption by plants

Find two groundsel plants of the same size. (Any other small plants may be substituted for this series of investigations.) Pull one out of the ground and carefully dig the other out, leaving a ball of soil round the roots. Place the plant with the ball of soil in a bucket of water and gently wash the soil away from the roots. Support the two plants in flasks so that their roots are covered with water, as shown in Figure 11.1.

A convenient way of supporting the plants is to bore a hole, large enough to take the stem, in a cork and then cut the cork so that the stem can be placed between the two halves. Cut a groove down the outside of the cork so that air can get into the flask, ensuring that the air pressure inside the flask is always the same as the air pressure outside the flask (see Figure 11.2).

Figure 11.2 Grooved cork for Investigation 11a

When the roots are in water and before fitting the corks, put a thin layer of oil on the surface of the water in each flask and mark the levels of the water. Place both flasks near a window.

Compare the appearance of the two plants and the water levels every day for two or three days. Which plant has absorbed more water? Which part of the root system is mainly responsible for the absorption of water?

Immediately behind the growing point of each root of a growing plant is a covering of fine, hair-like structures. These are called the **root hairs**. Each root hair is an extended single cell on the outer layer of cells of the root. The cell wall of the root hair has a large surface area which is in contact with soil particles and the water held between them (see Figure 11.3). Water can pass through the cell wall freely, but just inside the cell wall is a membrane through which water enters the cell. It passes through this membrane by **osmosis**.

Figure 11.3 Root hairs

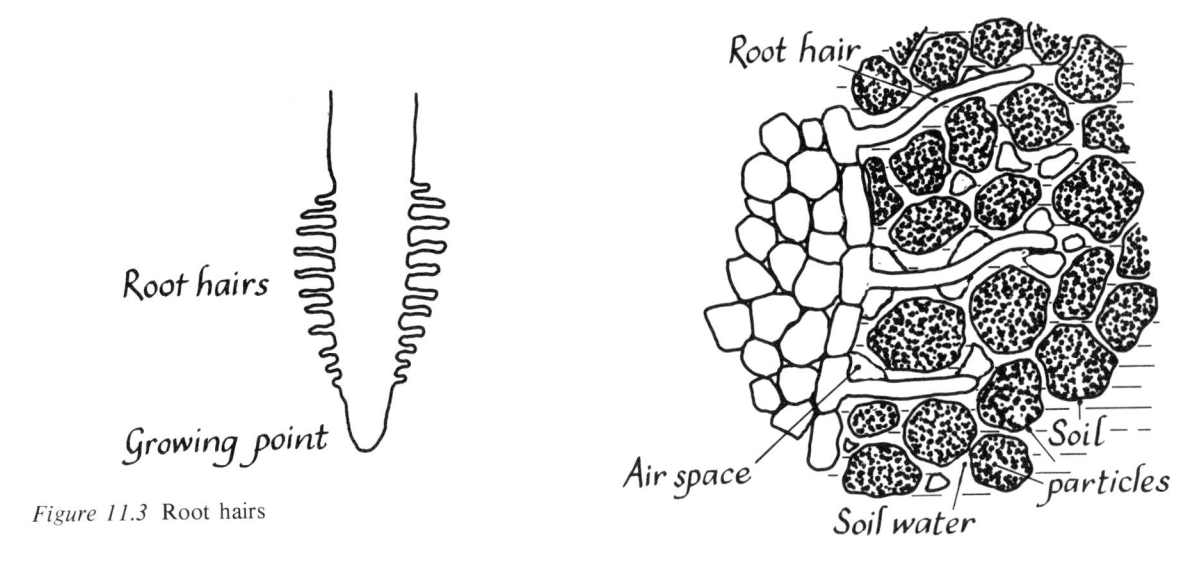

Investigation 11b. Osmosis

In order to investigate the process of osmosis, a membrane which has properties similar to those of the cell membranes must be used. Suitable membranes are a pig's bladder, parchment and some types of cellophane. Lightly smear the rim of a thistle funnel with 'Vaseline' and fix a piece of suitable membrane over the mouth of the thistle funnel with several turns of string. Set up the thistle funnel, as shown in Figure 11.4, so that the membrane is below the surface of the water in the beaker. Pour a syrup solution into the funnel with a pipette.

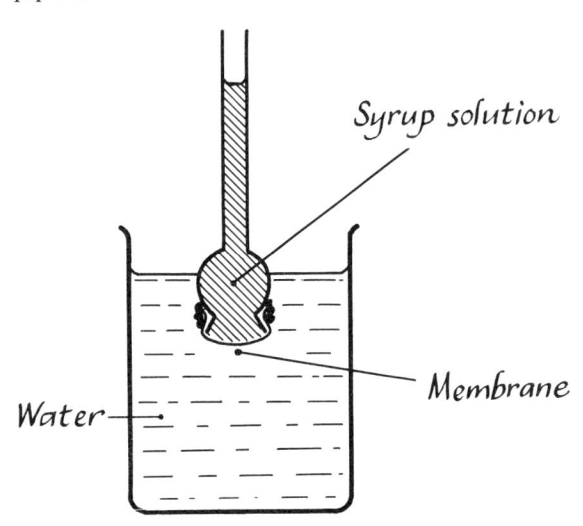

Figure 11.4 Osmosis

Fix a narrow strip of paper to the stem of the thistle funnel with 'Sellotape' and mark the level of the liquid. Leave the apparatus set up for several days and note any change in the level of the liquid in the thistle funnel. Any change in the level must have been caused by the movement of liquid through the membrane.

Cell membranes and the membrane used in this investigation are called **semi-permeable membranes**, because they allow the passage of the smaller solvent particles (the water) but not the larger solute particles (the syrup). If your investigation does not confirm this, you should suspect a leakage or that your membrane was not acting as a semi-permeable membrane.

Another way of considering osmosis is as a special type of diffusion. If you drop a large crystal of copper sulphate into a beaker of water, it will slowly dissolve to produce a dense blue layer of copper sulphate solution at the bottom of the beaker. In the course of a day or two, the blue colour will be equally distributed throughout the beaker. This indicates that particles of copper sulphate have diffused upwards, in spite of the fact that copper sulphate has a greater density than water. The direction of the movement of particles during diffusion is from the zone of greater concentration to the zone of less concentration. Diffusion continues until the concentration is the same throughout.

If the stopper of a bottle of ammonium hydroxide is removed at one end of the laboratory, the smell of ammonia will soon be noticeable at the other end, thus showing that diffusion also occurs in gases.

When two solutions of different concentration are separated by a semi-permeable membrane, diffusion of solvent particles takes place through the membrane from the zone of greater solvent concentration to the zone of less solvent concentration. This type of diffusion is osmosis (see Figure 11.5).

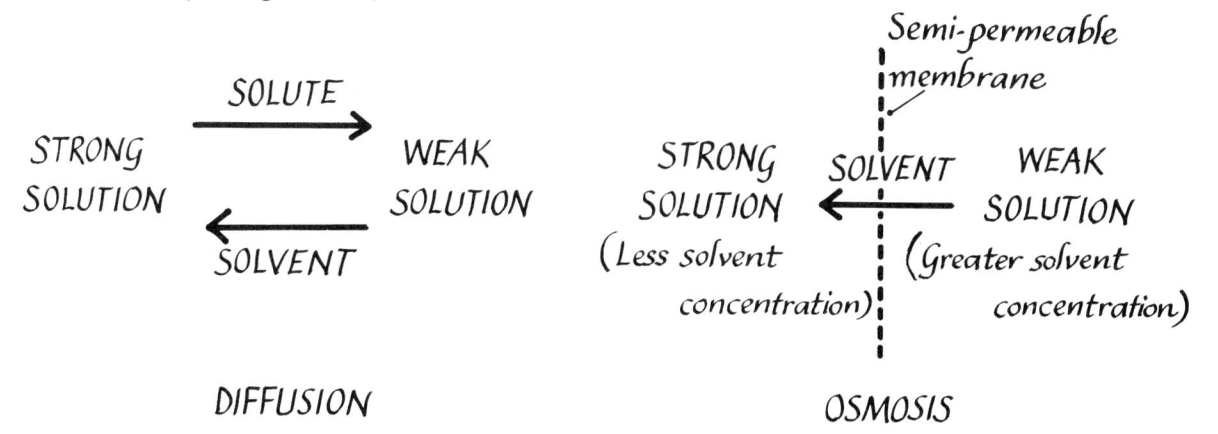

Figure 11.5 Diffusion and osmosis

The cell sap in the root hair is a stronger solution than the soil water, which is a very dilute solution of mineral salts. The soil water passes into the root hair by osmosis, thus making the cell sap weaker. The cell sap in the root hair is now more dilute than the cell sap of adjacent cells so the excess water passes, by osmosis, to these adjacent cells. From here, the water passes to **xylem vessels** which form a network through which water passes to all parts of the plant.

When a plant cell absorbs water by the process of osmosis, the pressure inside the cell increases and the cell swells. A cell in this state is said to be **turgid.** When the cells of a plant lose their turgidity, the plant wilts.

Investigation 11c. Turgidity

Using a large cork borer, cut two cylinders from a raw potato. Trim the ends so that both cylinders are of the same length. Note the length and diameter of the cylinders. If you have access to a sensitive balance, mop the surfaces of the cylinders of potato dry and weigh them separately (use watchglasses to protect the balance pans). Now immerse one of the cylinders in distilled water and the other cylinder in a strong solution of common salt. Leave the cylinders immersed for at least a day, remove them, mop them dry, measure their lengths and diameters, and weigh them. Have the cylinders altered in size or weight? Break each of the cylinders in half by bending it. What difference do you notice between the two broken cylinders? Can you explain the reasons for any differences?

Investigation 11d. Transpiration

Dig up two groundsel plants of the same size, each with a ball of soil round the roots, and gently wash the soil from the roots. Remove all the leaves from one of the plants.

Cut out two circles of thin card, each about 10 cm in diameter, and cut a slot from the outside to the centre of each. Fill two jam-jars to within 2 cm of the top with water, place the roots in the water and mark the water levels. Cover the water in each jam-jar with a thin layer of oil and support the stems with the slotted circles. If the stems tend to slip through the slots, a little modelling clay will hold them firmly, as shown in Figure 11.6.

Thoroughly dry the inside of two more jam-jars and fix a small piece of dry cobalt chloride paper to the inside of each with 'Sellotape'. Cover the plants with these jam-jars and seal any gaps with modelling clay. Place both plants near a window and, from time to time, note any colour change in the cobalt chloride paper. Investigation 10a will remind you of the significance of any colour change.

Which plant causes the colour change first? Which plant causes the water level to drop faster? What else can you deduce from the results of this investigation?

Cobalt chloride paper
Modelling clay
Layer of oil
Water

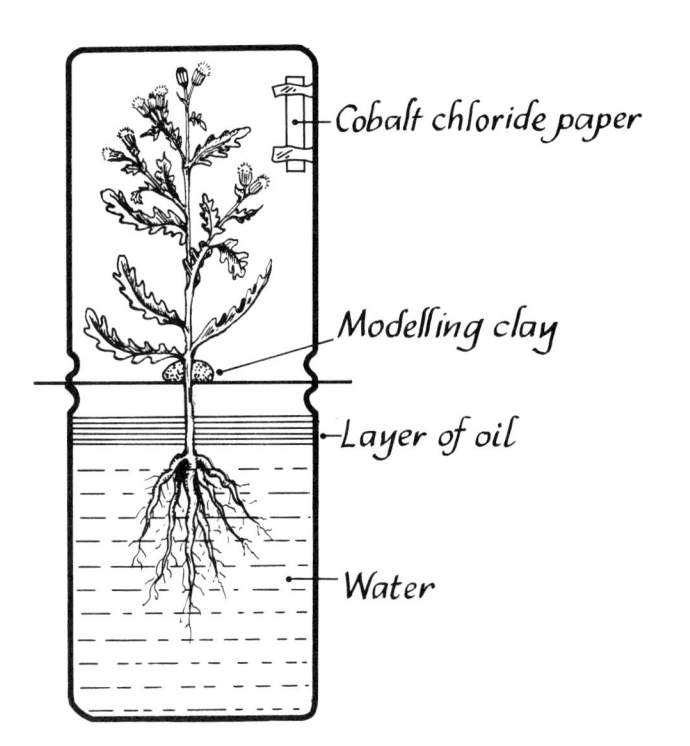

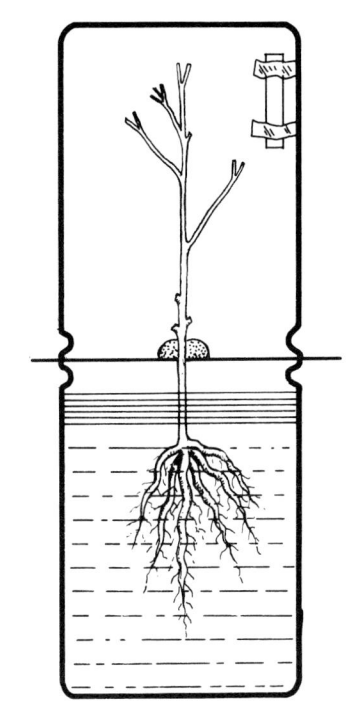

Figure 11.6 Apparatus for Investigation 11d

Figure 11.7 Apparatus for Investigation 11e

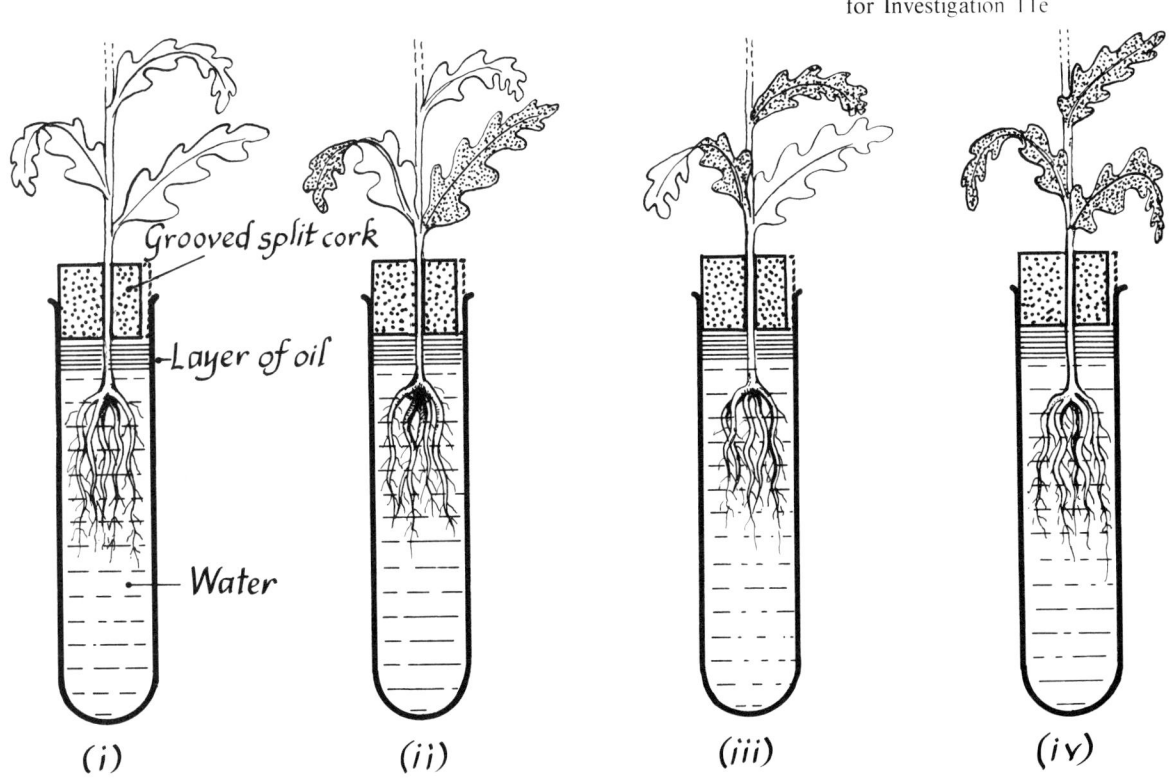

Grooved split cork
Layer of oil
Water

(i) (ii) (iii) (iv)

Investigation 11e. Transpiration and leaf surfaces

Dig up four groundsel plants of the same size, each with a ball of soil round the roots, and gently wash the soil from the roots. Fill four test-tubes to within 3 cm of the top with water, place the roots of the plants in the water and cover the water with a thin layer of oil. Support the stems in grooved split corks, as shown in Figure 11.7. Number the test-tubes and treat the leaves as follows:
1. No treatment. This plant will act as a control.
2. Cover the upper sides of the leaves with 'Vaseline'.
3. Cover the lower sides of the leaves with 'Vaseline'.
4. Cover both sides of the leaves with 'Vaseline'.

Weigh each of the test-tubes and place them in a rack near a window. Weigh each test-tube from time to time during the next few days. Which plant has given out most water? Which plant has given out least water? Which part of the plant is mainly responsible for giving out water?

Paint the upper and lower surfaces of a leaf with clear nail varnish. When the varnish is dry, peel it off and examine the impression with the low power of a microscope. Can you see any small bumps which indicate the position of small holes (or pores) through which the leaf gives out water? These holes are called **stomata** (singular: *stoma*) and the degree to which they open is controlled by two guard cells, one on each side.

The purpose of the oil layer in Investigations 11a, 11d and 11e is to prevent any drop in water level due to evaporation from the surface of the water. Can you explain why the oil layer is added *after* the roots have been put in the water?

The evaporation of water from the leaves of plants is called **transpiration**. The overall movement of water through plants is called the **transpiration stream**. A very considerable amount of water is transpired by growing crops. During a growing season, from sowing to harvesting, a grain crop can transpire as much as 250 litres of water per square metre—this is more than half of the average annual rainfall per square metre in the British Isles!

11.2. Oxygen and carbon dioxide in plants

Investigation 11f. Is carbon dioxide needed by plants?

Fill two test-tubes with a very weak solution of bromo-thymol blue, and bubble carbon dioxide through each until the colour changes to orange. Bromo-thymol blue is a chemical indicator which turns orange in an acid. Carbon dioxide dissolves in water, forming a weak acid—carbonic acid. When the carbon dioxide is removed from the water, the indicator will return to its original colour.

Put the same amount of elodea (or any similar water plant) into each test-tube, cork both test-tubes and completely wrap one of them with aluminium foil. Place both test-tubes near a bench lamp and, after several hours, examine both tubes, noting any colour change. Has any carbon dioxide been removed from the water in either of the tubes? Because the two tubes were left under different conditions, what further information can you deduce?

Investigation 11g. What gas do plants give out?

Fill two large beakers with water. Into one of them, bubble some carbon dioxide or, if carbon dioxide is not readily available, add a pinch of sodium hydrogen carbonate. Put the same amount of elodea (or any similar water plant) into each beaker and cover the elodea with an inverted filter funnel. Now fill two test-tubes with water, put your thumbs over the open ends, turn them upside down and, without allowing any of the water to escape, put them over the stems of the inverted funnels. To allow a free circulation of water round the plants, support the funnels on coins, as shown in Figure 11.8. Place both beakers near a sunny window and note what happens—you may have to wait for several days.

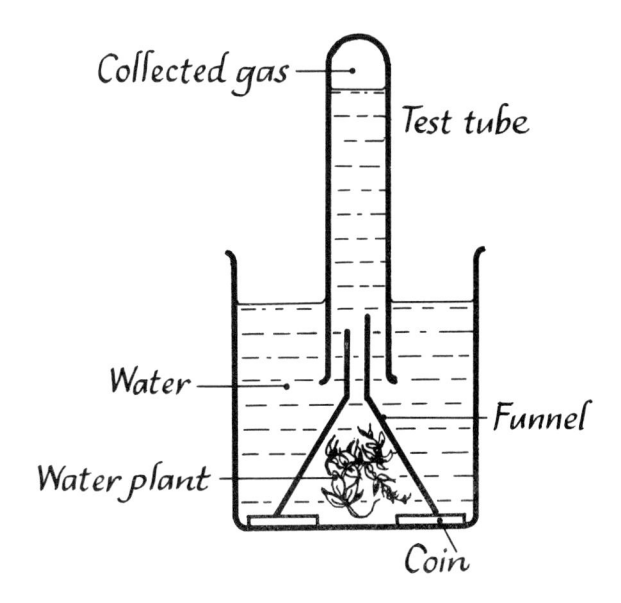

Figure 11.8 Apparatus for Investigation 11g

Has the amount of carbon dioxide in the water any effect on the amount of gas given out by the plant? What gas do you think is given out by the plants? When one of the test-tubes is full of gas, carry out the appropriate test to verify your deduction.

11.3. The leaf as a factory

In a factory raw materials are made into an end-product by using energy. In addition to the end-product there are also waste products, some of which can be converted into useful substances called by-products. In many ways, the functions of the leaves of a plant can be compared with the functions of a factory: the leaves convert raw materials into an end-product and useful by-products.

The leaves of most plants are green because the cells which make up the leaves contain **chloroplasts**. These chloroplasts contain a green pigment called **chlorophyll**, which is needed in the production of starch.

Investigation 11h. Removing chlorophyll from a leaf

1. Dip the leaf in boiling water for about thirty seconds. This coagulates and kills the protein in the cell membrane, making it fully permeable so that the next stage is more effective.

★ WARNING. *Do not heat the alcohol over a bunsen flame.*

2. Soak the leaf in hot alcohol until it has lost its green colour. (Since alcohol vapour is highly inflammable, the alcohol should be poured into a beaker and the beaker stood in hot water.) You will notice that the alcohol is now green. If you pour the alcohol into a shallow dish and allow it to evaporate, the chlorophyll will be left in the dish.

3. The alcohol has not only decolorized the leaf, but has also dehydrated it (removed the water from it). This makes the leaf brittle. The leaf should now be soaked in cold water to replace the water removed by the alcohol, thus making the leaf pliable. These three stages are shown in Figure 11.9.

Figure 11.9 Decolorizing a leaf

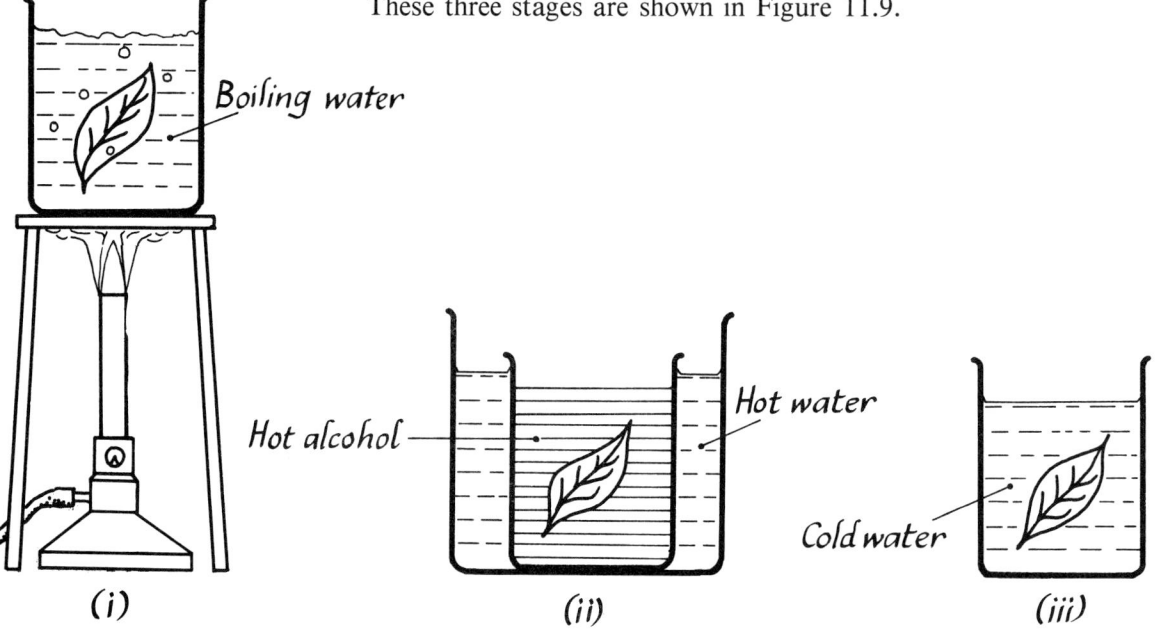

Investigation 11i. What does a green leaf produce?

On a sunny day, pick a green leaf from a growing plant and remove the chlorophyll. Put the decolorized leaf on a white tile or in a shallow dish and pour a little iodine solution over it. Note what happens. Does this tell you what a green leaf produces? If you are not certain, see Investigation 10d. What was the purpose of decolorizing the leaf first?

Investigation 11j. Is sunlight necessary for making starch?

1. Put a healthy, green pot plant in a dark cupboard for two or three days. Remove a leaf, decolorize it and test it for starch.
2. Remove the plant from the cupboard and immediately mask a part of one of the leaves by placing a piece of black paper or tinfoil on both sides, securing them with paper clips, as shown in Figure 11.10. Now place the plant in a sunny position for at least four hours. Remove the leaf from the plant, remove the masking, decolorize the leaf and test it for starch. What do you deduce from the results of this investigation?

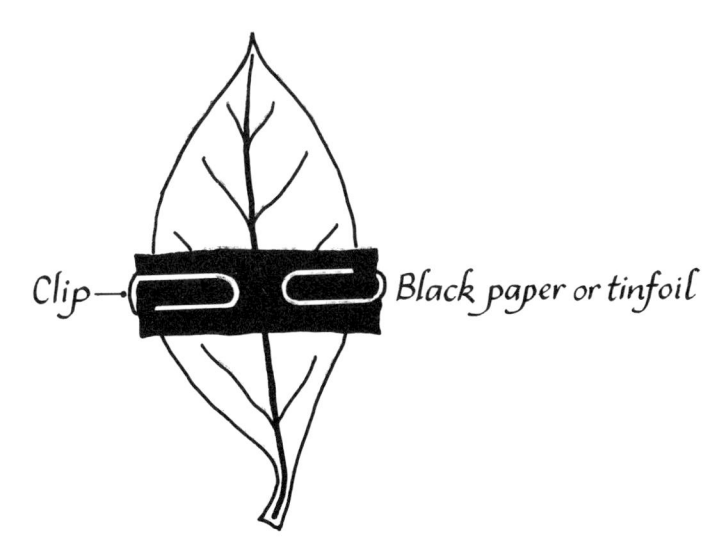

Figure 11.10 A masked leaf

Investigation 11k. Is carbon dioxide necessary?

Put a healthy pot plant (nasturtium is ideal) in a dark cupboard for two or three days. What does this do? Then place it in a sunny position for at least a day, with one of its leaves (still attached to the plant) in an airtight flask containing a little sodium hydroxide (caustic soda) solution. Sodium hydroxide will absorb all the carbon dioxide from the air in the flask. The best way to ensure that the flask is airtight is to use a split cork like the one shown in Figure 11.2, but without the groove. The cork should be thoroughly soaked in melted paraffin wax and, after the leaf has been put in the flask, sealed with modelling clay or more melted paraffin wax. The arrangement is shown in Figure 11.11.

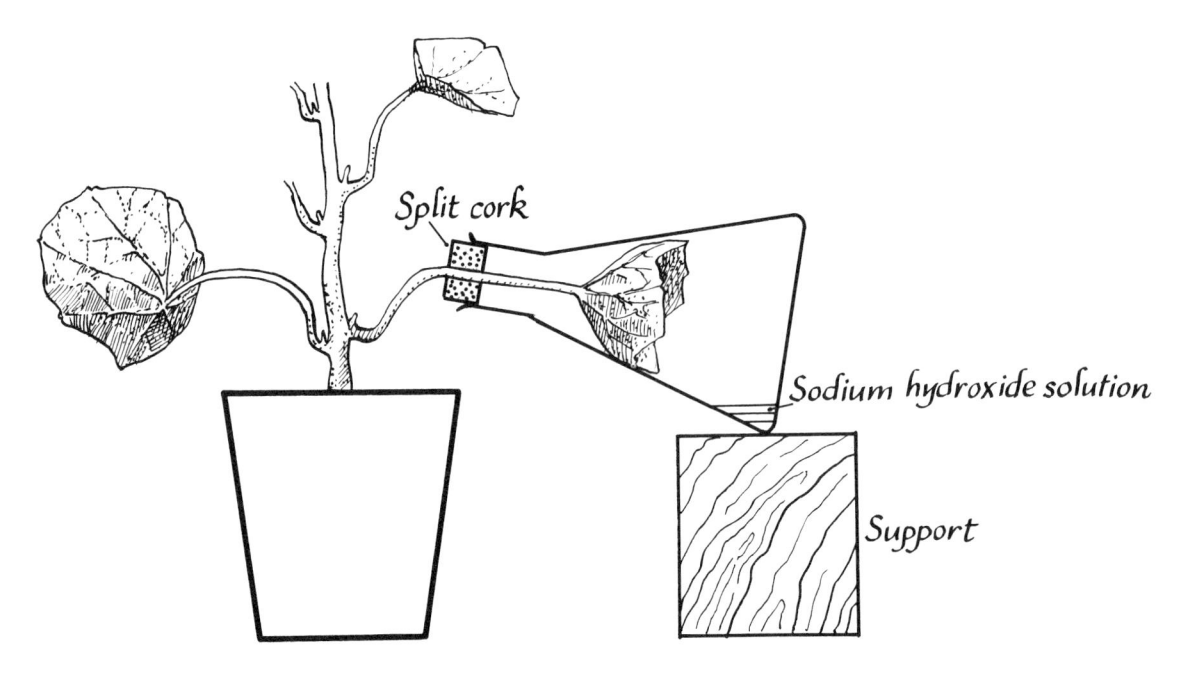

Figure 11.11 Apparatus for Investigation 11k

Remove the leaf which was deprived of carbon dioxide and another leaf (as a control) from the plant and test both for starch. Is carbon dioxide necessary for making starch?

Investigation 11l. Is chlorophyll necessary?

On a sunny day, remove a leaf from a plant which has variegated leaves (tradescantia, rose geranium, variegated privet, etc.). Place a gas jar cover, a sheet of ground glass or greaseproof paper over the leaf and trace the pattern of the green parts of the leaf. Decolourize the leaf and test it for starch. How does the leaf compare with your tracing? Is chlorophyll necessary for making starch?

11.4. Photosynthesis

The leaf of a green plant acts as a factory. Its raw materials are:
a. Water, which is absorbed from the soil by the roots and transported up the stem to the leaves.

b. Carbon dioxide, which enters the leaves by diffusion through the stomata.

In order to combine these raw materials, chlorophyll and the radiant energy of the sun are necessary. The process by which green plants produce starch from these raw materials is called **photosynthesis**. 'Photo' means 'light' and 'synthesis' means 'building up'. The main by-product of photosynthesis is oxygen; a large proportion of this is used by the plant itself for respiration, the remainder being given out by the stomata. Excess water is also given out by the leaves.

$$\text{Water} + \text{Carbon Dioxide} \xrightarrow{\frac{\text{Sunlight}}{\text{Chlorophyll}}} \text{Starch} + \text{Oxygen}$$

The process of photosynthesis is summarized in Figure 11.12.

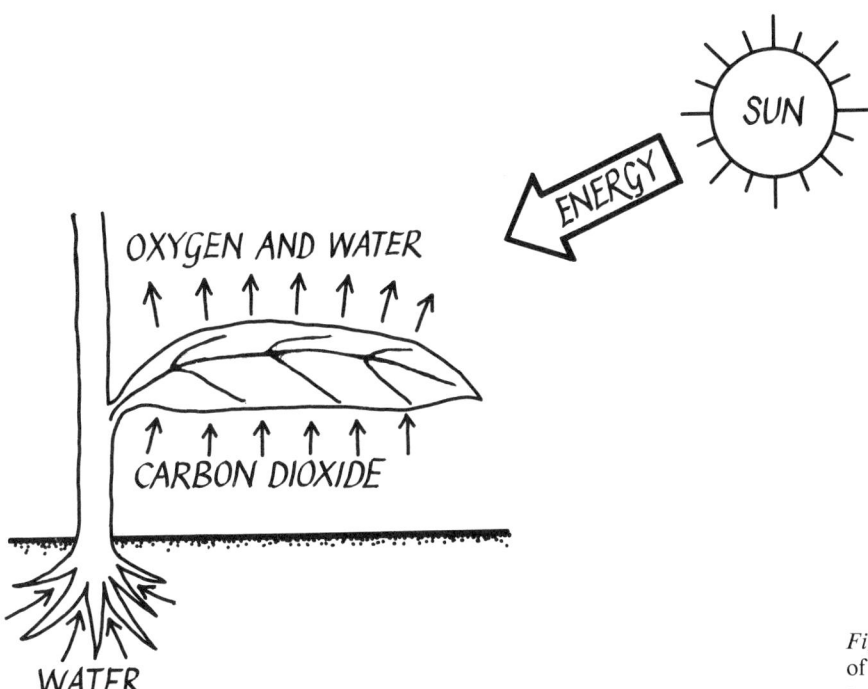

Figure 11.12 The transfer of materials in photosynthesis

In our investigations we have used the test for starch, but it is possible that the immediate product of photosynthesis is sugar in the form of glucose, which is converted into starch for storage in the leaves and elsewhere in the plant. Starch will not dissolve in water so, in order to transport it to other parts of the plant, either to be used by the plant itself for growth and other vital functions or to be stored, the starch is converted into sugar by the action of **diastase** (a plant enzyme).

Green plants are producers. Photosynthesis not only provides our carbohydrates but also maintains our oxygen supply. Vegetable fats and vegetable proteins could not be produced without photosynthesis because plants rely on this process for their very existence.

Animals are consumers. Since animals feed, either directly or indirectly on plants, animal fats and animal proteins could not be produced without the fundamental process of photosynthesis.

11.5. The oxygen cycle

One of the effects of photosynthesis is to absorb carbon dioxide from the atmosphere and replace it with oxygen. This is a very important effect as all forms of life (with the exception of certain types of bacteria) require large amounts of oxygen for the process of respiration, which is the liberation of energy by the oxidation of food. Plants respire day and night. Fortunately, during the day, they photosynthesize at a very much greater rate than they respire, so the overall effect is that they replace the oxygen used in all forms of oxidation and absorb the carbon dioxide which is a waste product of oxidation. The constant interchange of oxygen and carbon dioxide is called the oxygen cycle (see Figure 11.13).

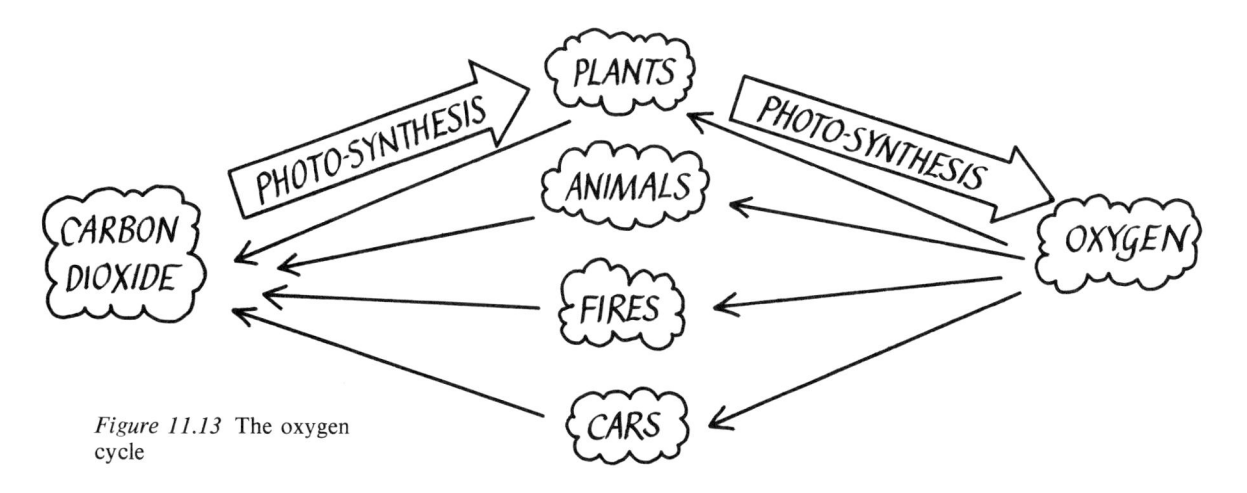

Figure 11.13 The oxygen cycle

11.6. The carbon cycle

The carbohydrates, which animals use as a fuel food, are obtained by eating plants or other animals which, in turn, have eaten plants. Carbohydrates are compounds which contain carbon. In prehistoric times, large plants died, became covered with sediment and, in the course of time, under great pressure, formed coal, from which we obtain coal-gas, coke and many by-products. These fuels contain carbon, and the burning (oxidation) of these fuels produces carbon dioxide. Petroleum (sometimes called crude oil) is formed from the remains of minute marine animals and plants. From petroleum we obtain petrol which is a carbon compound, together with oils and greases and many other by-products. The burning of petrol in a car engine produces carbon dioxide.

The constant interchange in the forms of carbon is called the carbon cycle (see Figure 11.14).

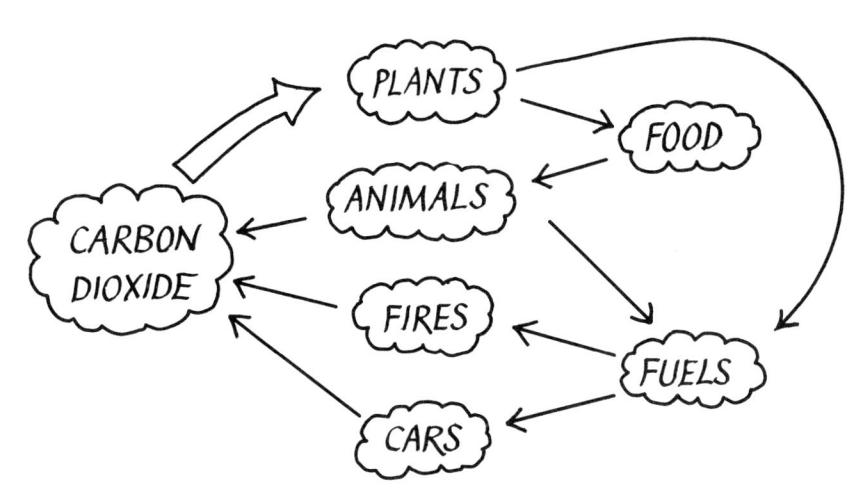

Figure 11.14 The carbon cycle

11.7. The nitrogen cycle

All living cells consist mainly of protoplasm, the main component of which is protein. Nitrogen is essential for protein synthesis.

Although the atmosphere is almost 80 per cent nitrogen, plants are unable to make use of it in this form. In order to be absorbed into the roots by diffusion, the nitrogen must be in the form of a dilute solution of nitrates (e.g. sodium nitrate, $NaNO_3$) or ammonium salts (e.g. ammonium sulphate, $(NH_4)_2SO_4$) in the soil.

There are several ways in which these salts are produced:

a. During a thunderstorm, electrical discharge (lightning) causes some of the nitrogen in the air to combine with oxygen to form nitric oxide (NO) which changes to nitrogen dioxide (NO_2) on cooling. Falling drops of rainwater dissolve the nitrogen dioxide to form nitrous acid (HNO_2) and nitric acid (HNO_3).

$$2NO_2 + H_2O \rightarrow HNO_2 + HNO_3$$

When these acids reach the soil, they form nitrites and nitrates. A comparatively small proportion of the nitrites and nitrates required by plants is produced in this way.

b. Dead leaves, dead plants, dead animals and animal wastes (urine and faeces) are broken down into ammonium compounds by decay bacteria and fungi. Some of these ammonium compounds are absorbed by living plants, while a small amount is converted into nitrites and nitrates by nitrifying bacteria.

c. Some nitrogen-fixing bacteria living in the soil are able to convert atmospheric nitrogen, which is present in the soil, into proteins. These proteins are released into the soil when they die and are treated by the nitrifying bacteria.

d. Other nitrogen-fixing bacteria live in the nodules on the roots of certain plants called legumes (clover, peas, beans, etc.). These bacteria feed on the plant but, in exchange, produce nitrates. This source of nitrates can be used by farmers in crop rotation, by growing a crop of legumes in a field every three or four years. The roots of this crop, together with the dead bacteria, replace the nitrates absorbed by the crops in the intermediate years. This system of crop rotation is not used as much now as it was in earlier times because of the greater use that is made of artificial fertilizers.

Although large amounts of nitrates and ammonium salts are produced in the soil in these ways or are added in the form of artificial fertilizers, some of them are unavoidably not available to plant life:

a. In the soil, there are denitrifying bacteria which change nitrates and ammonium salts back into nitrogen gas, which escapes into the atmosphere. These bacteria are **anaerobic** (live without free oxygen) and oxygen reduces their numbers. If a compost heap is turned over regularly, the loss of nitrates and ammonium salts through the action of these bacteria is reduced.

b. Some of the nitrates and ammonium salts are leached out (washed out) of the soil by water, and are carried too deep to be reached by the roots. This occurs to a greater extent in light, sandy soils.

The constant interchange in the forms of nitrogen is called the nitrogen cycle (see Figure 11.15).

11.8. The water cycle

We have seen, in Section 11.1, how water is constantly being passed through plants by the transpiration stream. Although this accounts for a large proportion of the total rainfall, there are other ways in which water is returned to the atmosphere:

a. Evaporation from the surface of oceans, seas, lakes, rivers, etc.
b. Evaporation from the surface of the soil.
c. Evaporation of animal perspiration.
d. Water vapour produced by animals during respiration.
e. Water vapour given off as a result of the oxidation of hydrogen compounds (the burning of fuels).

The atmosphere always contains water in varying quantities.

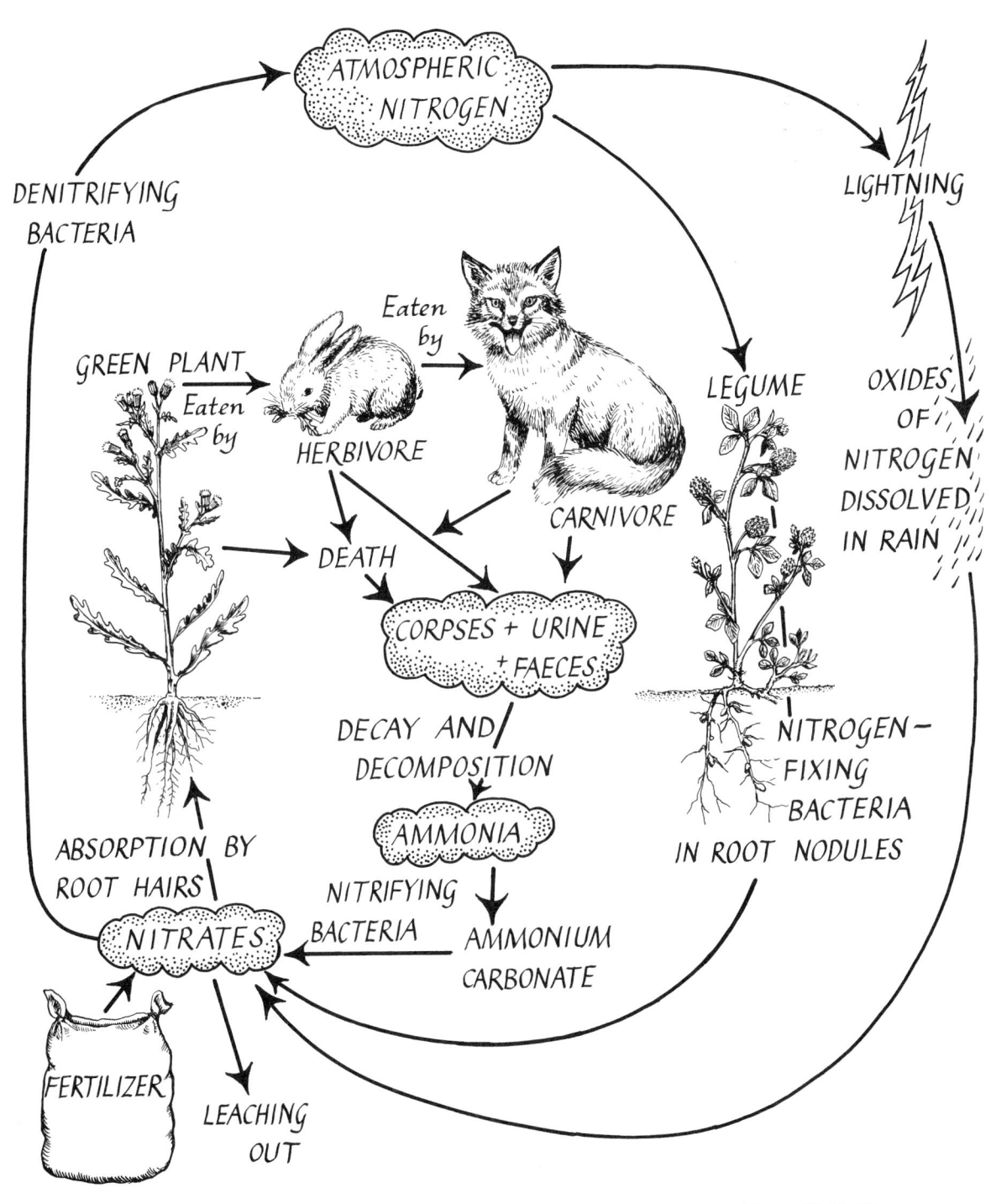

Figure 11.15 The nitrogen cycle

When the air is unable to absorb any more water, it is said to be saturated. Warm air is able to absorb more water than cold air. When saturated air is cooled, it is unable to retain as great a quantity of water and the excess is released in the form of rain, hail, snow or sleet, depending upon the rate of cooling and the temperature of the air. This cooling occurs in two stages:

a. The initial cooling, which causes the water vapour to change into minute droplets of water suspended in the air, forming clouds. This is called **condensation**.

b. The secondary cooling, which causes the droplets to join together into larger drops which are too heavy to be supported by the air, forming rain, etc. This is called **precipitation.**

When condensation and precipitation occur at temperatures below 0 °C, ice crystals instead of water droplets are formed. This produces snow.

The interchange in the forms of water through the stages of evaporation, condensation, precipitation and its return to the earth to begin over again, is called the **rain cycle** (see Figure 11.16). The radiant energy of the sun is an essential factor in the rain cycle.

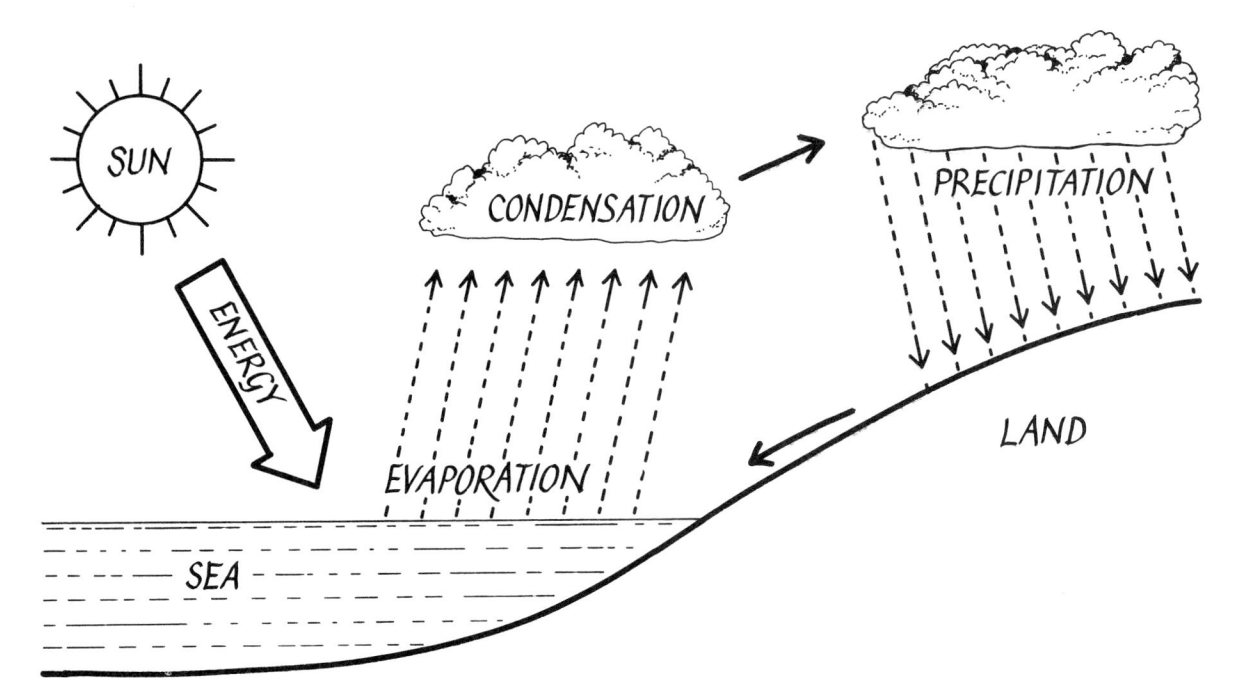

Figure 11.16 The rain cycle

At night, when the earth's surface is losing heat by radiation, the temperature drops. When moist air comes into contact with a surface at a lower temperature than the air, water is condensed on the surface. This water is called **dew**. The temperature at which this occurs is called the **dew point** and depends upon the temperature of the moist air and the amount of water vapour in it. If the dew point is below 0 °C, then ice crystals instead of water droplets are formed. This is called **frost**.

Dew and frost are more likely to occur on a still, cloudless night, because a cloud layer acts as an insulating layer which reduces heat loss.

Investigation 11m. Finding the dew point

Take a highly polished copper calorimeter and half fill it with water at room temperature. Support a sheet of glass or perspex to shield the calorimeter from your breath. This precaution is necessary because the air which you breathe out is warmer and wetter than the surrounding air. Place a thermometer in the water and add a small ice cube. Stir the water until the ice cube melts. Continue adding ice cubes, one at a time and stirring until each one dissolves, until you see a fine film of dew forming on the outside of the calorimeter. Note the temperature of the water when this happens. Do not add any more ice cubes and allow the calorimeter to warm up. When you see the last trace of dew disappearing, note the temperature. An average of these two temperatures will give you the dew point.

A combination of all the ways in which water is interchanged from the atmosphere to the earth and back to the atmosphere is called the water cycle (see Figure 11.17).

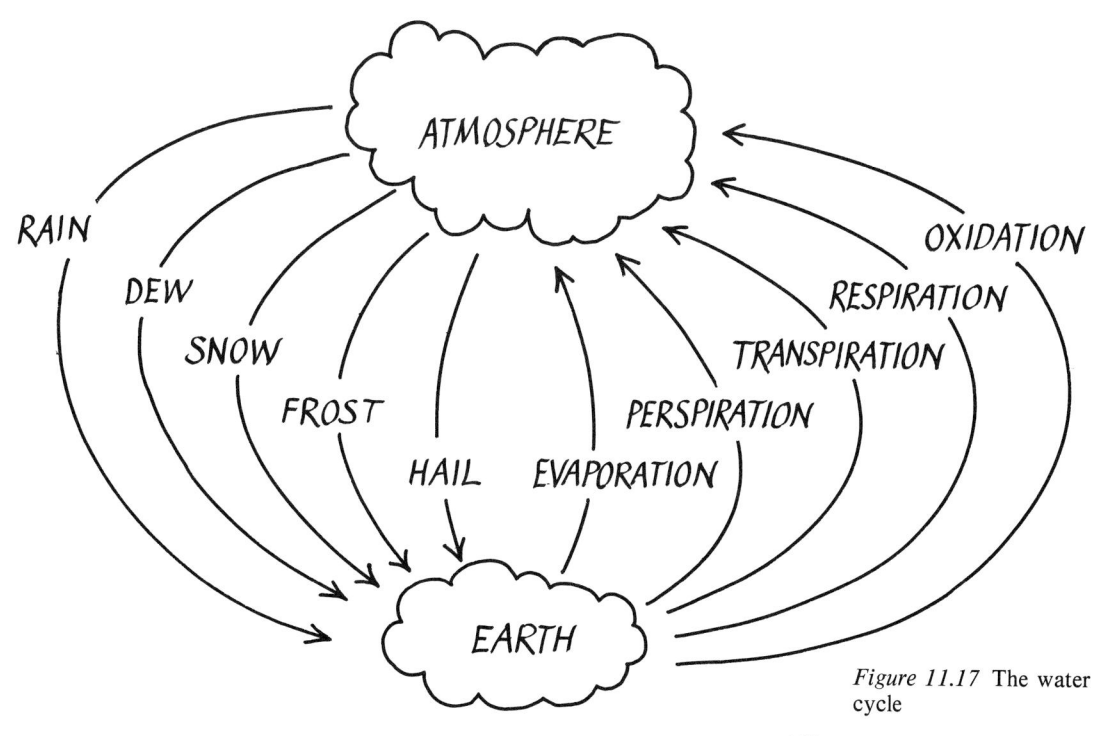

Figure 11.17 The water cycle

11.9. Water supply

Since water is a part of our diet (and a very important part), it is necessary to have a regular supply of safe drinking water. It has been estimated that, in a civilized society such as ours, the daily water requirements amount to between 100 and 200 litres per person. This includes water for drinking, cooking, washing (our clothes and ourselves), toilet flushing and the vast quantities used by industry. Our domestic supply can be obtained from a number of sources:

    a. Rainwater falling on high ground which is collected in reservoirs.
    b. River water.
    c. Spring water.
    d. Water from shallow wells.
    e. Water from deep wells.

All of these sources will contain impurities, some of which are harmless or even beneficial, and others which could be very dangerous. These impurities can be divided into five groups:

    a. Solid matter floating on the surface (this type of impurity is found mainly in river water).
    b. Solid matter suspended in the water.
    c. Dissolved solids.
    d. Dissolved gases.
    f. Bacteria and other living organisms.

To make water safe for drinking, it must be purified. Floating matter is removed by passing the water through fine wire-mesh screens. Suspended matter is removed in two ways:

    a. By allowing the water to stand in tanks, so that the suspended solids sink to the bottom. This process is called **sedimentation** and can be speeded up by the addition of alum. This forms small jelly-like crystals in the water which attract the suspended solids and then sink to the bottom.
    b. By passing the water through sand filters.

Investigation 11n. Sedimentation

Take two gas jars of equal size and put 5 g of fine soil into each of them. Fill both gas jars with water, shake them, place them side by side, and into one of them sprinkle 1 g of alum. Compare the time taken for the water in the two gas jars to clear. Repeat the investigation, using different amounts of alum, to find out if clearing is faster with greater amounts.

Investigation 11o. Sand filtration

Using a nail, punch a large number of small holes in the bottom of a tall tin can. Thoroughly wash some small stones, gravel and sand in running water and put them in the tin. Arrange them with a thin layer of stones in the bottom of the tin, then a layer of gravel and, finally, fill the tin almost to the top with sand, as shown in Figure 11.18. Now fill two flasks with water, add equal quantities of fine soil to each and stir them. Support the tin over a dish and pour the contents of one of the flasks into the top of the tin. When the water has stopped dripping out of the tin, wash out the empty flask and pour the filtered water from the dish into it. Now compare the appearance of the water in the two flasks.

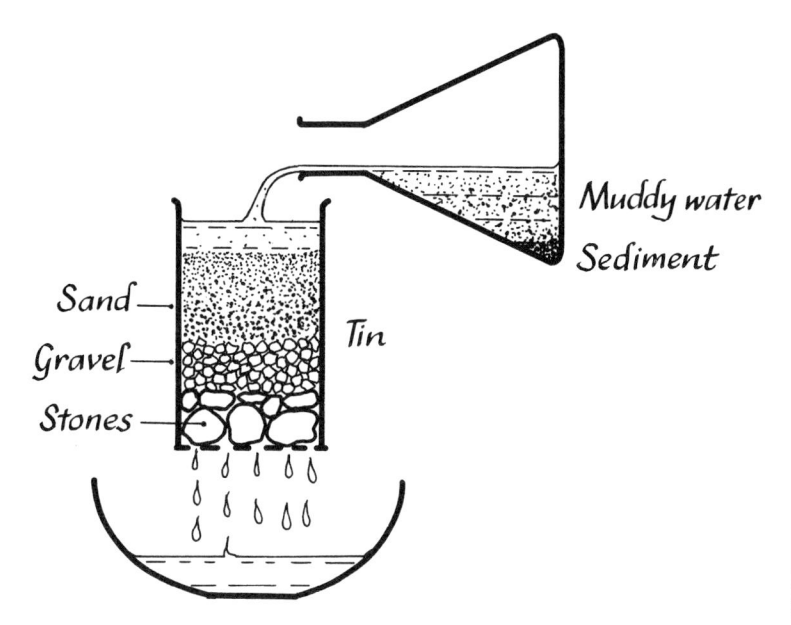

Figure 11.18 Sand filtration

Dissolved gases, such as sulphur dioxide and hydrogen sulphide, can be removed by passing the water over a series of small waterfalls. This allows a large surface area of water to be in contact with the air, enabling the gases to escape. The dissolved solids are likely to be mainly calcium and magnesium salts which, although they make the water hard, are not harmful to health. In fact, the calcium salts are an essential part of our diet.

Bacteria and other minute living organisms are killed as the last process before the water is fed into the mains. This is usually done by adding a small amount of chlorine to the water; the proportion is about one part of chlorine per million parts of water. This proportion of chlorine does not give the water any appreciable taste. In time of war or during epidemics of water-borne disease, the proportion of chlorine is increased as a precaution against bacteria entering the supply. This stronger solution has a definite taste.

Some water supply authorities now add a small amount of fluoride to the water, as this has been shown to greatly reduce decay in the teeth of young children.

Test your understanding

1. The label on a bottle of a liquid fertilizer gives the following directions: 'Put one cap-full of the concentrated fertilizer in four litres of water, stir and apply to the soil round the roots of your plants.' An inexperienced gardener applied a solution of four cap-fulls in one litre of water to his plants and found that, on the next day, they had withered. Can you explain why this happened?
2. If a rose bush has to be transplanted during the summer months instead of the winter, some of the leaves should be removed. Explain the reason for this.
3. Name the raw materials used by a green plant in the production of starch, and explain how they enter the plant.
4. Apart from the production of food, what is the other important function performed by green plants which enables animals to live?
5. How can a thunderstorm save a farmer money?
6. How is dew formed?
7. What is the purpose of planting water plants in an aquarium?

Chapter 12

Digestion

Digestion is the method by which animals obtain nourishment from the food they eat. A simple animal, such as the amoeba (a single-celled animal which lives in water), obtains its nourishment from food particles floating in the water. These food particles may consist of even smaller living organisms. The amoeba engulfs the food particles by flowing round them and, after digestion, it absorbs the substances it requires and disposes of the unwanted part by flowing away from it, as shown in Figure 12.1.

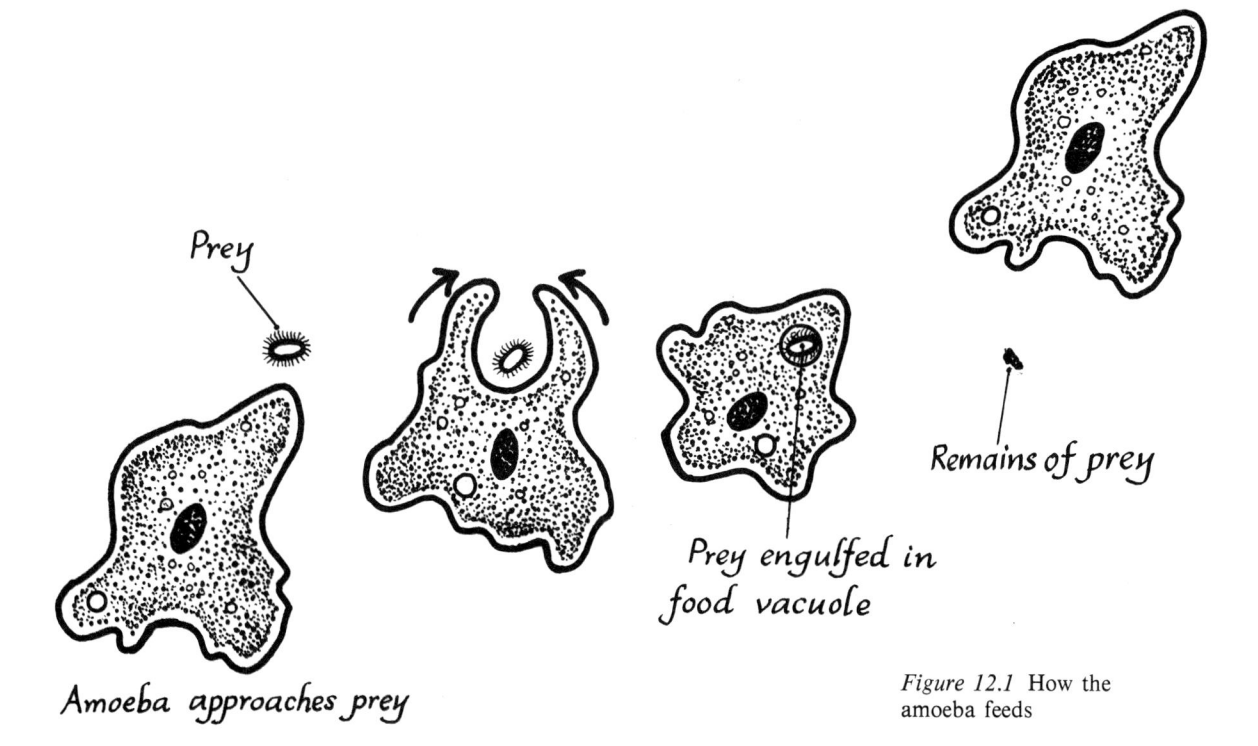

Figure 12.1 How the amoeba feeds

Whilst the amoeba can take in food at any point on its surface, more complicated animals can only take in food at one point (the mouth) and, from there, it follows a definite route. This route is called the **alimentary canal**.

12.1. Digestion in the earthworm

The alimentary canal of the earthworm is a straight tube from the mouth at one end to the anus at the other end. This tube widens behind the mouth to form a crop and then a gizzard. The earthworm's food, consisting of decayed animals and plants, is taken in through the mouth, together with soil. This mixture is acted upon by digestive juices in the **oesophagus** (between the mouth and the crop) and in the crop itself. These digestive juices react with the food and produce soluble substances. From the crop, the mixture of dissolved food and soil passes to the gizzard, where it is ground into fine particles. From there, it passes along the intestine, where the dissolved food is absorbed through the walls of the intestine into the bloodstream. The remainder, consisting of soil and undigested matter, is excreted through the anus in the form of worm casts.

Investigation 12a. Making a wormery

Make a wooden box (30 cm × 20 cm × 5 cm is a suitable size) with a glass side. Fill the box to within 5 cm of the top with alternate layers of soil and sand. Put some earthworms and a few leaves on top, cover the glass side and leave the wormery undisturbed for a week. Figure 12.2 shows a wormery. At the end of the week, uncover the glass side and examine the wormery. Can you still see the

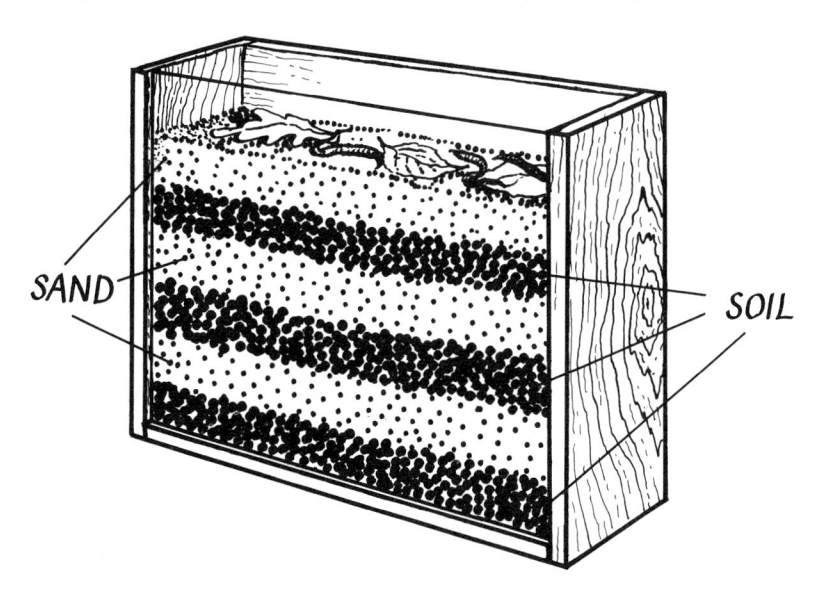

Figure 12.2 A wormery

alternate layers of soil and sand? Can you still see the leaves? Compare some of the original soil and sand with the worm casts, using a hand lens or a low-power microscope. What differences can you see?

12.2. The alimentary canal in man

Man's food is much more complicated than that of the amoeba or the earthworm. In order that it may be broken down and absorbed, food must remain in the alimentary canal for a comparatively long time. The alimentary canal in man is between 8 and 9 metres long, and obviously cannot be a straight tube. The canal consists of the following sections: mouth, gullet (or oesophagus), stomach, duodenum, small intestine, large intestine (or colon), rectum and anus, as shown in Figure 12.3.

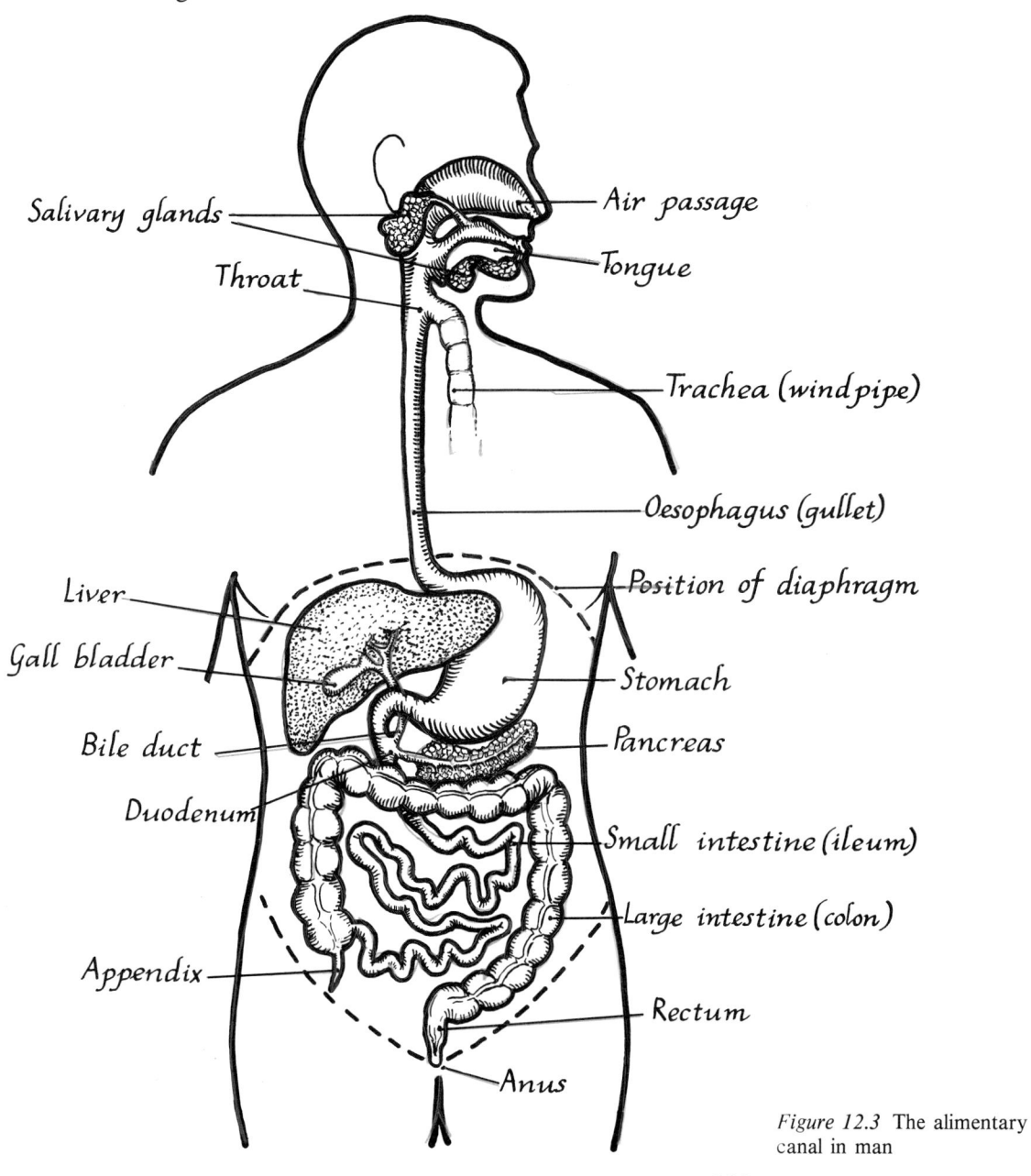

Figure 12.3 The alimentary canal in man

12.3. Digestion in man

There are two types of digestive process, mechanical and chemical. Mechanical digestion consists of breaking up the food into small pieces and mixing it with various substances along the alimentary canal. Chemical digestion is the process of breaking down substances so that they can be absorbed by the blood. This involves the action of chemical substances, which are contained in digestive juices produced by the digestive organs of the body. These chemical substances are called **enzymes**. The action of an enzyme is specific—it will react only with one particular type of food and has no effect on any other.

The process of digestion begins before you start eating. It starts when the saliva begins to flow in the mouth. This happens at the sight of food, the smell of food or even at the thought of food. There are three salivary glands on each side of the mouth (the parotid gland, the submaxillary gland and the sublingual gland). Between them, they produce about a litre of saliva per day. Saliva performs two main functions:

a. It moistens the food so that it is more easily tasted, and lubricates it so that it can be swallowed.

b. It converts cooked starch into **maltose** (a form of sugar), by the action of the enzyme **ptyalin**, which it contains.

The mechanical breaking up of food and the enzymic action of ptyalin are the best safeguards against indigestion. Even rice pudding should be well chewed to mix it with ptyalin.

Investigation 12b. How surface area affects dissolving rate

For this investigation you will need two small beakers (100 cm^3 would be suitable), two measuring cylinders, a balance, a watch or stop-clock, some small pieces of marble, some powdered marble and some dilute hydrochloric acid.

Put about 2 g of marble pieces on one side of the balance and slowly add powdered marble to the other side until it balances. Put the marble pieces into one beaker and the powdered marble into the other beaker. Measure 25 cm^3 of dilute hydrochloric acid into each measuring cylinder and pour this amount from the measuring cylinders into each beaker. Note how long it takes to dissolve the marble in each beaker. In which beaker was the action more vigorous? Which form of marble had the greater surface area, the pieces or the powder? Which form of marble dissolved faster?

How do the results of this investigation show the advantages of the thorough chewing of food?

Investigation 12c. The action of ptyalin

Put a small piece of dry bread in your mouth and chew it for about five minutes. Is there any change in its taste? Now put the piece of chewed bread in a test-tube and test it for sugar (see Investigation 10f). As a control, try the same test on a piece of unchewed bread.

When food has been chewed and mixed with saliva, it is swallowed by being squeezed against the roof of the mouth by the tongue and passed to the gullet. At this stage, the food is a pulpy mass called a **bolus**.

Along the length of the gullet and the small intestine is a series of ring muscles (**sphincter muscles**). By contracting in sequence, the sphincter muscles produce a wave-like movement along the length of the gullet and the small intestine. This wave-like movement is called **peristalsis**.

Investigation 12d. To demonstrate peristalsis

Lubricate the inside of a length of rubber tubing by pouring some soap solution or washing-up liquid through it. Lay the rubber tubing on a flat surface and place an orange pip just inside one end of it. Now make the orange pip pass through the rubber tubing to the other end by squeezing just behind the pip.

Figure 12.4 shows the effect of peristalsis.

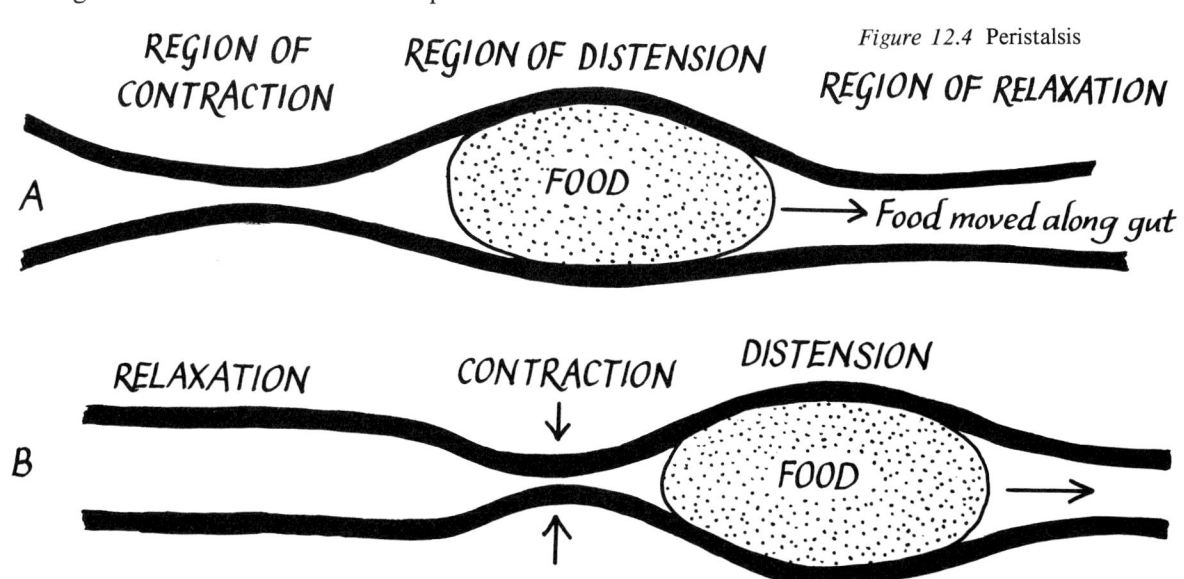

Figure 12.4 Peristalsis

During the day, peristaltic contractions occur about ten to twelve times per minute. At night, when the body is resting, contractions slow down to about seven or eight times per minute. As soon as some food is eaten, after a long rest, peristalsis speeds up beyond its normal rate to between fourteen and sixteen waves per minute. Because of this speed-up, the most effective time to empty the bowel is shortly after breakfast. The habit of emptying the bowel at this time will do much to prevent the most common complaint of modern civilized society—constipation.

The stomach is a stretchable bag with a pitted lining. Because these pits occur in the stomach, they are called gastric pits (see Figure 12.5). Through the large surface area of the gastric pits, dilute

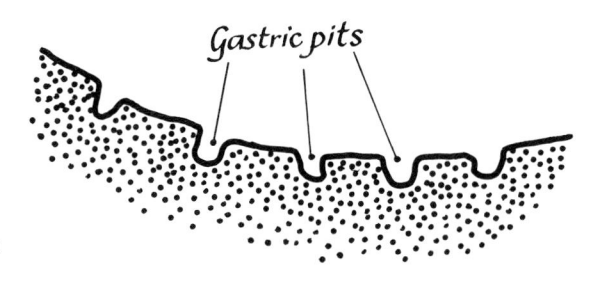

Figure 12.5 The stomach lining

hydrochloric acid and gastric juice are secreted. Gastric juice contains two enzymes: **rennin** and **pepsin**. Rennin is the enzyme which curdles milk by its action on casein (milk protein). Pepsin changes protein into soluble **peptones**. The hydrochloric acid is essential because pepsin acts faster when acid is present. The acid also dissolves mineral salts.

The length of time that food remains in the stomach depends on the nature of the food. Carbohydrates stay in the stomach for a relatively short time because enzymic action has already started in the mouth. Proteins remain in the stomach longer than carbohydrates and fats remain longest of all. When the food has been thoroughly mixed with gastric juice in the stomach, it becomes a thick liquid called **chyme**.

At the lower end of the stomach, just under the navel, is a sphincter muscle which is normally tightly closed. This sphincter muscle is called the **pylorus**, which comes from the Greek word meaning 'gatekeeper'. When the chyme in the stomach has the correct amount of gastric juice in it, the pylorus opens and the contents of the stomach pass into the **duodenum**, the first 30 cm of the small intestine. Some time after a meal, you may hear your stomach rumbling. This is caused when the pylorus opens and the stomach empties.

In the duodenum, the chyme is mixed with two more liquids, **liver bile** and **pancreatic juice**. Liver bile, which is made in the liver but is stored in the gall bladder, is not an enzyme. One of its functions is to split fat into small droplets, thus increasing the surface area which can be acted on. Pancreatic juice, produced by the pancreas, contains three enzymes: **trypsin**, **amylase** and **lipase**. Trypsin changes the peptones (from the action of pepsin on protein in the stomach) into simpler substances called proteoses. Amylase changes any starch (cooked or uncooked) which has not already been converted (by the action of ptyalin) into maltose. Lipase acts upon the fat droplets, already broken up by the liver bile, and converts them into fatty acids and glycerine.

By the time the chyme reaches the end of the duodenum, it is much thinner because of the addition of more liquid; it is now called **chyle**. The small intestine is about 6.5 m long. It is called the 'small' intestine because it is only about 2 cm in diameter. Its inside lining is covered with long pimple-like structures called **villi** (singular: *villus*), as shown in Figure 12.6.

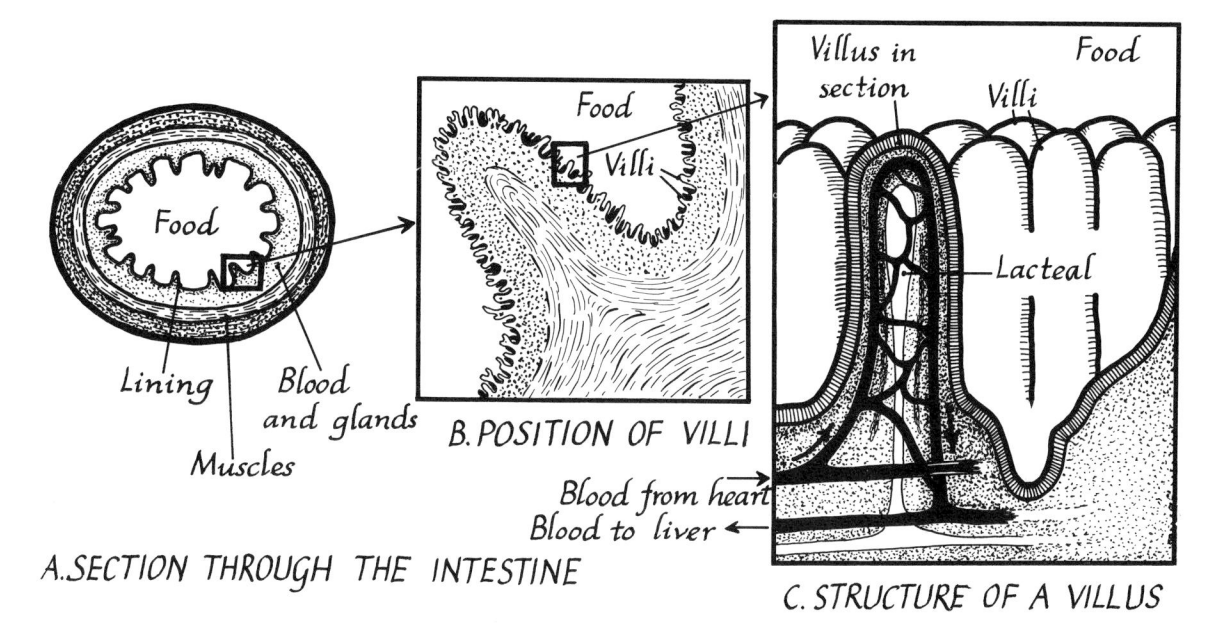

Figure 12.6 The structure of the small intestine

Glands which open between the villi of the small intestine produce four enzymes: **maltase, lactase, sucrase** and **erepsin**. Maltase changes maltose into glucose. Lactase changes lactose (milk sugar) into glucose and galactose. Sucrase changes sucrose (cane sugar) into glucose and fructose. Erepsin converts partly digested protein into soluble amino acids. All of this enzymic action takes place in the upper part of the small intestine.

A summary of the action of enzymes is shown in Table 12.1.

The products of digestion are absorbed through the intestinal wall. Most of the absorption takes place in the lower part of the small intestine, but some absorption occurs in the upper part and, to a more limited extent, in the stomach. Glucose from carbohydrate digestion and the amino acids from protein digestion are absorbed directly into the bloodstream. Some of the fatty acids and glycerine

TABLE 12.1. ENZYMES AND THEIR ACTION

Enzyme	Part of Alimentary Canal	Converts	Into
Ptyalin	Mouth	Starch	Maltose
Rennin	Stomach	Milk	Simpler proteins
Pepsin	Stomach	Proteins	Peptones
Trypsin	Duodenum	Peptones	Proteoses
Amylase	Duodenum	Starch	Maltose
Lipase	Duodenum	Fats	Fatty acids and glycerine
Maltase	Small intestine	Maltose	Glucose
Lactase	Small intestine	Lactose	Glucose and galactose
Sucrase	Small intestine	Sucrose	Glucose and fructose
Erepsin	Small intestine	Proteins	Amino acids

from the fats are absorbed into **lymph vessels** (or **lacteals**) which eventually connect with the bloodstream through the thoracic duct near the neck.

At the lower end of the small intestine is a valve which regulates the flow of chyle into the large intestine. Below this valve is a wide blind tube called the **caecum** (pronounced 'see come'). In herbivores (plant-eating animals), the caecum is comparatively long. Cellulose in the diet of herbivores is digested in the caecum. The caecum in man and other carnivores (meat-eating animals) is very short. At the lower end of the caecum is a thin blind tube called the **appendix**, which serves no useful purpose in man.

The large intestine is about 1·5 m long and about 6 cm in diameter. There is no peristalsis in the large intestine. As the chyle passes through the large intestine, excess water is absorbed through the walls into the bloodstream. This excess water, together with other waste products, is extracted from the blood by the kidneys and passed to the bladder as urine. When the content of the large intestine reaches the rectum, it is in a semi-solid state and consists mainly of the indigestible part of the food. This is now retained in the **rectum**, which acts as a reservoir until the residue is passed through the **anus** as faeces.

The indigestible part of our food consists of cellulose, which is the material of the cell walls of plants. The human digestive system does not produce an enzyme to act on this, so it passes through the alimentary canal unchanged. Although it does not provide any nourishment, it is a very important part of our diet, because it provides the 'bulk' which is necessary for peristalsis. Without this bulk, peristalsis would have very little effect in pushing food along the alimentary canal. This essential part of our diet is called **roughage** and is present in foods such as breakfast cereals, wholemeal bread, celery, raw fruit, etc. The inclusion of roughage in our diet is the best method of avoiding constipation.

Test your understanding

1. What happens to the food of an earthworm when it is in the gizzard?
2. Can you explain why the earthworm is sometimes called 'the farmers' friend'?
3. What are the reasons for chewing food?
4. How could indigestion be caused by toothache?
5. What two precautions could you take in order to prevent constipation?
6. 'Jack Spratt could eat no fat, his wife could eat no lean.' Can you suggest what might have been lacking in the digestive systems of Mr and Mrs Spratt?
7. What is the function of the caecum in herbivores?
8. Why is it unwise to eat a heavy meal just before going to bed?

Chapter 13

Cooking

In their natural surroundings, animals of all dietary types, herbivores, carnivores and omnivores, eat their rood raw. Man, on the other hand, cooks much of his food before he eats it. In this chapter you will learn the reasons for cooking some foods, and some of the methods of cooking food.

13.1. Why do we cook food?

From the point of view of nutrition, cooking usually reduces the food value. Consequently, foods that can be eaten either raw or cooked (fruits and some vegetables) are better eaten raw. This loss of nutritional value can be reduced, in the case of vegetables, if the water in which they are cooked is used as the basis for a soup or gravy. Cooking also breaks down the physical structure of many foods, softens them and reduces the amount of roughage (see Chapter 12).

So far, it would seem that the cooking of food has no advantages; this is true for many fruits and vegetables. It is only when we consider meats and cereals that we can see that the advantages of cooking far outweigh the disadvantages.

The advantages of cooking food can be classified as follows:

a. Cooking softens most foods, thus making them easier to chew; raw meat, for example, would be rather tough. Cooking swells and bursts the starch granules in foods such as potatoes and rice, thus enabling the enzymic action of ptyalin in the saliva to take place (see Chapter 12).

b. The appearance and taste of food is often more attractive when it has been cooked. This, in itself, increases the enjoyment of one of the greatest pleasures in life—eating. It also enables us to eat foods which, in their raw state, we would be unable or unwilling to eat. The flavour of a food can be altered by using different cooking methods. This introduces further variety into our diet and variety is 'the spice of life'.

c. Raw meat is always likely to contain bacteria and, although very rare nowadays, the eggs of parasites such as the tapeworm.

These can be rendered harmless by cooking the meat although, once cooked, the meat becomes highly susceptible to re-infection by bacteria from its surroundings. The risk of re-infection can be reduced by taking sensible precautions: covering the meat or wrapping it in foil and keeping it in a refrigerator or, safest of all, eating it as soon as possible after cooking, before re-infection can occur.

d. Many foods can be preserved for considerable periods of time if they are cooked and sealed in suitable containers. Most fruits, for example, can be preserved by bottling in the following way. The fruit is washed or peeled and then sliced (according to the type of fruit), packed into wide-necked preserving jars and covered with water or syrup (a solution of sugar in water). Glass lids fitted with rubber sealing rings are placed on top of the jars. Brass caps are lightly screwed down so that the lids are held in position loosely enough for steam to escape during the next stage. The jars are then placed in a wide saucepan, or preserving pan, with a little water, which is heated very slowly to a temperature of 72 °C for plums and other stone fruits or 82 °C for apples, pears, etc., and kept at this temperature for five minutes. This process sterilizes the jars and their contents. As the jars and their contents cool down, a partial vacuum is produced inside the jars and air pressure acts on the glass lids, forcing them down on the rubber sealing rings, thus preventing air (and, consequently, bacteria) from entering. The brass caps are now tightened. The effectiveness of the seal can be tested by removing the brass cap and lifting the jar by its lid. If the lid comes off in this test, the whole process should be repeated. The bottling and the canning of food both use the same principle: kill the existing bacteria and then seal to prevent more bacteria from entering.

e. Most people prefer to eat hot food in cold weather because it gives them an immediate heating sensation. In fact, the amount of heat transferred to the body due to the temperature difference between the food and the body is very small, compared with the heat transferred to the body by the digestion of food and subsequent oxidation in the body cells.

13.2. Low-temperature cooking

In low-temperature cooking, water is used as a medium to transfer heat to the food being cooked. The use of water limits the temperature and ensures an even distribution of heat throughout the food.

There are three ways of cooking food by using water: simmering, steaming and boiling.

a. *Simmering*. Food is simmered at a temperature of about 85 °C, bubbles appearing only occasionally at the surface. Stewing (meat and fruit) and poaching (eggs) are terms used when foods are cooked by simmering. Meat is made more tender when it is cooked by this method.

b. *Steaming*. Food being steamed is maintained at 100 °C. The food may be supported in a perforated container above the surface of boiling water in a saucepan so that it is in direct contact with the steam, or it may be in an enclosed basin or dish placed in the saucepan. Steaming is a method suitable for cooking fish, vegetables and puddings.

c. *Boiling*. Food being boiled is surrounded by boiling water. Often a little common salt is added to the water to improve the flavour and alter the boiling point (see Investigation 13a).

Investigation 13a. The effect of salt on the boiling point

Half fill a beaker with water, place it on a tripod and gauze and heat it. Support a thermometer so that the bulb is surrounded by the water. Make a note of the highest temperature registered by the thermometer. Now, add a little common salt to the water and, again, note the highest temperature registered by the thermometer. Continue adding salt, a little at a time, noting the temperature, until the addition of salt has no further effect on the temperature. How does the addition of common salt affect the boiling point of water?

Investigation 13b. The effect of adding salt when boiling potatoes

Half fill two beakers with water and, to one of them, add a little common salt. Place the beakers on tripods and gauzes and heat them. While the beakers are being heated, peel a potato and cut two pieces of about the same shape and size from it. The size and shape of the pieces must be such that, when they are put into the beakers, there will be a free flow of water round them and the beakers will not overflow. When the water in both beakers is boiling, put one piece of potato into each. Can you explain why the water stops boiling when the potato is put in? Continue heating both beakers and, from time to time, prod the pieces of potato. Do they both feel the same or does one feel softer than the other? When both pieces are cooked (from fifteen to twenty minutes), allow them to cool and taste them. Do you notice any difference? Now, sprinkle a little salt on the remainder of the piece which was cooked in the unsalted water and compare its taste with the remainder of the piece which was cooked in the salted water. What difference do you notice?

An alternative method of cooking with water is to use a pressure cooker. At normal atmospheric pressure, water boils at 100 °C. By increasing the pressure on the surface of water, it can be made to boil at a higher temperature. An ordinary saucepan has a loose-fitting lid so that the steam can escape easily. The lid of a pressure cooker fits tightly and has a pressure-regulating valve, which can be set to control the steam pressure which is produced when the water inside the pressure cooker boils. When this predetermined pressure has been reached, steam slowly escapes from this valve and the pressure inside the pressure cooker can be maintained with a low setting of the heat source. In case the regulating valve becomes blocked, a safety valve is also fitted to the lid. When the cooking time has elapsed, the cooker is removed from the stove and is either allowed to cool normally or is cooled rapidly in cold water. As the pressure cooker cools the steam inside it condenses, thus reducing the pressure so that the lid can be removed safely.

Pressure cooking is not only faster, consequently saving fuel (potatoes are normally boiled for fifteen to twenty minutes but can be pressure cooked in only six to eight minutes), but it also preserves the flavour, texture and vitamin content of foods better than conventional cooking.

13.3. High-temperature cooking

There are basically four different methods of cooking food at high temperature :

a. *Oven cooking.* When food is cooked in an oven, air is used as a medium to transfer heat to the food. In oven cooking, the name given to the method depends on the type of dish being cooked:
1. **Baking** is the term used when cooking foods which contain a high proportion of flour, such as bread, cakes, pastries, buns and biscuits.
2. **Roasting** is the term used when cooking foods which contain a high proportion of fat, such as meat and poultry, or food which has been coated with fat.
3. **Braising** is the term used when meat (and sometimes poultry) is cooked in a covered container in an oven, either by itself or with vegetables and water. Because the container for this method of cooking is called a casserole, the method is sometimes referred to as casseroling.

The temperature of an oven is normally controlled by a thermostat which can be pre-set. Gas ovens are controlled by turning a knob to a **Regulo setting** while electric ovens are set to a given temperature. The comparison between Regulo settings and temperature is shown in Table 13.1.

TABLE 13.1. OVEN TEMPERATURES

Regulo Setting	Temperature (°C)	Description
$\frac{1}{4}$	115	
$\frac{1}{2}$	130	Very cool
1	145	
2	155	Cool
3	170	Warm
4	180	Moderate
5	195	
6	205	Fairly hot
7	220	Hot
8	230	
9	245	Very hot

b. *Frying*. When a food is fried, heat is transferred to it by an oil or a fat. When a fat is heated, it melts and, as the temperature reaches the boiling point of water, any water present in the fat will boil, forming bubbles of steam which escape through the surface. On further heating, after the water content of the fat has been removed, a temperature will be reached when the fat begins to smoke and burn. This temperature is known as the smoking temperature of the fat. If a fat is heated beyond its smoking temperature, it decomposes, darkens in colour and becomes unfit for further use. A good frying oil or fat should have a smoking temperature of 180 °C or higher. Particles of food which are allowed to remain in a frying fat will, on repeated heating, decompose and leave particles of carbon in the fat. The presence of these particles has the effect of reducing the smoking temperature of the fat. Such particles can be removed by putting the fat into a saucepan with some water, bringing it to the boil and pouring the contents into a clean dish. When cold, the fat will have formed a solid crust on top and the particles will be in the water or on the lower surface of the crust, from which they can be easily removed. The fat should then be heated until it stops bubbling, in order to remove any water. This process is known as **clarifying**.

Shallow frying is used for foods which contain a high proportion of fat. Such foods may be put directly into a heated dry frying-pan or into a frying-pan with just sufficient hot fat to cover the bottom.

In deep frying, the food is entirely immersed in hot fat. Some foods which are to be deep fried are given a coating of egg and breadcrumbs or batter before being fried. This coating quickly absorbs fat and partially seals the food, thus reducing the loss of water by evaporation and the absorption of fat into the food.

c. *Grilling.* When food is grilled, it is exposed to radiant heat from a high-temperature source. Grilling is a suitable method for cooking foods such as tomatoes, mushrooms, fish, chops, steaks, cutlets, hamburgers, rissoles, etc. The food is placed on a greased grill rack which is supported just below the heat source. The surface of non-fatty foods should be basted from time to time, but foods with a high fat content, such as bacon and herrings, do not need basting to prevent them from becoming too dry. Grilled foods are less fatty than fried foods and are therefore easier to digest.

Because only one side of the food is exposed to the heat source, it is necessary to turn the food several times during the grilling, to ensure that not only both sides but also the centre of the food is fully cooked. In some modern cookers, the food is constantly rotated on a spit which is driven by an electric motor. This part of the cooker is called a rotary grill or a rotisserie. This method of grilling may be compared with the mediaeval method of spit-roasting an entire carcass over a charcoal fire.

d. *Dielectric or high-frequency cooking.* In most electronic circuits (radio, television, etc.) there are components called **capacitors**. Basically, a capacitor consists of two metal plates separated by an insulating material. The plates are called the electrodes and the insulating material is called the dielectric.

When a direct current supply is connected to the electrodes, lines of electrical stress are set up in the dielectric, producing an electrostatic field in much the same way as a magnetic field is set up in the space between two magnets (see Figure 13.1).

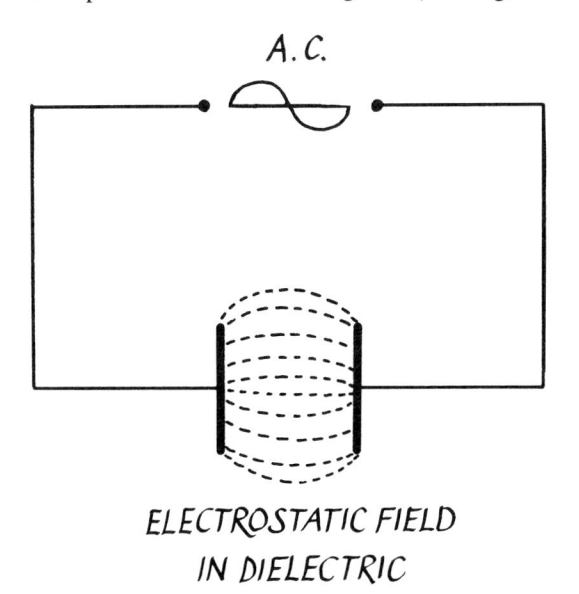

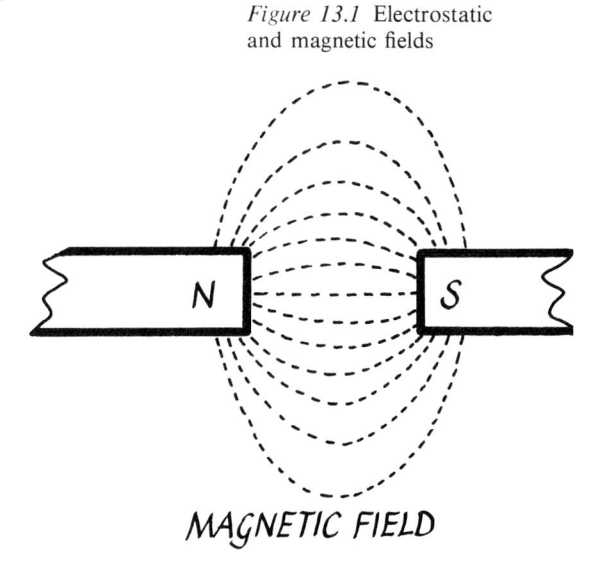

Figure 13.1 Electrostatic and magnetic fields

When an alternating current supply is connected to the electrodes, the strength and direction of the electrostatic field is constantly varying. A dielectric oven designed for cooking food operates on an alternating current supply with a frequency of about 2 000 megahertz (2 000 000 000 hertz).

In a dielectric oven the electrodes form the floor and roof of the oven. The particles of food placed between the electrodes are agitated by the varying electrostatic field. This agitation has the effect of heating the food at the same rate throughout, as opposed to the conventional oven in which the outside of the food is heated first and the inside of the food is then heated by conduction.

Dielectric cooking has sometimes been wrongly called 'cooking from the inside out'. The reason for this inaccurate description is that, in the dielectric oven, the electrodes and the air between them are comparatively cool; consequently, the outside surface of the food cools more rapidly than the inside. This surface cooling can be overcome by the use of normal heating elements in the walls of the oven.

Because any solid material placed between the electrodes will become heated when the current is switched on, an automatic cut-out switch is operated when the door of the oven is opened. No doubt you can see the reason for this precaution.

13.4. Making bread rise

The main constituents of flour are starch and the proteins **gliadin** and **glutenin**. When water is added to flour, the proteins combine to form **gluten** which is a pliable, tacky substance. When gluten is heated, it coagulates (hardens and sets).

Bread is produced by baking a dough made from flour, water, a little common salt and a raising agent. If the raising agent is omitted, the product is unleavened bread, which is rather like hard tack or ship's biscuits. The raising agent normally used in making bread is yeast.

Investigation 13c. The action of yeast

Label four test-tubes and half fill them with the following:

Tube A: ⎱ 4 g of yeast shaken up in 50 cm^3 of 10 per cent
Tube B: ⎰ glucose solution
Tube C: 10 per cent glucose solution
Tube D: 4 g of yeast shaken up in 50 cm^3 of water

Fit a toy balloon over the neck of each tube. Leave tube A in a test-tube rack and place tubes B, C and D in a water bath, maintaining the temperature at about 37 °C for about twenty minutes.

Any reaction which produces a gas will have the effect of inflating a balloon. Has any gas been produced in any of the tubes? If so, pinch the neck of the balloon, remove it from the tube and fit a piece of glass tubing to it. Put the end of the glass tubing into some limewater and allow the gas to bubble through. What happens to the limewater? What does this indicate? What was the purpose of performing this investigation with more than one tube?

Yeast is a single-celled plant which feeds on organic substances, mainly solutions of simple sugars, producing ethanol and carbon dioxide. In a bread dough, the sugar on which the yeast feeds is produced by the action of diastase (a plant enzyme present in the flour which reacts in a similar way to ptyalin in saliva) on starch to form maltose.

The stages in bread-making are as follows:

a. The flour and salt are thoroughly mixed in a warm bowl.

b. The yeast is mixed with warm water and added to the flour and the salt.

c. The mixture is kneaded to form a dough.

d. The dough is covered with a cloth and put in a warm place. The purpose of the cloth is to maintain the temperature of the dough. During this stage, which is called **proving**, the dough will rise to about double its original size in an hour and a half. This increase in size is produced by the release of carbon dioxide.

e. The dough is placed on a flour-dusted board and cut into pieces. Because the dough will lose about 10 per cent of its weight through evaporation during baking, the pieces are cut to allow for this loss. This process is called **scaling**. When the dough is being scaled, there is a small loss of carbon dioxide.

f. The scaled dough is placed in greased tins, is slightly kneaded, to bring fresh supplies of food into contact with the yeast, and is proved a second time, to allow for the loss of carbon dioxide which occurred during the scaling.

g. The tins of dough are placed in an oven to bake until the outer crust is brown. During the baking, the carbon dioxide expands and, although some escapes, most of it is trapped by the gluten as it coagulates, thus producing a light-textured bread.

An alternative method of making bread rise is to put the mixture of flour and salt into an enclosed vessel fitted with a mechanical mixer. Water with carbon dioxide dissolved in it is then added to the flour and salt mixture. The ingredients are then mixed together and the dough is forced out of the vessel into greased tins by the force of the carbon dioxide. The tins are placed in the oven where the expansion of the carbon dioxide and the coagulation of the gluten produces bread with a spongy texture. Bread made by this method is called *aerated* bread and needs no yeast. Because yeast is not used, proving is not necessary.

13.5. Making cakes rise

There are two main methods used to make cakes rise: mechanical and chemical. In the mechanical method, air is incorporated into the mixture and is trapped by the coagulation of proteins during baking. In Yorkshire pudding, air is beaten into a thin batter of flour, eggs, salt and milk. In cake mixtures, the air is trapped by the white of the eggs.

The chemical method relies on the carbon dioxide produced either by heating sodium hydrogen carbonate (**baking soda**) or by the action of an acid on the sodium hydrogen carbonate.

Investigation 13d. The action of heat on sodium hydrogen carbonate

Put about 2 g of sodium hydrogen carbonate in the bottom of a test-tube. Fit a cork and delivery tube and heat the test-tube over a bunsen burner. While heating, put the free end of the delivery tube into some limewater and note any change. What does this change tell you about the gas being given off by the heated sodium hydrogen carbonate? While you were heating the test-tube, did you notice any condensation on the cold part of the test-tube? If not, repeat the investigation and test any condensation that forms with cobalt chloride paper. If you are not certain of what this indicates, refer to Investigation 10a.

Investigation 13e. The action of acids on carbonates

Take six test-tubes, label them, and add the following mixtures.

Tube A: 1 g of sodium carbonate and 1 g of citric acid
Tube B: 1 g of sodium carbonate and 1 g of tartaric acid
Tube C: 1 g of sodium carbonate and 1 g of potassium tartrate (cream of tartar)
Tube D: 1 g of sodium hydrogen carbonate and 1 g of citric acid
Tube E: 1 g of sodium hydrogen carbonate and 1 g of tartaric acid
Tube F: 1 g of sodium hydrogen carbonate and 1 g of potassium tartrate

Thoroughly mix the contents of each tube. Make a copy of Table 13.2 and treat each of the test-tubes, one at a time, as follows: add water, a drop at a time, and note if there is any effervescence (fizzing); test any gas given off by dipping a glass rod into limewater and holding the rod in the mouth of the test-tube; gently heat each test-tube, note if there is any increase in effervescence and re-test the gas. Enter your results in the table.

TABLE 13.2. RESULTS OF INVESTIGATION 13e

Acid	Sodium carbonate		Sodium hydrogen carbonate	
	Cold	Warm	Cold	Warm
Citric acid				
Tartaric acid				
Potassium tartrate (cream of tartar)				

In Investigation 13d, sodium hydrogen carbonate was decomposed into sodium carbonate, water and carbon dioxide:

$$2NaHCO_3 \rightarrow Na_2CO_3 + H_2O + CO_2$$

When sodium hydrogen carbonate alone is used as a raising agent, the cake, scone or bun may have some sodium carbonate remaining in it after baking. This will give it an unpleasant taste and may cause discoloration. In many recipes, an ingredient containing a mild acid is included. Such ingredients include lemon juice, sour milk, syrup, jam, honey and treacle. The acid in these ingredients has the effect of reducing the amount of sodium carbonate remaining in the final product by reacting with it in the same way as the acids in Investigation 13e, producing additional carbon dioxide.

A more reliable raising agent is **baking powder**. This is a mixture of sodium hydrogen carbonate (baking soda), an acid substance such as potassium tartrate (cream of tartar) and a suitable filler. The proportions of baking soda and the acid are calculated in such a way that they react completely, leaving neither of the original substances in the finished cake. You will have noticed from Investigation 13e that no reaction took place until water was added to the mixture. Even a damp atmosphere will allow the reaction to take place, thus reducing the amount of carbon dioxide produced when the baking powder is used, and upsetting the balance of the recipe. Because of this, it is important that baking powder should be stored in dry conditions. The function of the filler is to absorb any water that may be present in the storage container. Rice starch or cornflour may be used as a filler. A typical baking powder can be made by mixing 4 g of baking soda with 9 g of cream of tartar and 2 g of rice starch.

Because cream of tartar is only slightly soluble in cold water and becomes more soluble as the temperature of the water rises, most of the carbon dioxide is released during the baking. The products of the reaction between cream of tartar and sodium hydrogen carbonate (baking soda) are Rochelle salt, water and carbon dioxide:

$$\underset{\text{(cream of tartar)}}{KHC_4H_4O_6} + \underset{\text{(baking soda)}}{NaHCO_3} \rightarrow \underset{\text{(Rochelle salt)}}{NaKC_4H_4O_6} + \underset{\text{(water)}}{H_2O} + \underset{\text{(carbon dioxide)}}{CO_2}$$

The Rochelle salt remaining in the finished cake has no unpleasant taste and does not produce any discoloration.

The most convenient way of making a cake is to use a self-raising flour. This is plain flour with the correct amount of raising agent added and thoroughly mixed.

13.6. Cooking utensils

Containers in which food is cooked should be made from a material which fulfils the following conditions:
a. It must be able to withstand high temperatures without burning, buckling or becoming unserviceable.
b. It must be easily cleaned.
c. It must conduct heat evenly from the heat source to the food.
d. It must not be poisonous.
e. It must be chemically resistant to attack by food being cooked.

Materials used for cooking utensils include:
a. **Iron** utensils are difficult to keep clean and may cause discoloration of the food. Acidic foods, such as fruit, may react with the iron, but traces of iron compounds in the food are not harmful.
b. **Tinned iron** trays and tins for use in an oven should be cleaned carefully to avoid damage, otherwise the iron beneath will rust.
c. **Enamelled iron** utensils are easy to keep clean but are quickly ruined if they are accidentally left empty on a hot stove, because the enamel will flake off. Tinned iron and enamelled iron utensils are called *hollow-ware*.
d. **Copper** utensils, when kept polished, are not attacked by most foods, but prolonged exposure to acidic foods will cause oxidation. Copper is the best conductor of heat of all the materials used for cooking utensils.
e. **Aluminium** is a comparatively good conductor of heat and resists corrosion. Aluminium utensils may become blackened with time, but this blackened surface can be cleaned with steel wool, a scouring powder or by cooking rhubarb in the utensil. Aluminium utensils should be thoroughly dried after cleaning, otherwise the surface will become pitted.
f. **Stainless steel** is an ideal material. It is easy to keep clean and does not easily become corroded.

g. **Earthenware** dishes for pies, casseroles, etc., are satisfactory so long as the glaze is not a lead glaze and they are not subjected to rapid changes in temperature which would crack them.

h. **Boro-silicate glass** ('Pyrex', 'Phoenix', etc.) is an ideal material for ovenware. It is easy to keep clean, will withstand scouring and high temperature but, in common with earthenware, has the disadvantage that it is breakable.

The base of a good cooking utensil for use on a stove should be thick. A thin base is more likely to buckle, and this will cause uneven heating. This is particularly important in a frying-pan where the fat layer between the food and the bottom of the pan should be of an even depth, otherwise burning of the food may occur. When a saucepan or frying-pan is to be used on an electric hot-plate, where heat is conducted from the surface of the hot-plate directly to the base of the utensil in contact with it, any buckling of the base will reduce the area in contact (see Figure 13.2), and this leads to a wastage of heat.

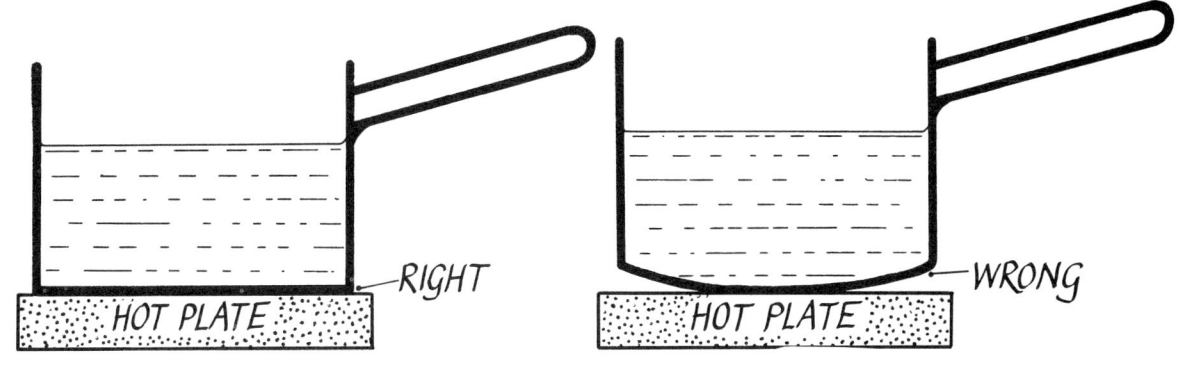

Figure 13.2 Contact areas of flat-based and buckled saucepans

Test your understanding

1. Why does the cooking of starch make it more digestible?
2. What is the main danger of keeping cooked meat for too long?
3. What causes the hissing sound when a bottle of preserved fruit is opened?
4. What is the best method of cooking a tough piece of meat?
5. Why does a pressure cooker cook food in a shorter time than it would take in an ordinary saucepan?
6. What causes the 'spitting' when chipped potatoes are put into hot fat?
7. Why is an automatic cut-out switch fitted to a dielectric cooker?
8. Explain the meanings of 'proving' and 'scaling'.
9. What happens when baking soda is heated?
10. Why should baking powder be stored in an airtight tin rather than a paper bag?

11. What is the difference between plain flour and self-raising flour?
12. What would be the effect of using plain flour in a recipe which required self-raising flour?
13. What is hollow-ware?
14. In what ways is a thick-based saucepan better than a thin-based saucepan?

Chapter 14

Atomic Structure

14.1. Early ideas

For more than 2 000 years men have puzzled about the way in which matter is made up. Greek philosophers, about the year 500 B.C., argued rather like this: 'Suppose you take a piece of metal and break it in halves, and then take one of the halves and break it again. If this is done repeatedly you will eventually have a piece which cannot be broken into anything smaller. This would be the smallest possible piece of metal (an **atom** of metal, i.e. according to the ancient Greeks, that which cannot be split).

14.2. Dalton's atomic theory

Early in the nineteenth century, John Dalton suggested his atomic theory, which contained the following main points:
  a. Chemical elements are made up of very small pieces of matter called atoms.
  b. Atoms cannot be divided, created or destroyed.
  c. Atoms of the same element are all alike, but are different from atoms of other elements.
  d. Atoms of different elements have different weights.
  e. When chemical combination takes place, small whole numbers of the atoms of the combining elements are involved.
  f. The 'compound atoms' so produced are all alike.

14.3. Chemical combination

Dalton based his theory on the observed laws of chemical combination, which include:
  a. **The law of conservation of mass.** This states that matter cannot be created or destroyed.
  b. **The law of constant composition.** This states that the same pure compound always contains the same elements combined together in the same proportion by weight.

14.4. Twentieth-century ideas

Dalton regarded an atom as a solid lump of matter which could not be broken into anything simpler. By the end of the nineteenth century, it was clear that this idea could not be correct.

In 1898, Becquerel discovered that some atoms broke up spontaneously. If a piece of uranium or uranium compound was placed near a photographic plate wrapped in black paper, the plate became fogged, as if it had been exposed to light. This was the discovery of radioactivity.

A year before, J. J. Thompson had produced a stream of cathode rays and had shown that these behaved like a stream of fast-moving particles, each having about $\frac{1}{1840}$ of the mass of a hydrogen atom. Each of these particles carried a negative charge of electricity. These particles are called **electrons**.

About this time, the work of M. and Mme Curie led to the discovery of two other radioactive elements, radium and polonium. It was found that radium atoms broke up spontaneously, forming another element, radon, and at the same time producing energy and positively charged helium atoms.

14.5. Electrons and protons

Matter is usually electrically neutral (it displays neither a negative nor a positive charge). Therefore, if atoms possess electrons which are negatively charged, they must also possess positively charged particles. These positively charged particles are called **protons**. They are very much heavier than electrons, but the positive charge of a proton is equal in strength to the negative charge of an electron. The simplest atom is that of hydrogen (the lightest element), which contains one electron and one proton.

14.6. Neutrons

The second lightest atom is that of helium, which contains two electrons and two protons. This suggests that the weight of a helium atom would be twice that of a hydrogen atom but, in fact, it is found to be nearly four times as heavy. From this we may deduce that, in addition to the proton and the electron, there is a third particle, which has no electrical charge. This particle is the **neutron** and has the same mass as a proton.

14.7. Atomic structure: the Rutherford–Bohr model

Work done by Rutherford, Bohr, Chadwick and others, at the beginning of the twentieth century, led to the idea that the atom might consist of a central nucleus containing protons and (except for the hydrogen atom) neutrons, with electrons moving around it.

Rutherford pictured the atom as being rather like a miniature solar system having a nucleus with the electrons moving around it in orbits, in much the same way as the planets orbit the sun. Bohr stated that the position of the orbit in which an electron happened to be depended on the energy of rotation possessed by it. This quantity of energy must be one of a definite set of possible values. An electron

Figure 14.1 Atomic structure

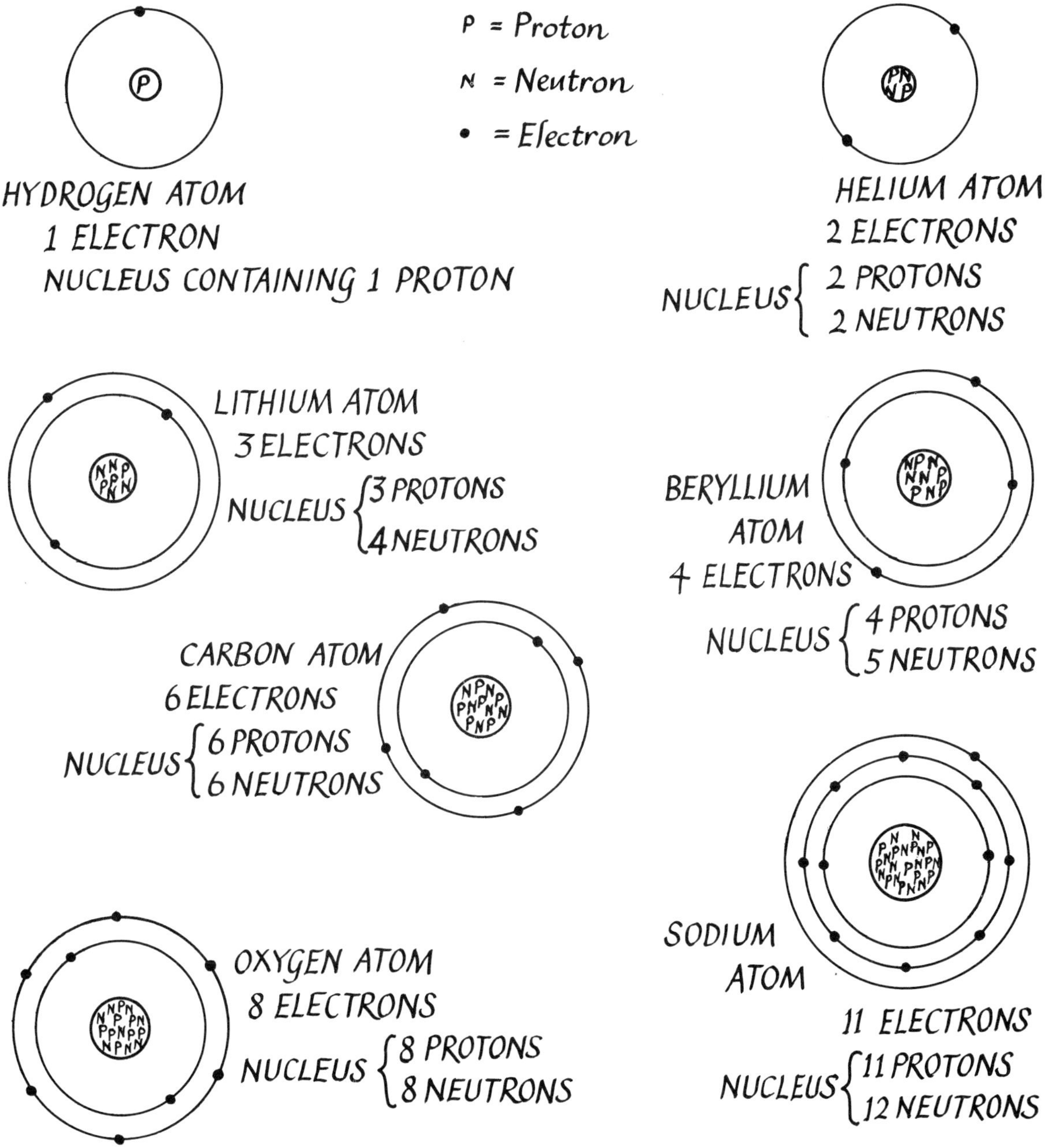

P = Proton
N = Neutron
• = Electron

HYDROGEN ATOM
1 ELECTRON
NUCLEUS CONTAINING 1 PROTON

HELIUM ATOM
2 ELECTRONS
NUCLEUS { 2 PROTONS, 2 NEUTRONS

LITHIUM ATOM
3 ELECTRONS
NUCLEUS { 3 PROTONS, 4 NEUTRONS

BERYLLIUM ATOM
4 ELECTRONS
NUCLEUS { 4 PROTONS, 5 NEUTRONS

CARBON ATOM
6 ELECTRONS
NUCLEUS { 6 PROTONS, 6 NEUTRONS

OXYGEN ATOM
8 ELECTRONS
NUCLEUS { 8 PROTONS, 8 NEUTRONS

SODIUM ATOM
11 ELECTRONS
NUCLEUS { 11 PROTONS, 12 NEUTRONS

at the lowest possible energy level was said to possess one **quantum** of energy. An electron at a higher energy level would have a greater, but whole, number of quanta of energy. This meant that there were definite orbits in which electrons might be found.

Figure 14.1 represents the arrangement of electrons, protons and neutrons for the atoms of various elements. You will notice that the electrons have been represented in definite 'shells'. The innermost shell cannot contain more than two electrons, the second shell may contain up to eight and the third may have as many as eighteen electrons. We shall see shortly that the arrangement of electrons in the outermost shell can be used to explain many chemical reactions.

14.8. Atomic number

The atomic number of an element is the number of protons which its nucleus contains. The atomic number of hydrogen is 1, that of helium is 2, and that of lithium is 3. Write the atomic numbers of beryllium, carbon, oxygen and sodium.

14.9 Mass number

The mass number of an element is the total number of protons and neutrons possessed by the nucleus of its atom. The mass number of hydrogen is 1, that of helium is 4, and that of lithium is 7. Write the mass numbers of beryllium, carbon, oxygen and sodium.

14.10. Isotopes

Contrary to Dalton's belief, we now know that it is possible for an element to have more than one kind of atom. An atom of chlorine, for example, may have a mass number of 35 or 37. Although both types of atom possess the same number of protons, the former has eighteen neutrons, while the latter has twenty. These different kinds of atom are called **isotopes**. We may distinguish between them by writing them thus: ^{35}Cl; ^{37}Cl. In nuclear physics we may represent a certain isotope of uranium by writing $^{238}_{92}U$. This represents an isotope of mass number 238 and of atomic number 92 (i.e. it has a nucleus containing 92 protons and 146 neutrons). You may have heard people speak of 'uranium two-three-five' and 'uranium two-three-eight'. These are the two isotopes of uranium.

14.11. Ions

It is possible for an atom to lose or gain electrons. When this happens, it is no longer called an atom, but an **ion** (pronounced 'eye-on'). Unlike atoms, ions possess an electric charge. A chlorine ion is an atom of chlorine which has gained an electron and, since it has one more electron than it has protons, it is negatively charged. A sodium ion is an atom of sodium which has lost an electron. It is, therefore, positively charged.

14.12. Formation of compounds

Chemical compounds may be formed by the electrical attraction between ions or by the sharing of electrons.

A compound such as sodium chloride, which is formed by the attraction between oppositely charged ions, is said to be **electrovalent** (see Figure 14.2).

A crystal of sodium chloride is a group of ions. The pattern taken up by these ions determines the shape of the crystal (see Figure 14.3).

Figure 14.2 Formation of sodium chloride

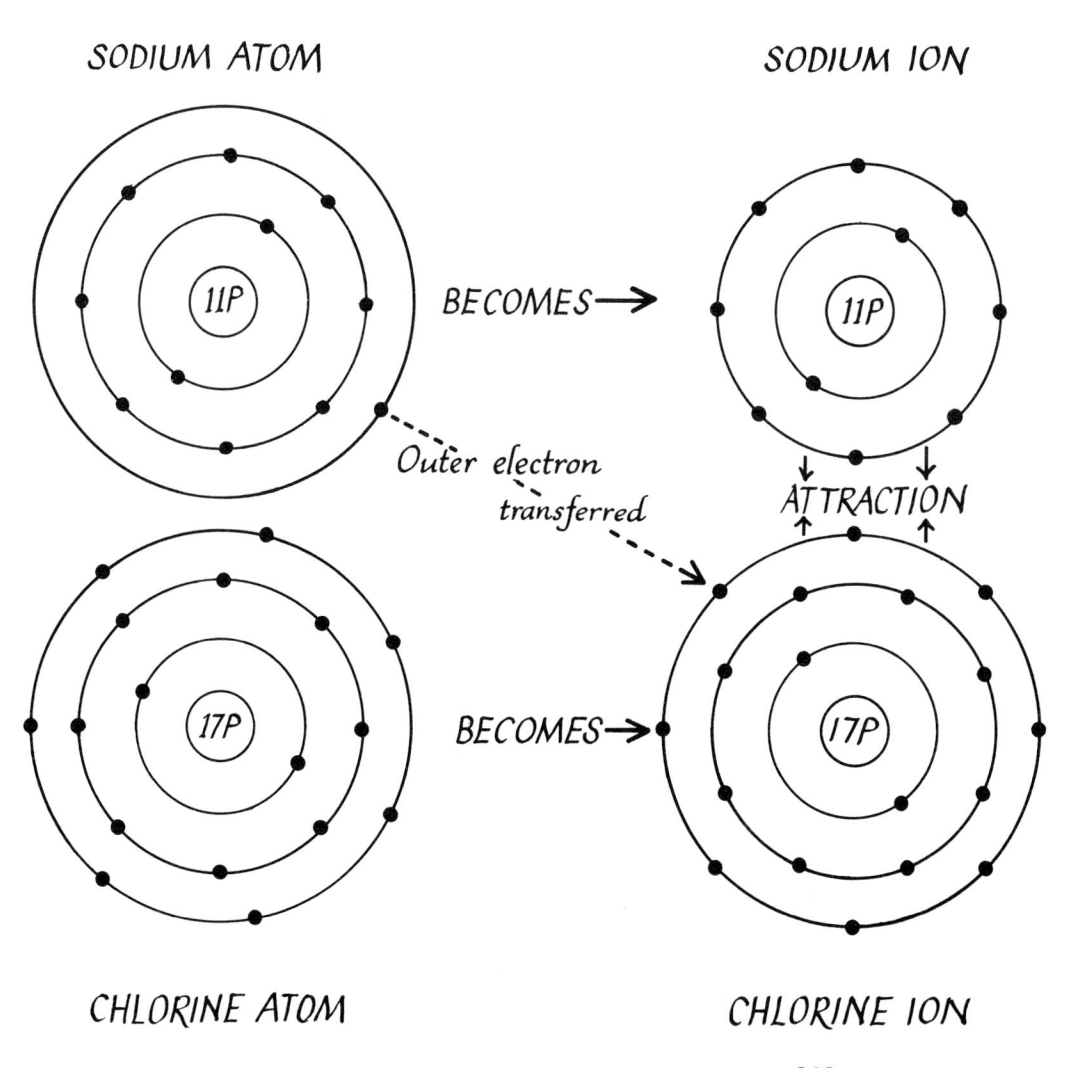

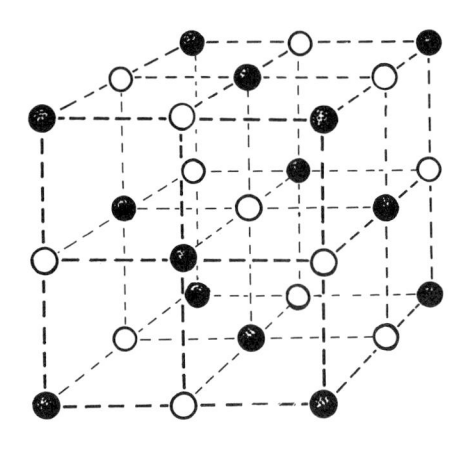

Figure 14.3 Arrangement of ions in a crystal of sodium chloride

● = Centre of sodium ion

○ = Centre of chlorine ion

Figure 14.4 Sharing of electrons in carbon dioxide

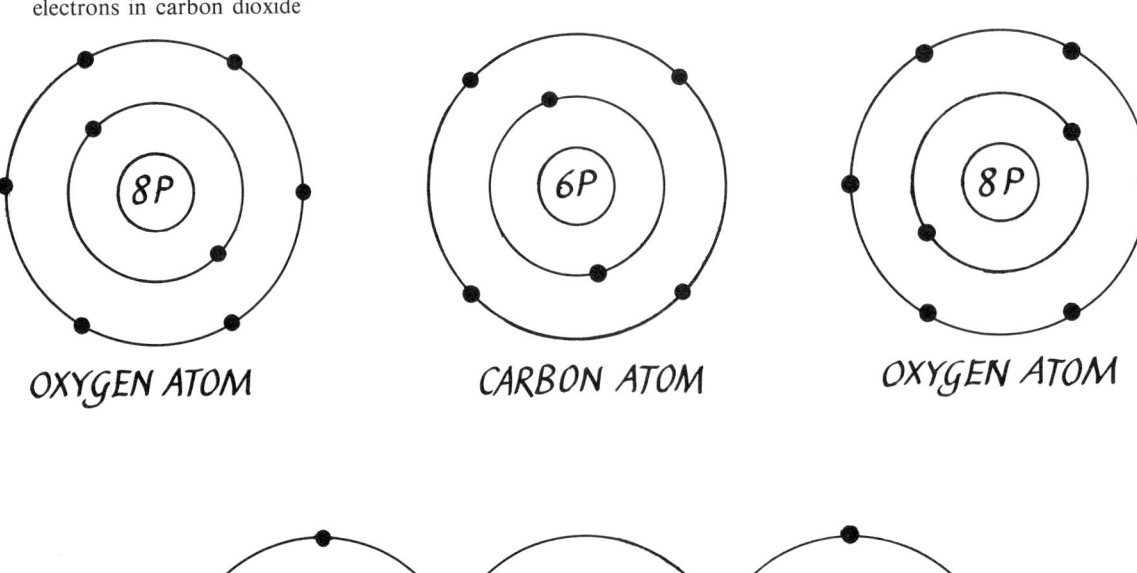

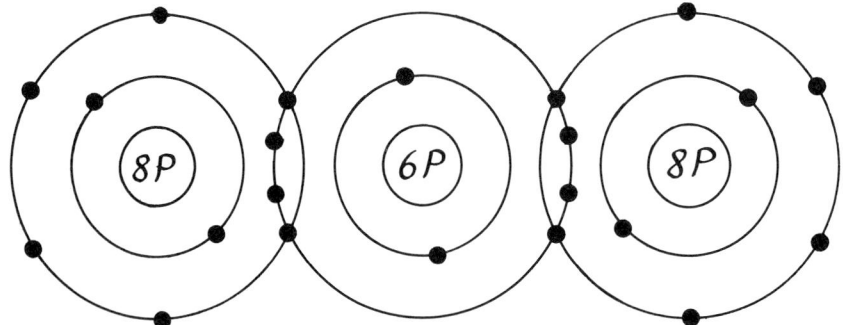

OXYGEN ATOM   CARBON ATOM   OXYGEN ATOM

MOLECULE OF CARBON DIOXIDE

Some compounds are formed by the sharing of electrons between the atoms making up the compound. Such a compound is said to be **covalent**. Carbon dioxide is an example of such a compound (see Figure 14.4). When electrons are shared to form molecules of covalent substances, the shared electrons enable the outside shells to be filled.

Elements having atoms in which the outer electron shell is full do not usually combine with other elements. These are called the **inert** elements and include helium, neon, argon, krypton and xenon. All these are gases which are found, in small quantities, in the atmosphere.

14.13. Electrolytes

Investigation 14a. Do liquids conduct electricity?

Obtain two carbon rods from used dry cells and solder a short length of connecting wire to the cap of each. Make two holes in a large cork and fit the carbon rods into these. Connect the rods in a circuit with a flashlamp bulb and a 4·5-volt battery, as shown in Figure 14.5.

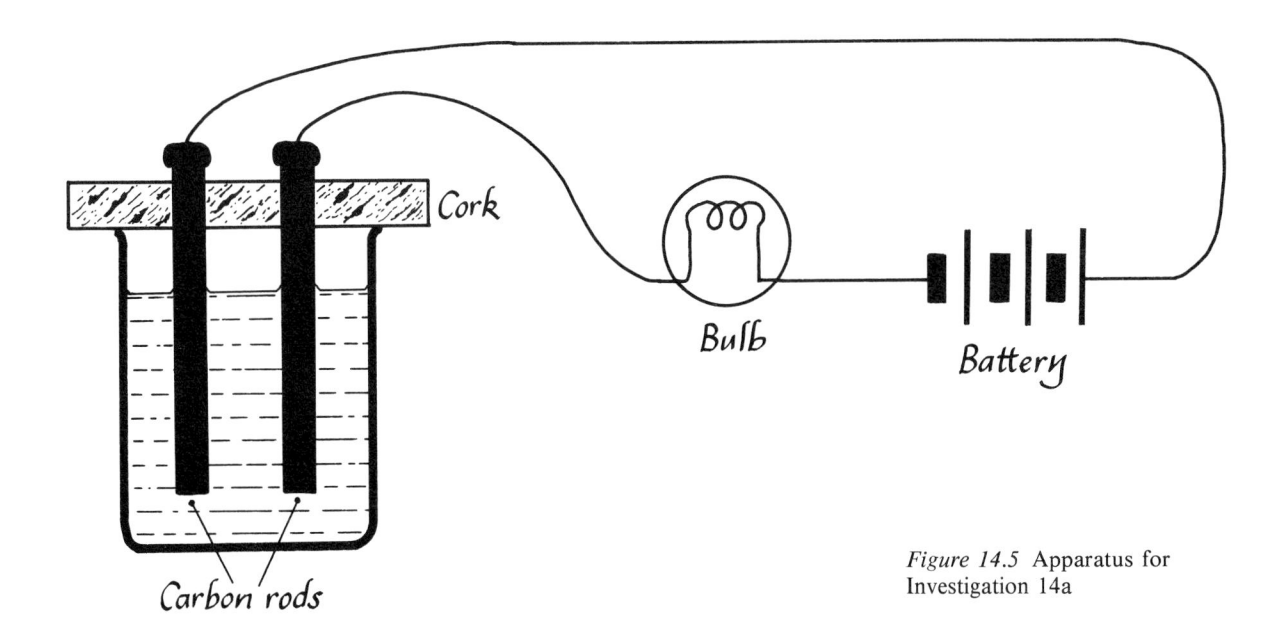

Figure 14.5 Apparatus for Investigation 14a

Take a beaker containing a little water and place the rods in it. Dissolve some sodium chloride in the water and replace the rods. Repeat the investigation, using the following substances in place of the sodium chloride solution: (1) sodium carbonate solution, (2) ammonium chloride solution, (3) copper sulphate solution, (4) sodium hydroxide solution, (5) dilute sulphuric acid, (6) sugar solution, (7) paraffin oil. Which of the liquids conduct electricity?

Investigation 14b. Electrolysis

For this investigation you will need the carbon rods fitted into a cork as in Investigation 14a. You will also require a battery (a lead–acid or nickel–iron battery would be very suitable), a switch, a beaker, some copper sulphate solution and some dilute sulphuric acid.

Set up the apparatus, as shown in Figure 14.6, with copper sulphate solution in the beaker. Switch on and allow the current to flow for about five minutes. Switch off and examine the rods.

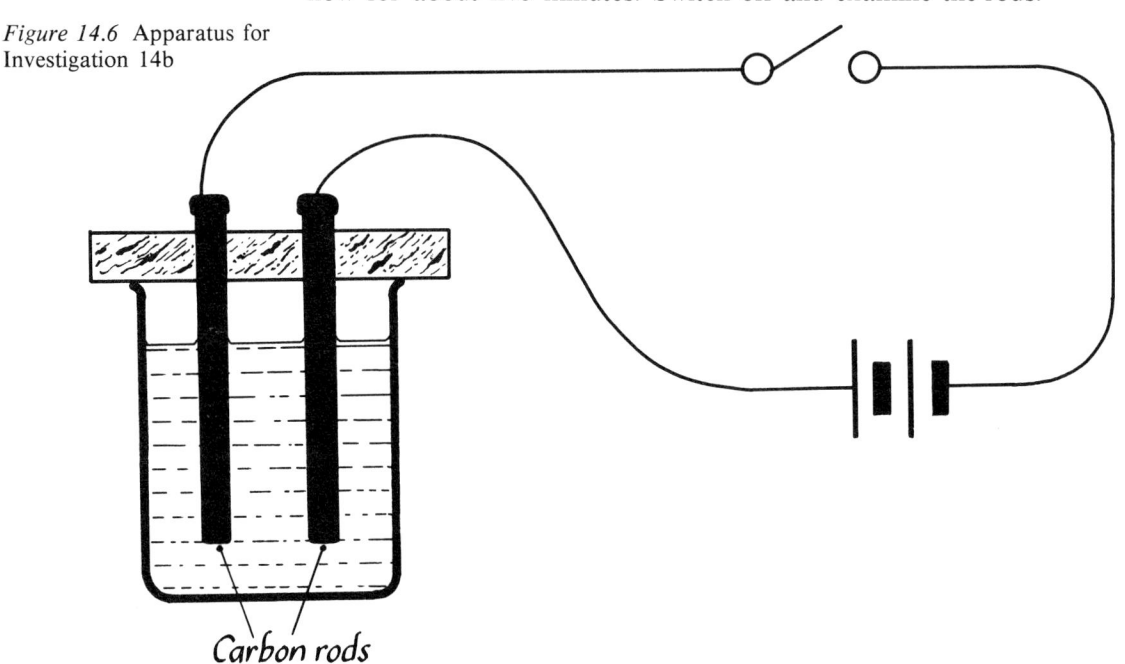

Figure 14.6 Apparatus for Investigation 14b

Carbon rods

Reverse the battery connections and repeat the investigation, noting all your observations.

Switch off, change the copper sulphate solution for dilute sulphuric acid, switch on and observe what happens.

An electrolyte is a compound which, when molten or when in solution, will conduct electricity and at the same time will be chemically decomposed by the passage of the current.

14.14. Electrolysis

The process you carried out in Investigation 14b is called **electrolysis**.

Copper sulphate consists of positively charged copper ions and negatively charged sulphate ions. Water contains some positively charged hydroxonium ions and negatively charged hydroxyl ions.

$$CuSO_4 \rightarrow Cu^{++} + SO_4^{--}$$
(copper sulphate) (copper ion) (sulphate ion)

$$2H_2O \rightarrow H_3O^+ + OH^-$$
(water) (hydroxonium ion) (hydroxyl ion)

The positive ions are attracted by the rod connected to the negative terminal of the battery. This rod is called the **cathode**. The negative ions are attracted by the rod which is connected to the positive terminal of the battery. This rod is called the **anode**.

At the cathode, the copper ions gain electrons and are deposited as copper atoms on the rod. The hydroxonium ions gain electrons and they form water and hydrogen. The hydrogen is given off as a gas.

$$Cu^{++} + 2 \text{ electrons} \rightarrow Cu$$
$$2H_3O^+ + 2 \text{ electrons} \rightarrow 2H_2O + H_2 \uparrow$$

At the anode, the hydroxyl ions lose electrons to form water and oxygen. The oxygen is given off as a gas.

$$4OH^- - 4 \text{ electrons} \rightarrow 2H_2O + O_2 \uparrow$$

During electrolysis, some ions gain or lose electrons more readily than others. At the anode, the hydroxyl ions 'win' in the competition to lose electrons and so are discharged in preference to the sulphate ions. At the cathode, copper ions gain electrons more readily than do the hydroxonium ions. Thus, in the electrolysis of copper sulphate solution, only the copper ions and the hydroxyl ions are discharged.

Dilute sulphuric acid contains hydroxonium ions (positive), hydroxyl ions (negative) and sulphate ions (negative). At the cathode, the positive ions gain electrons to form water and hydrogen, while at the anode, the hydroxyl ions lose electrons to form water and oxygen.

14.15. Matter and energy

In 1905, Albert Einstein stated that matter and energy are equivalent to one another; thus, if it were possible to convert matter into energy, it would be possible to calculate the amount of energy which would be obtained. He gave the formula, $E = mc^2$, in which E represents the amount of energy obtainable from m kilogrammes of matter and c is the velocity of light (300 million metres per second). This means that one kilogramme of matter is equivalent to approximately 90 thousand million, million joules of energy.

In nuclear changes, Einstein's equation is obeyed. We must now include the law of conservation of matter in the more general law of conservation of energy, which states that energy cannot be created or destroyed. It must be emphasized, however, that in ordinary chemical reactions matter is not converted into energy, so, in these conditions, the law of conservation of matter is still obeyed.

14.16. Radioactivity

Some elements have atoms having nuclei which are said to be **unstable**. An unstable nucleus may break up at any time, causing **radioactive radiation** to occur. Sometimes such a break-up results in the formation of a new element. If this happens, **transmutation** is said to take place.

Radioactive radiation is of three kinds.

a. Alpha (α) radiation

This consists of a stream of particles. Each particle is a helium nucleus, comprising two neutrons and two protons. Alpha radiation can be stopped by a few centimetres of air or by a thin sheet of card.

b. Beta (β) radiation

This is also a discharge of particles. They may be electrons, which are negatively charged, or **positrons**, which are particles of similar mass to an electron but have a positive charge. Beta radiation has a range of about a metre in air, but can be stopped by a fairly thin sheet of metal.

c. Gamma (γ) radiation

This is energy transmitted in the form of electromagnetic waves of very short wavelength. They can never be completely stopped, but a shield of thick lead or concrete will considerably reduce the amount of radiation.

★ WARNING. *This must be a demonstration carried out only by a qualified science teacher.*

Investigation 14c. The penetrating power of radiation

A geiger counter is set up, as shown in Figure 14.7, with the tube mounted on a baseboard. You will notice that the counter starts counting the moment it is set up. This is because of **background radiation**, which is always present. If the luminous dial of a watch

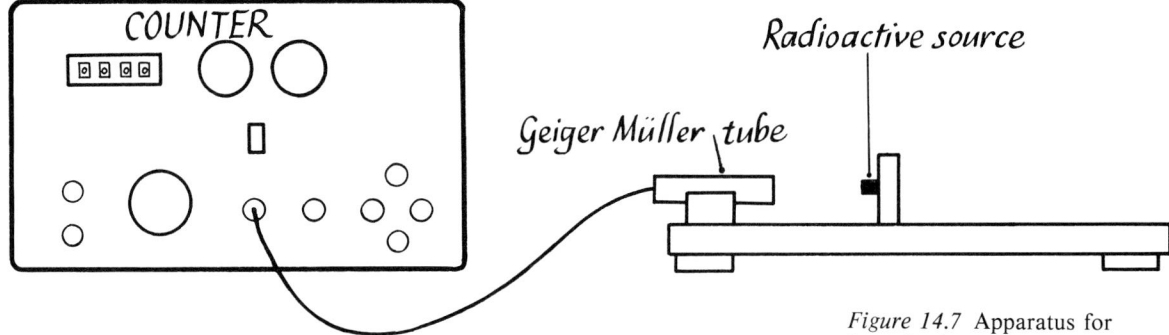

Figure 14.7 Apparatus for Investigation 14c

See D.E.S. Administrative Memorandum 1/65 concerning radioactive substances.

is brought near to the window of the tube, you will notice that the count becomes very much more rapid. This indicates that the dial is radioactive.

A closed source of alpha radiation is now mounted on the baseboard, a few centimetres from the tube. Note the count. Note how the count changes as the source is moved away from the tube. With the source mounted close to the tube, note what happens when a piece of thin card is placed between the source and the tube.

The investigation is repeated using a beta source. Does the thin card stop the beta radiation? Note what happens when a thin aluminium sheet is placed between the source and the tube.

The investigation is now repeated using the gamma source. Does the card stop gamma radiation? Does the thin aluminium sheet stop it? With the source placed about 6 cm from the tube, note the effect of interposing lead plates of various thicknesses.

14.17. Radioactive decay

Although we cannot tell when any particular unstable nucleus will break up, we are always dealing with very large numbers of unstable nuclei and we can measure, or forecast, the time it will take for half of the unstable nuclei to disintegrate. This time is called the **half-life** of the material.

After this time, the specimen will still contain a very large number of unstable nuclei (half the original number) and will still be radioactive. After two half-lives, there will remain one-quarter of the original number of unstable nuclei and after three half-lives, one-eighth of the original number will remain.

14.18. The harmful effects of radiation

Radiation destroys living cells. This may be immediately obvious, as in the case of burns, but the symptoms may appear at any time from six to twenty years after the exposure. If the affected cells are of a type which the body can replace, the symptoms are likely to be temporary. If there are few of that kind of cell in the body, or if the cells are of a type which are irreplaceable, then death may ensue.

Some sources of radiation tend to settle in certain parts of the body; for example, uranium salts tend to settle in the bone marrow, preventing the formation of red blood cells.

If the sex cells are affected, mutations will occur.

14.19. Uses of radiation

a. Tracer techniques

★ WARNING. *Although this could be carried out as an investigation, use of open sources is not normally permitted. See D.E.S. Administrative Memorandum 1/65 concerning radioactive substances.*

Radioactive isotopes may be used to trace the path of a substance through a plant or an animal body. To trace carbon through a plant, it is only necessary to put the plant in a sealed chamber containing some radioactive carbon dioxide. This is used by the plant, in the same way as normal carbon dioxide, and the radioactive carbon follows the same course through the plant as ordinary carbon. By testing the plant for radioactivity, it is possible to measure the rate of flow and to estimate the concentration in various parts of the plant. Similar methods can be used with radioactive isotopes of any element taken up by the plant.

This technique may also be used in medicine; for example, to check the circulatory system, a small quantity of sodium chloride containing a little radioactive sodium may be injected into a patient's bloodstream. The rate of flow may then be checked using a geiger counter. If this is normal, all is well, but if the rate is slow, then there may be an obstruction, such as a blood clot, causing the trouble, and this can then be located.

b. Sterilization of instruments

Bacteria may be destroyed by exposure to radiation. This method is often used in hospitals in order to sterilize instruments and bandages. A chamber is set up containing a strong radioactive source, and boxes containing the material to be sterilized are passed slowly through the chamber. The bacteria are killed without the boxes or their contents becoming radioactive.

c. Thickness gauging

In many industries, it is important to know the thickness of a material accurately. If a radioactive source is placed on one side of the material and a counting tube on the other, it is possible to estimate changes in the thickness. Some of the radiation is absorbed by the material; an increase in thickness will cause more radiation to be absorbed, so the tube gives a lower count. This technique is used in the manufacture of paper.

d. Photographic techniques

X-rays and gamma rays have long been known to 'fog' photographic film. Use is made of this fact in medicine in examination of the body for fractures. If it is suspected that a patient has a fractured arm, a source of X-radiation is set up on one side of the limb and a photographic film is placed on the other. The radiation penetrates the softer tissue more easily than it does bone, so when the film is developed the outline of the bone can easily be seen.

In industry, a similar method is used to test the quality of welded joints. For this purpose, a source of gamma radiation is placed on one side of the joint and a photographic film on the other. When the film is developed, any flaws in the welding, such as cracks or air bubbles, will show up as dark spots on the film.

e. Nuclear power stations

There is a growing use of nuclear energy to produce electricity. A controlled nuclear reaction is used to produce heat, which, in turn, is used to produce steam to drive turbines. These turbines drive the generators. It should be noted that the only way in which a nuclear power station differs from a conventional power station is in the source of heat used to produce steam. A piece of uranium 235 the size of a pin-head can produce as much heat as the burning of about five million kilogrammes of coal. In the next section we shall discuss more fully the kind of reaction which takes place in a nuclear power station.

The above list gives only a few of the many ways in which radioactive isotopes are used today.

14.20. Nuclear fission

Some of the heavy radioactive nuclei can be made to split, forming new elements as well as several nuclear particles. This process is called **fission**.

The nucleus of an atom of uranium 235 will behave in this way. If it is struck by a neutron, the nucleus splits into two smaller nuclei and releases two or three more neutrons. These neutrons could strike other uranium 235 nuclei causing further fission. If this process continues, **a chain reaction** is said to occur. Figure 14.8 shows the principle of the chain reaction.

For a chain reaction to take place, a sufficiently large mass of the radioactive material must be present. If the mass is too small the emitted neutrons may leave the piece of material without striking other nuclei. When there is just enough material to support a chain reaction it is said to have **critical mass and shape**.

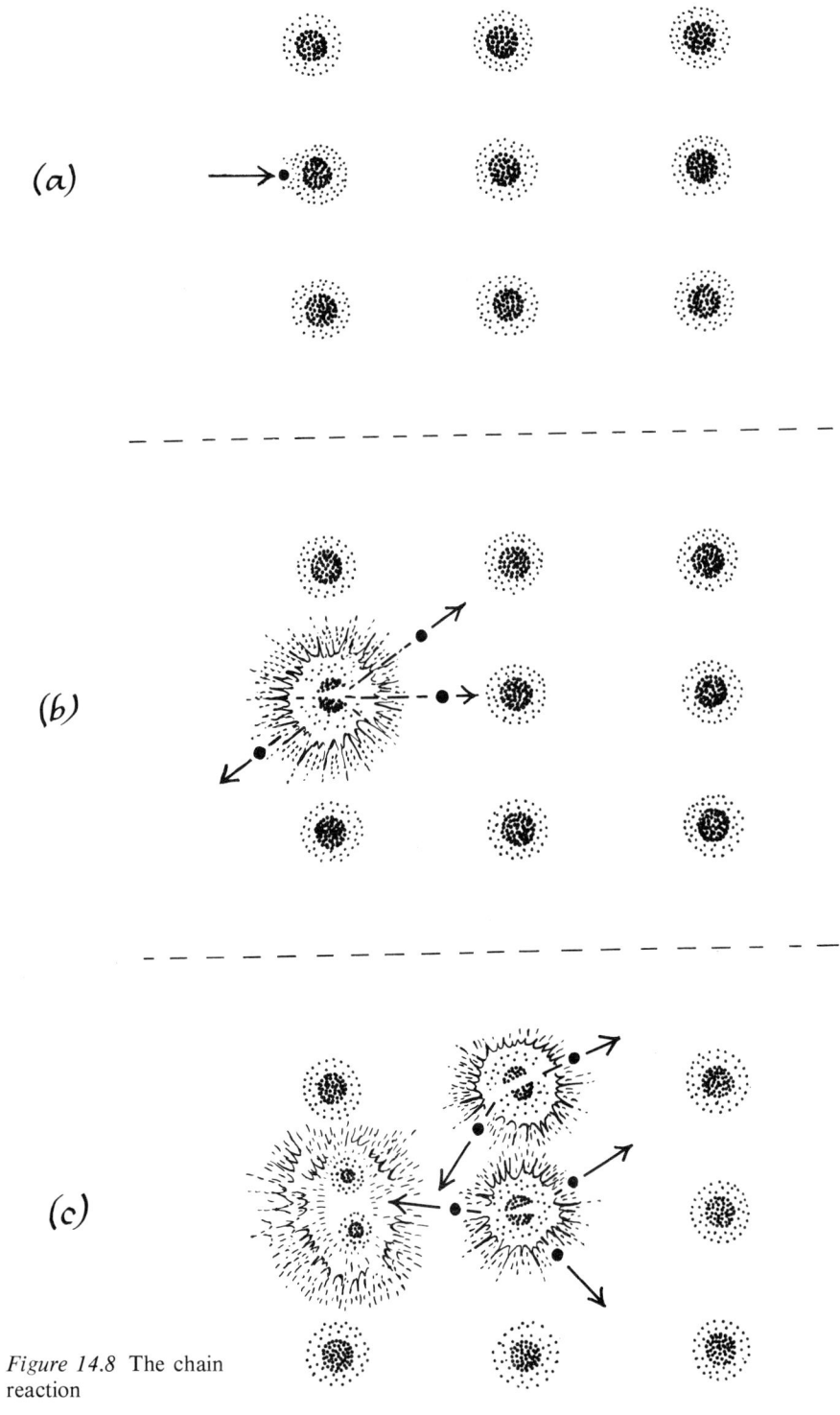

Figure 14.8 The chain reaction

If a chain reaction is not controlled, it proceeds rapidly, liberating vast amounts of energy. This is what takes place in the atomic bomb. In order to use a chain reaction in a power station, it is necessary to control the rate at which the reaction proceeds.

A nuclear reactor consists of a graphite **core** having a number of cylindrical holes in it. Some of these holes contain the fuel elements (rods of uranium) and the others contain boron rods which can be raised or lowered. The graphite core slows down the emitted neutrons, which increases the likelihood of their colliding with uranium nuclei (the graphite is said to **moderate** the reaction). The purpose of the boron rods is to absorb (soak up) neutrons. Each fuel element

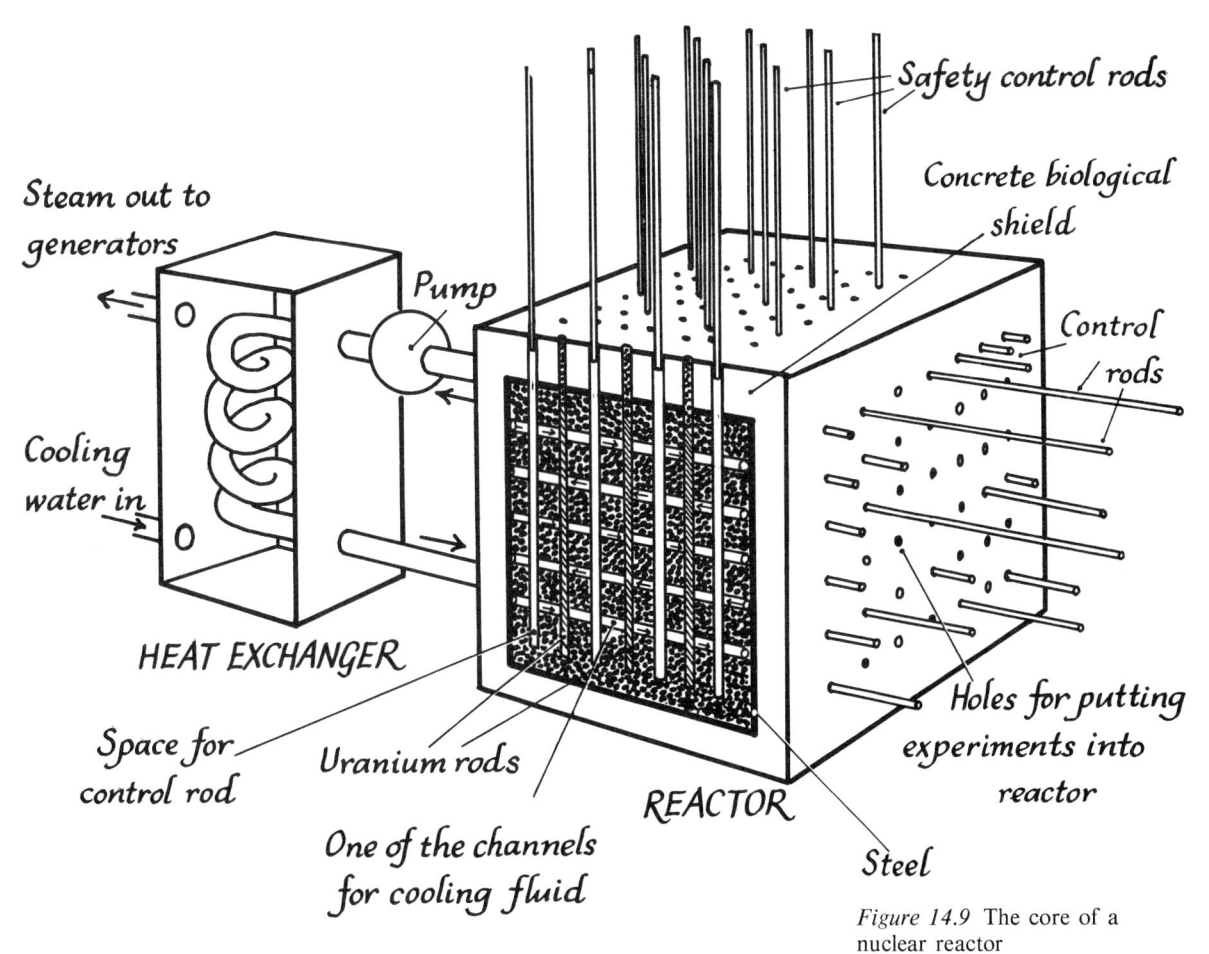

Figure 14.9 The core of a nuclear reactor

contains less than the critical mass of uranium and when the boron rods are lowered they soak up the neutrons, so that a chain reaction cannot take place. When the boron rods are raised, the chain reaction begins and the rate at which this proceeds is controlled by raising and lowering the rods. Figure 14.9 represents this arrangement of moderator, fuel rods and control rods.

The heat energy produced by the reaction must be removed from the reactor and used to produce steam. Usually, carbon dioxide gas under pressure is used to convey the heat from the reactor. The pipes carrying the hot gas are then used to heat the boilers, which produce the steam to drive the turbines.

Precautions have to be taken to protect personnel against radiation from the reactor. The reactor is enclosed by a thick steel container which is surrounded by a very thick layer of concrete. This is called a **biological shield**.

14.21. Nuclear fusion

Nuclear energy is released if hydrogen atoms are made to combine to form a helium atom. This can only occur at very high temperatures and is the reaction which is taking place on the sun (and the other stars).

Test your understanding

1. Draw a diagram representing a named atom. Label three different particles.
2. What is an isotope?
3. What do you understand by $^{238}_{92}U$?
4. What are ions?
5. What would you expect to happen when a direct current is passed between carbon electrodes through (a) a solution of copper sulphate, (b) dilute sulphuric acid?
6. What is an electrolyte?
7. The number of protons in the nucleus of an atom is known as the ..... number.
8. The total number of protons and neutrons in the nucleus of an atom is known as the ..... number.
9. What are the kinds of radioactive radiation? Which kind is most easily stopped? Which kind can never be completely stopped?
10. What is meant by 'transmutation'?
11. What is meant by 'half-life'?
12. A radioactive isotope has a half-life of ten minutes. What fraction of its original energy (of radiation) will remain after forty minutes have passed?
13. The splitting of an atomic nucleus is known as ......
14. What is a chain reaction?
15. What is the purpose of the graphite core in a nuclear reactor?
16. How is the rate of reaction of a nuclear reactor controlled?
17. List some uses of radioactive radiation.
18. Explain briefly the possible harmful effects of radioactive radiation.
19. What is meant by 'nuclear fusion'?

Chapter 15

Electricity

15.1. Electric charges

Investigation 15a. Charging

1. Blow up a balloon, tie it and then rub it vigorously with a clean, dry duster. Place the balloon on the wall or the ceiling.
2. Place a few very small pieces of paper on the bench. Rub the balloon with a duster and bring the balloon slowly towards the pieces of paper.
3. Blow up two balloons and tie a length of thread to each. Rub one balloon while your partner rubs the other. Hold the balloons suspended by their threads and try to bring them together.
4. Rub a balloon as before and hold it close to your partner's head.
5. Rub a balloon and bring it slowly towards your nose.
6. Rub a polythene rod with a duster and see what effect the rod has on small pieces of paper. Try bringing the rod close to your partner's hair.
7. Repeat procedure 6, using a glass rod in place of the polythene rod and rubbing it with a piece of silk.
8. Try using rods of other materials, such as vulcanite, sealing wax, copper, iron, carbon and perspex, in place of the glass rod.
Make a note of all your observations.

When a body, having been rubbed, has the property of attracting small, light objects to it, the body is said to be **electrically charged**. The word 'electricity' comes from the Greek word for 'amber'. When a piece of amber is rubbed with a dry cloth it becomes charged.

15.2. Positive and negative charges

When a balloon is rubbed with a duster, electrons (see Chapter 14) are transferred from the duster to the balloon. The balloon, having an excess of electrons, is said to be **negatively** charged. The duster, having lost electrons, is said to be **positively** charged.

We saw in Investigation 15a that two negatively charged balloons repelled each other.

Investigation 15b. Opposite charges

Charge a balloon and suspend it by a thread. Rub a glass rod with a piece of dry silk and bring the rod close to the balloon. Is the balloon repelled?

When a glass rod is rubbed with silk, electrons are lost by the rod and so it becomes positively charged.

The law of electric charges is that **like charges repel** one another, while **unlike charges attract** one another.

15.3. Conductors and insulators

Investigation 15c. Charging an electrophorus

In this investigation you will be using an electrophorus and a gold leaf electroscope.

An electrophorus is a metal disc fitted with an insulating handle. A pad made from an insulator is used with the electrophorus (see Figure 15.1).

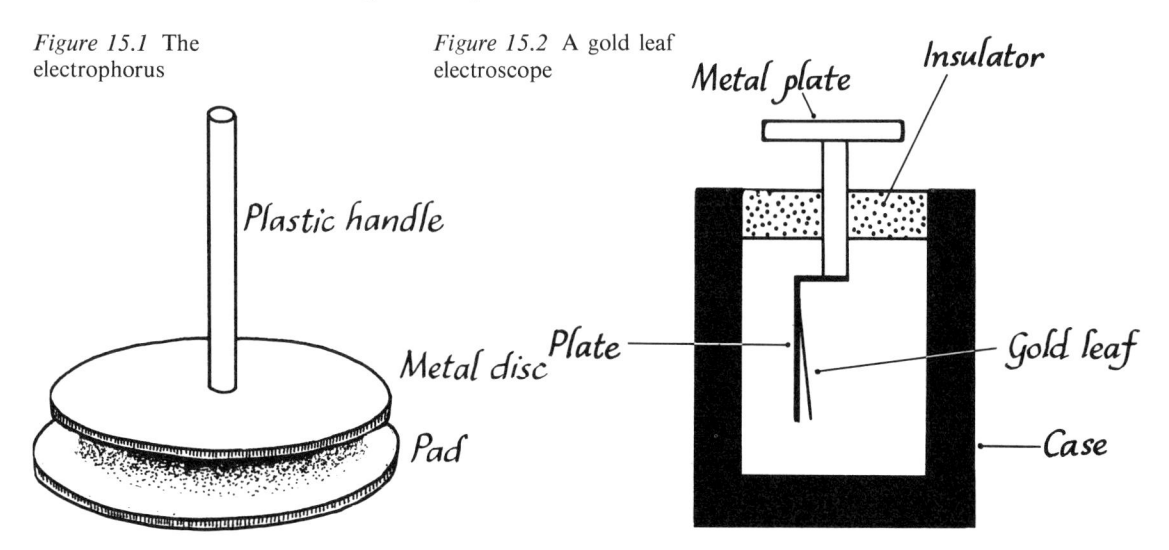

Figure 15.1 The electrophorus

Figure 15.2 A gold leaf electroscope

The electroscope consists of a metal disc attached by a short metal rod to a rigid metal plate. Suspended alongside the plate is a strip of gold leaf. A case surrounds this arrangement, as shown in Figure 15.2.

When a charged body is brought near to the metal disc of the gold leaf electroscope, the gold leaf diverges from the metal plate. Why do you think this happens? Try to bring a charged balloon near the disc.

Holding it by the handle, charge the electrophorus by rubbing it on its pad. Take care not to touch the metal disc of the electrophorus. Bring the disc near to the head of the electroscope. Is the electroscope charged?

Keeping it near the electroscope, touch the metal part of the electroscope with your finger. What happens?

Materials such as metals allow a ready movement of electrons through them. Such materials are good **conductors** of electricity. Substances which do not readily allow a flow of electrons through them are bad conductors, or **insulators**. When a charge is built up on an insulated body so that electrons cannot flow, the charge is said to be **static**.

When the electrophorus was rubbed, a charge built up on the metal plate. This could not flow through the insulated handle, but it was able to flow through your body when you touched the plate.

15.4. Current electricity

When electrons flow through a conductor, this constitutes a **current** of electricity. The path along which such a current flows is called an electrical **circuit**. When a battery is connected in a circuit it acts as an 'electron pump', driving electrons through the circuit. A battery is said to provide an **electromotive force**. The **volt** is the unit in which electromotive force is measured.

Investigation 15d. Simple circuits

For this investigation you will need a cell, two flashlamp bulbs (1·25 V) and holders, a switch, some connecting wire, two crocodile clips, strips of wood, rubber and carbon, and one-metre lengths of copper wire, iron wire and nichrome wire. These one-metre lengths of wire should be of the same diameter. If Worcester circuit boards are available, the circuits may be conveniently set up on these.

The making of circuit diagrams is greatly simplified by using special symbols. Some of these are given in Figure 15.3. Learn them. You will be meeting others in later chapters.

Figure 15.3 Some electrical symbols

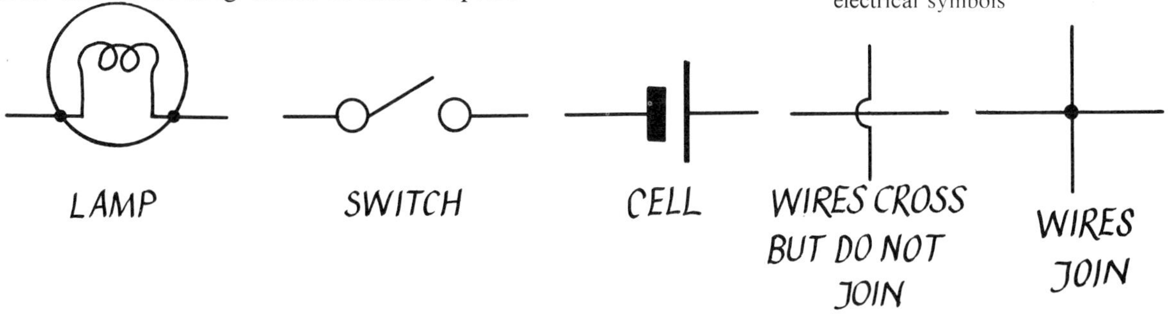

LAMP   SWITCH   CELL   WIRES CROSS BUT DO NOT JOIN   WIRES JOIN

Set up each of the circuits shown in Figure 15.4. In each case, note what happens when the switch is closed. (Make sure that the bulbs are screwed tightly into their holders.) In Circuits 15.4b and 15.4c, note what happens when one bulb is removed from its holder.

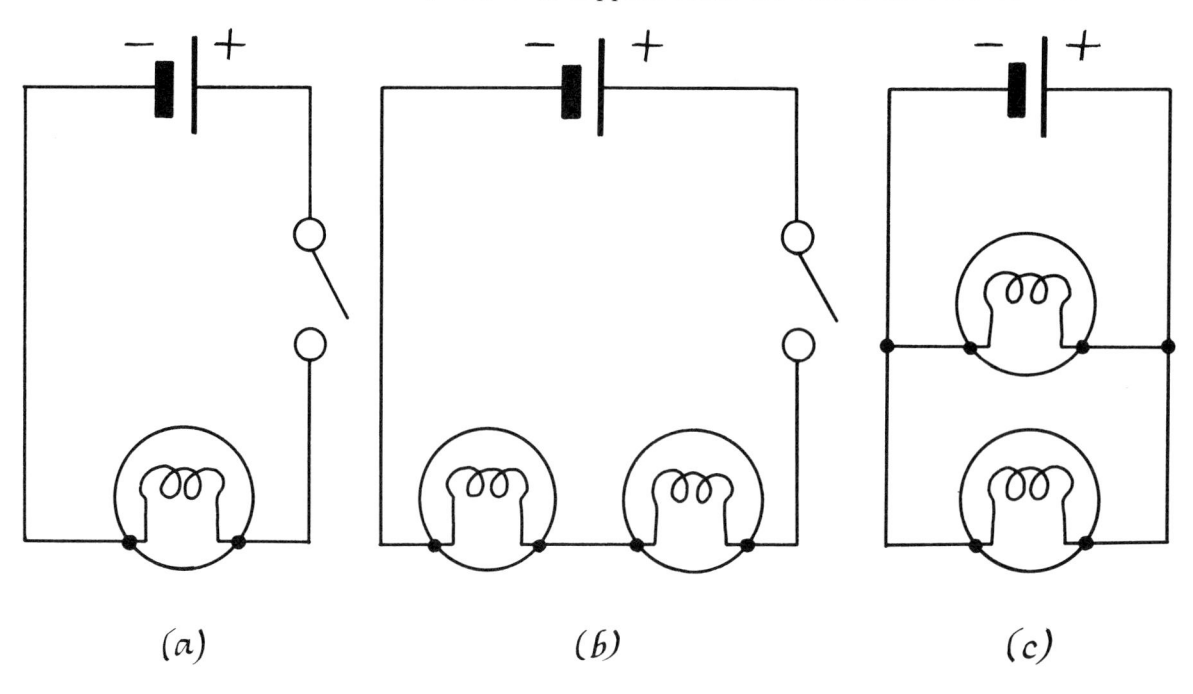

(a) (b) (c)

Figure 15.4 Some simple circuits

In Circuit 15.4b, the lamps are said to be connected in **series**. Christmas tree lights are often connected in this way. Circuit 15.4c shows lamps connected in **parallel**. The lamps in your home are connected in this way.

Now set up the circuit shown in Figure 15.5. In turn, connect each of the one-metre lengths of wire between the crocodile clips. Then

Figure 15.5 Good and bad conductors

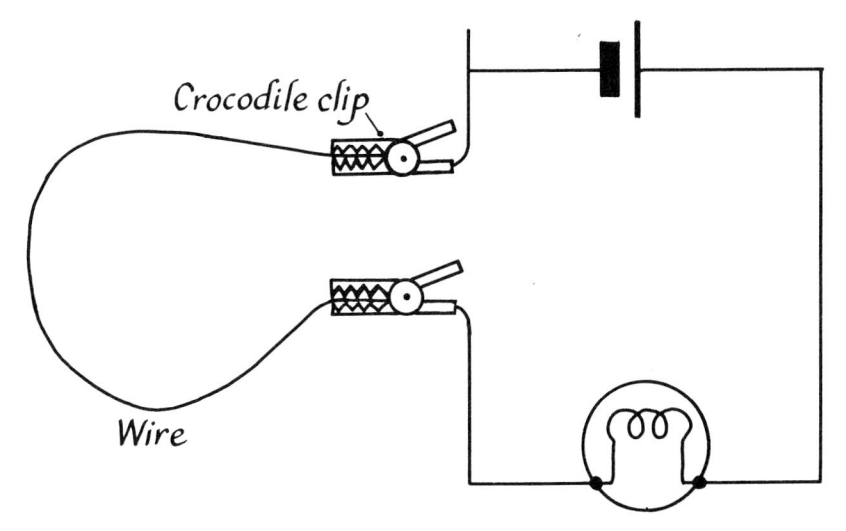

connect the strips of wood, rubber and carbon, in turn, between the crocodile clips.

From your observations it should be possible for you to make lists of conductors and insulators. You should also have noticed that some conductors are better than others.

15.5. Generating electricity

Investigation 15e. Simple voltaic cells

For this investigation you will need dilute solutions (about 2 M) of sulphuric acid, sodium hydroxide and ammonium chloride, and some ordinary table vinegar. You will also require two zinc plates, two copper plates, some connecting wire, a flashlamp bulb (1·25 V) and holder, and a voltmeter (range 0–3 V).

Pour some dilute sulphuric acid in a beaker and place a copper plate and a zinc plate in the acid, making sure that they do not touch. First connect the voltmeter and then the flashlamp bulb across the plates, as shown in Figure 15.6. Note what happens and then repeat the investigation, using first two copper plates and then two zinc plates.

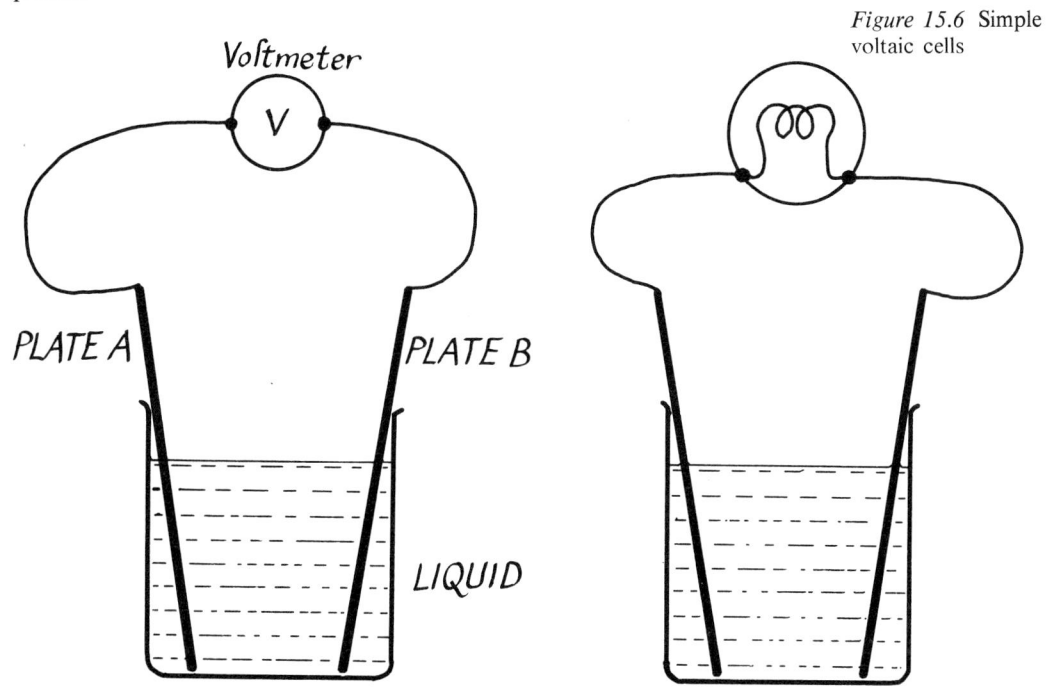

Figure 15.6 Simple voltaic cells

Repeat the whole investigation, using each of the other liquids, in turn, in place of the sulphuric acid. Every time you set up the apparatus rinse the plates thoroughly. Draw up a table of results, as shown in Table 15.1.

TABLE 15.1. SIMPLE CELLS

Liquid	Plate A	Plate B	Observations	
			Voltmeter	Lamp
Dilute sulphuric acid	Copper Copper Zinc	Zinc Copper Zinc		
Sodium hydroxide solution	Copper	Zinc		

Electricity may be produced by placing two plates of different conducting materials in a solution of an electrolyte. The plates must not touch within the liquid, and an electric current is produced only when the plates are connected externally by an electrical circuit. A device of this kind is called a **voltaic cell**. It derives its name from Alessandro Volta, an eighteenth-century Italian scientist, who carried out work in this sphere.

Investigation 15f. Polarization

Set up a simple voltaic cell consisting of copper and zinc plates immersed in dilute sulphuric acid. Connect a flashlamp bulb (1·25 V) across the plates and observe what happens.

After a short while, the lamp will grow dim and go out. When this happens, remove the plates, rinse them and wipe them with a dry cloth. When they are immersed again the bulb should light up.

When the lamp failed, did you notice that bubbles of gas had formed on the plates? In any practical form of cell, it is necessary to prevent the formation of such gas bubbles. This is usually achieved by making use of a suitable chemical compound, known as a **depolarizing agent**. (If bubbles of gas form on the positive plate of a cell it is said to be **polarized**.)

The most important type of voltaic cell is the Leclanché cell. This is shown in Figure 15.7 (a) and (b). Figure 15.7 (a) shows a wet Leclanché cell. This is an efficient cell, but it is relatively heavy and the electrolyte is easily spilled. If such a cell is available, examine it and connect a flashlamp bulb across its terminals. Notice that it does not become polarized. Figure 15.7 (b) shows a dry cell. In this, the container is made of zinc and also serves as one of the plates. The ammonium chloride solution is made into a paste. As this is moist the cell is not really dry.

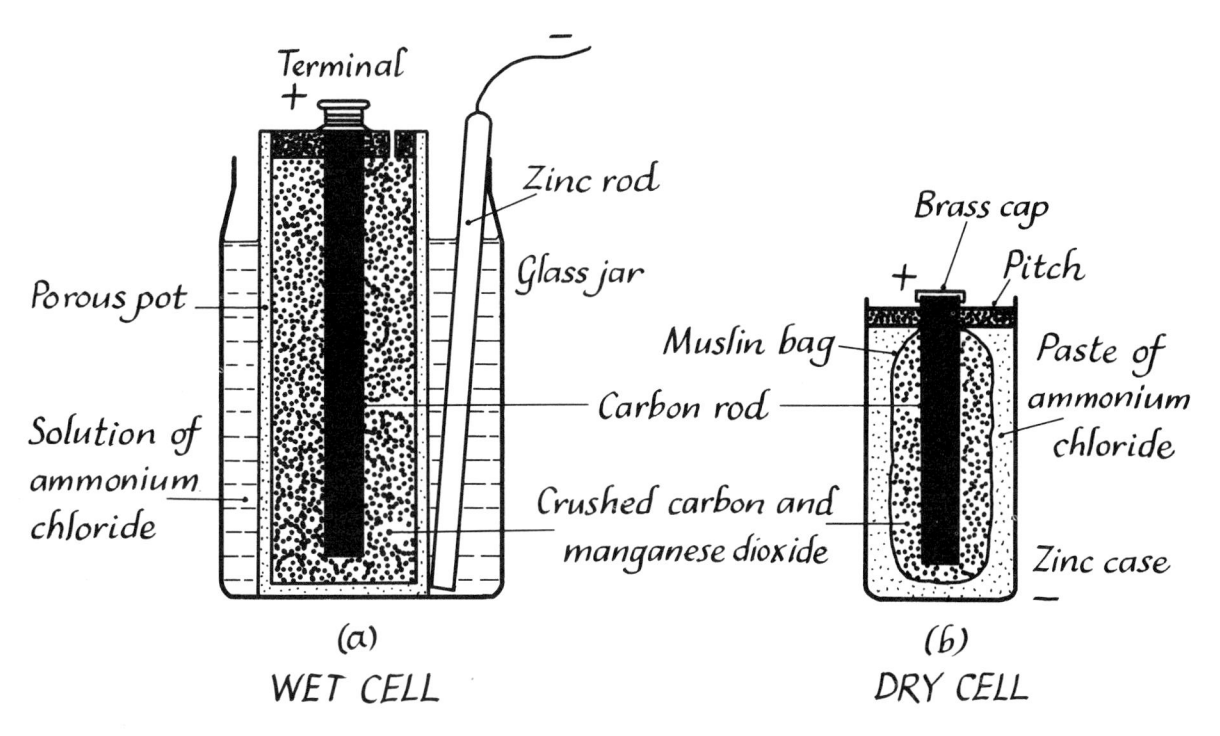

Figure 15.7 Wet and dry Leclanché cells

Investigation 15g. The photovoltaic cell

Obtain two copper plates and clean them thoroughly, using fine emery cloth. Support one of these plates well above a lighted bunsen burner and leave it there for about ten minutes. Allow it to cool and then examine it. It should be coated with a thin brown layer of copper oxide (copper (I) oxide).

Make up a solution of ammonium carbonate by dissolving about 20 g in 200 cm³ of water. Add a little ammonia solution to this. Pour the solution into a beaker, place the two plates in the beaker, making sure that they do not touch, and connect a centre-zero galvanometer to the plates, as shown in Figure 15.8. Shine a bright light on the plates in the cell. What happens?

Because this type of cell uses light energy in order to produce an electric current, it is called a **photovoltaic cell**. Photovoltaic cells are

Figure 15.8 Making a photovoltaic cell

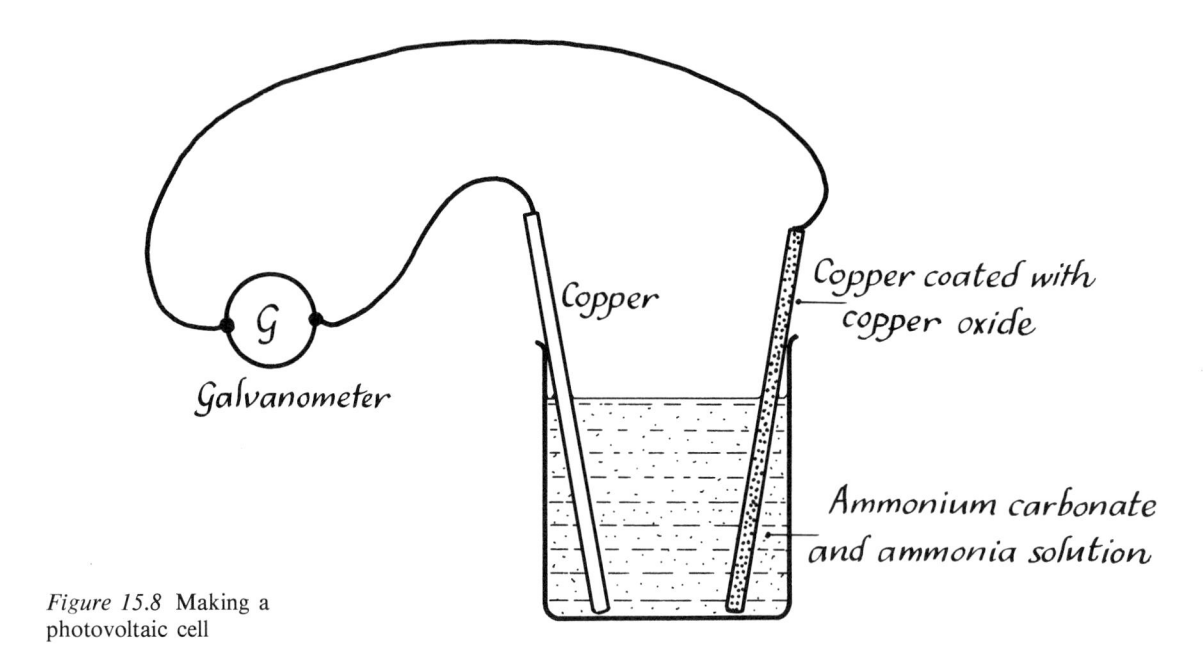

used in photographic light meters, and some types, known as solar cells, have been developed in order to operate apparatus used in man-made satellites. Figure 15.9 shows the construction of a practical photovoltaic cell.

In Chapter 4 a thermocouple was used as a method of measuring temperature. You will realize that, in order to read the temperature,

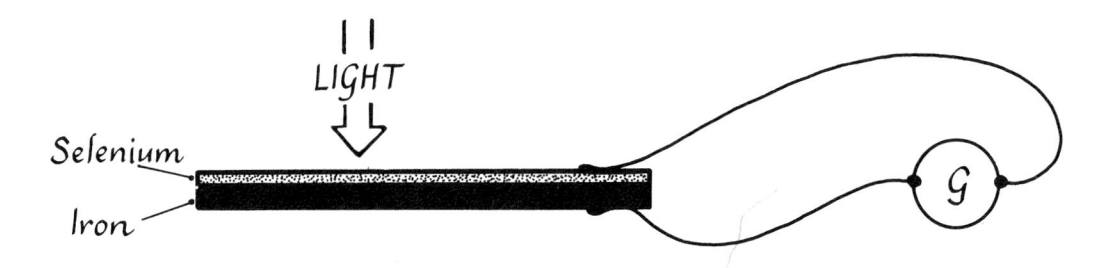

Figure 15.9 A photovoltaic cell

an electric meter of some sort had to be used. This device converts heat energy into electrical energy.

Investigation 15h. The dynamo principle

Obtain a solenoid and connect it to the terminals of a centre-zero galvanometer, as shown in Figure 15.10. Move a bar magnet slowly in and out of the solenoid. Note how the galvanometer pointer is deflected.

Then try the effect of moving the magnet more quickly. Repeat the experiment, using a more powerful magnet. Repeat the investigation, using a solenoid having a much larger, or a much smaller, number of turns.

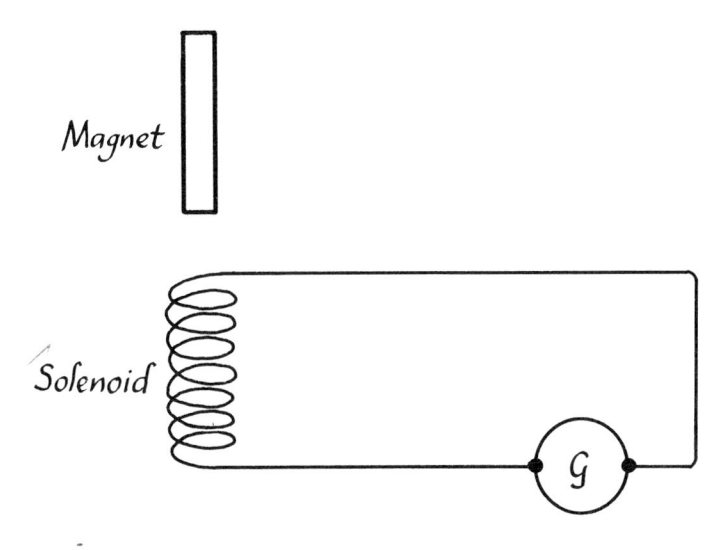

Figure 15.10 The dynamo principle

When a magnetic field cuts across a conductor, electricity is generated. In this way, mechanical energy is converted into electrical energy. The dynamo works on this principle.

One of the simplest forms of dynamo is the cycle dynamo. This has a solenoid wound on to the middle of a U-shaped iron core. A magnet rotates between the arms of the core. Figure 15.11 illustrates the general arrangement.

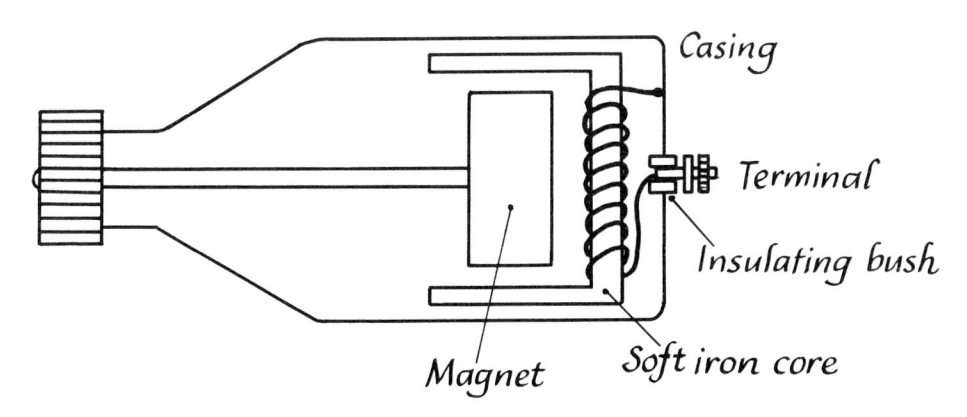

Figure 15.11 The cycle dynamo

Test your understanding

1. What is the law of charges?
2. Sketch a gold leaf electroscope. How is the electroscope affected by the presence of a charged body?
3. Make circuit diagrams showing (a) three lamps connected in parallel, (b) three cells connected in series.

4. Draw a labelled diagram of a dry cell.
5. How would you obtain (a) electrical energy from light energy, (b) electrical energy from heat energy?
6. Describe what you would expect to happen when a solenoid connected to a centre-zero galvanometer has a bar magnet moved slowly up and down inside it.

Chapter 16

Effects of Electricity

In Chapter 15 we saw how other forms of energy could be converted into electricity. Electricity is a very useful source of power because it can easily be converted into other forms of energy.

16.1. The magnetic effect

Investigation 16a. Some properties of magnets

For this investigation you will need two small bar magnets, strips of iron and as many other different metals as you can obtain, a large sheet of paper, a pepper pot containing iron filings, some thread and a wooden retort stand.

1. Using a piece of thread about 30 cm long, suspend one of the bar magnets from the stand. Place this as far as possible from the other apparatus and allow the suspended magnet to come to rest. When this happens, mark an arrow on one end of the magnet, and another on the bench, pointing in the same direction as the arrow on the magnet (see Figure 16.1). Repeat, using the other magnet.
2. Suspend one of the magnets as before. Take the other magnet and try to bring its marked end near to the marked end of the suspended magnet. Now try bringing the unmarked end of the magnet you are holding near to the marked end of the suspended magnet.
3. Place all the strips of metal on the bench and bring one of the bar magnets near to each strip in turn.
4. Place a bar magnet on the bench and cover it with a large sheet of paper. Holding the pepper pot well above the paper, sprinkle iron filings on to the paper. Tap the paper with your finger. Make a note of everything you observe.
5. Repeat experiment 4, using nickel powder in place of iron filings.

When a magnet is suspended, it comes to rest pointing in a north–south direction. The end of the magnet pointing north is called the north (or north-seeking) pole of the magnet and the other end is the south pole of the magnet.

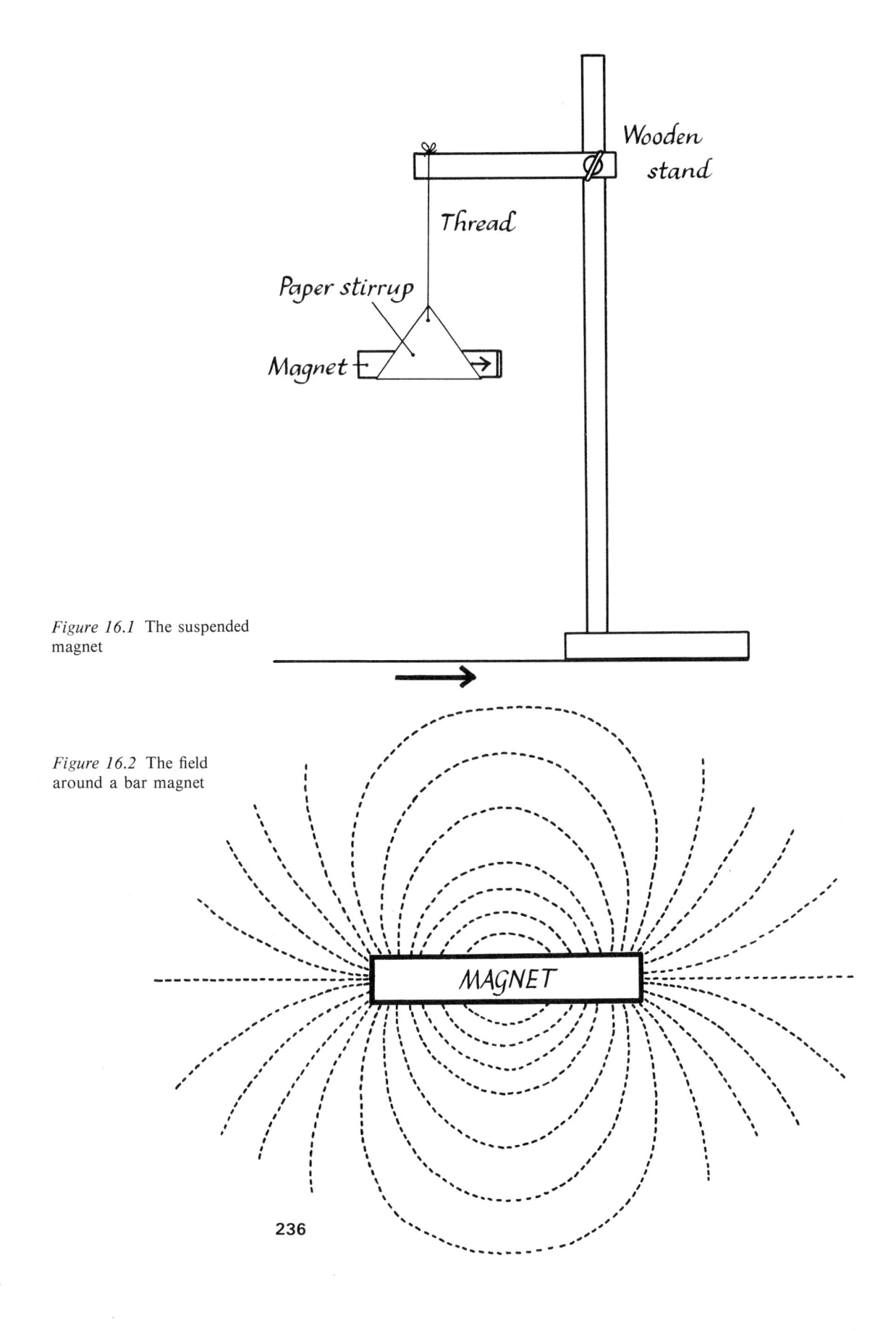

Figure 16.1 The suspended magnet

Figure 16.2 The field around a bar magnet

236

When two magnets are brought together, **like poles repel** one another, but **unlike poles attract** one another.

A magnet will attract to itself unmagnetized pieces of the following metals: iron (and steel), nickel and cobalt.

The space around a magnet where its influence can be detected is called the **field** of the magnet. Figure 16.2 shows the shape taken up by the iron filings in Investigation 16a (4).

Investigation 16b. The magnetic effect of a current

Support a piece of stiff copper wire vertically, so that it passes through a piece of stiff cardboard, held horizontally. Complete the circuit, as shown in Figure 16.3, using a lead–acid cell and a bell-push type of switch. Place several plotting compasses on the card.

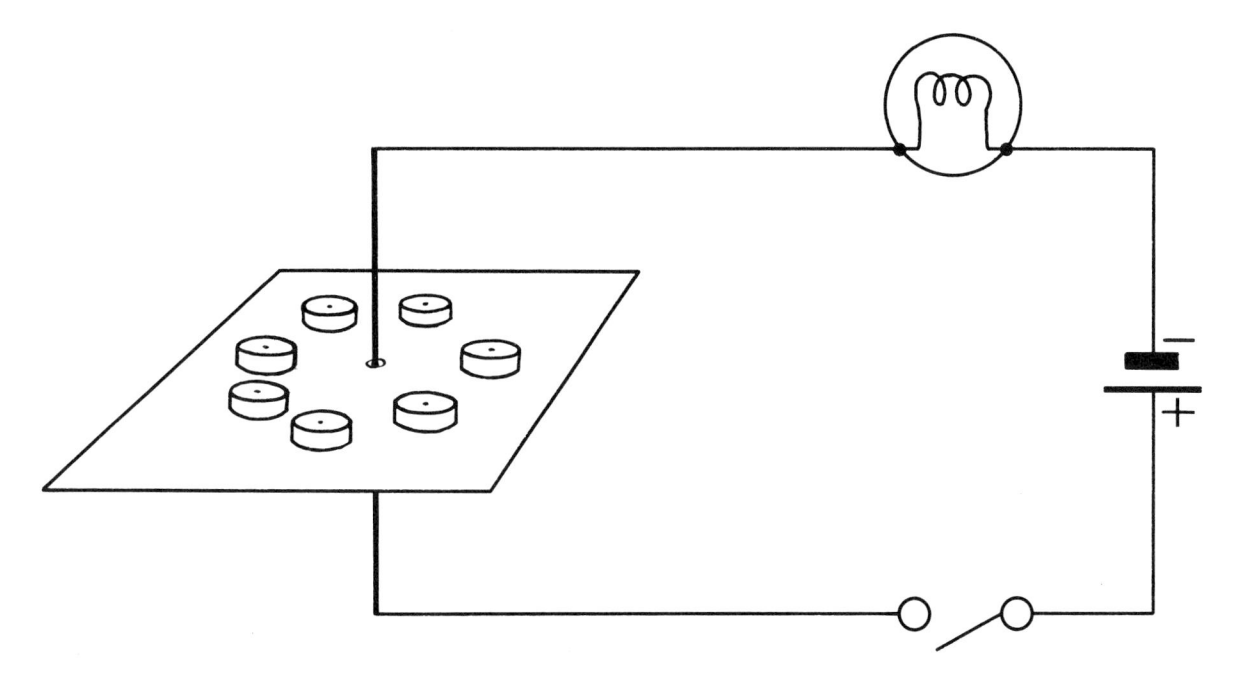

Switch on the current, briefly, and note what happens to the plotting compasses.

Reverse the direction of the current by reversing the connections to the cell and repeat.

Figure 16.3 The magnetic effect of a current

Investigation 16c. The effect of current in a solenoid

Obtain a piece of cardboard tubing (a diameter of from 1 to 2 cm is suitable) and wind on it about a hundred turns of insulated (enamelled or cotton covered) copper wire of diameter 0·56 mm. Secure the ends of the coil with strips of 'Sellotape'. This is a solenoid. Connect it in series with a cell and a bell-push, as shown in

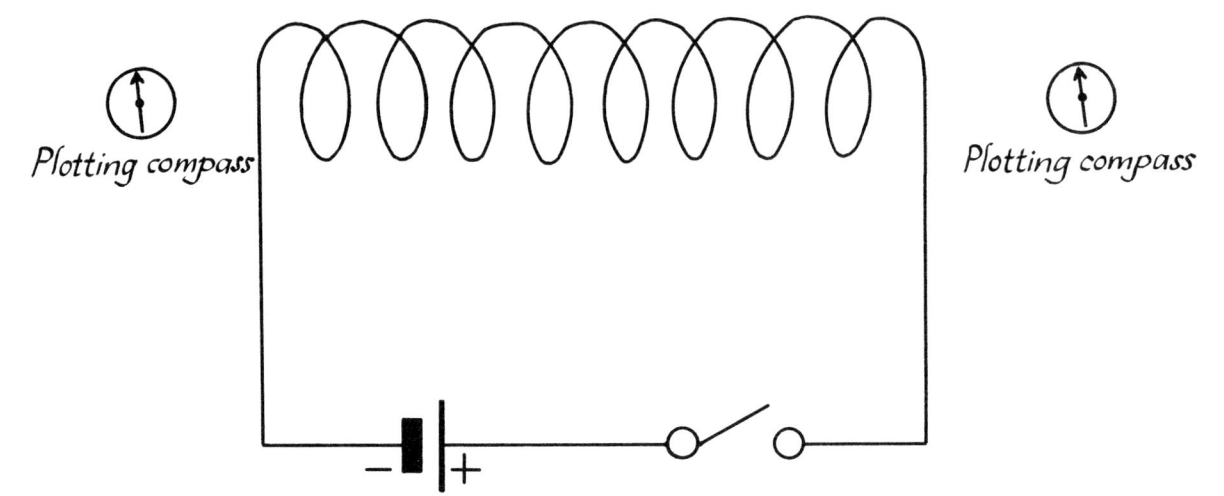

Figure 16.4 The effect of current in a solenoid

Figure 16.4. Place a plotting compass near to each end of the solenoid and briefly switch on the current, noting what happens to the plotting compasses.

Repeat with the current direction reversed.

Place a piece of iron rod in the solenoid and repeat the investigation.

When an electric current flows along a conductor, magnetism is produced. Figure 16.5 represents the magnetic fields produced in Investigations 16b and c. The arrows show the direction in which the north pole of a plotting compass would point, if one were placed in the field.

Figure 16.5 Magnetic fields produced by electric currents

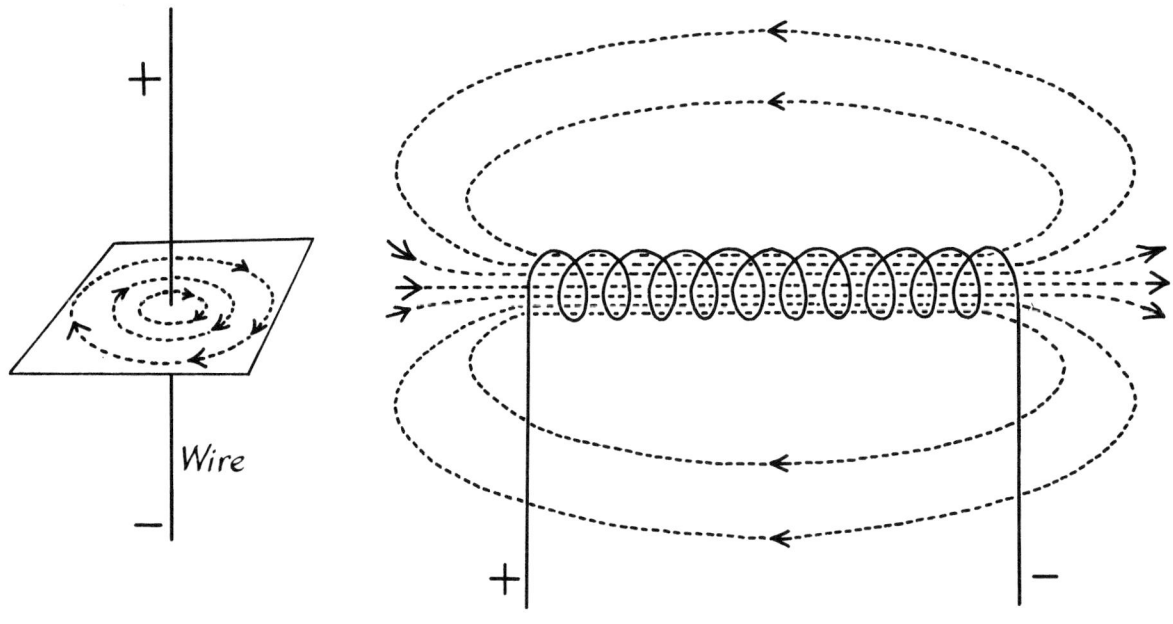

The field produced by a current flowing through a solenoid is like a magnetic field around a bar magnet. The strength of such a field is concentrated by placing an iron rod in the solenoid.

16.2. The heating effect

Investigation 16d. The heating coil

Make a heating coil by winding 300 mm of bare nichrome wire of diameter 0·71 mm on to a piece of glass tubing about 15 cm long. Attach flexible insulated wires to the ends by means of barrel connectors. This arrangement is illustrated in Figure 16.6.

Figure 16.6 The heating coil

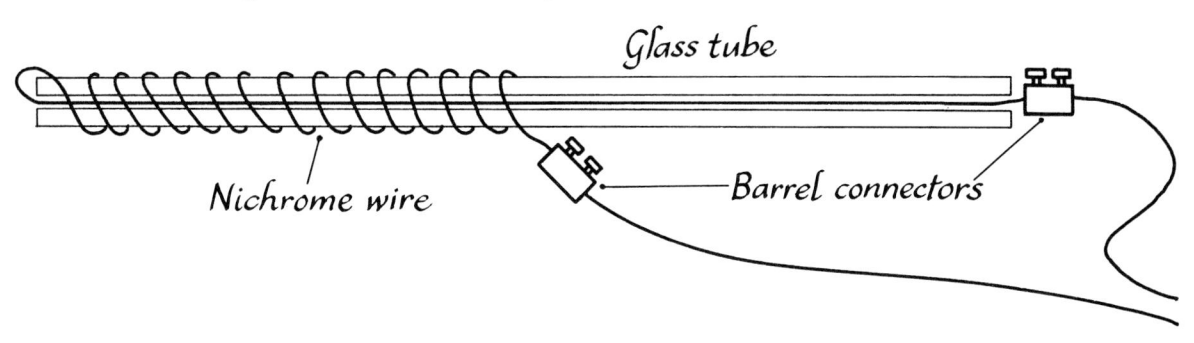

Pour 200 cm³ of water into a thin plastic beaker, place the heating coil in the water, making sure that it is completely immersed, and complete the circuit, as shown in Figure 16.7, by connecting a switch and a single lead–acid cell in series with the coil.

Note the temperature of the water, then switch on the current, and

Figure 16.7 The heating coil circuit

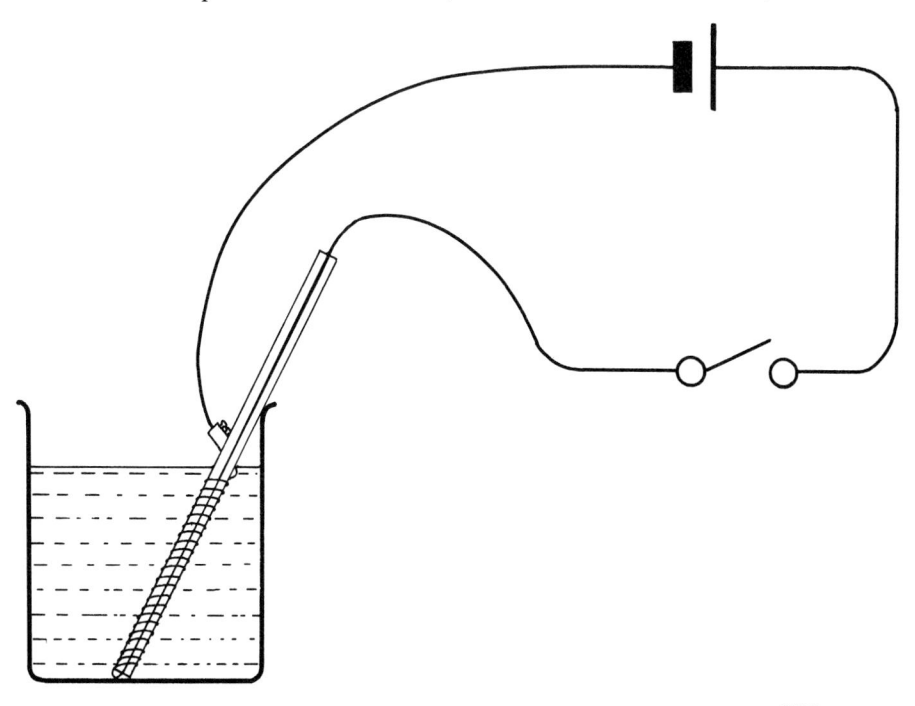

note the temperature of the water at two-minute intervals until a rise in temperature of about 10 kelvin is reached. Tabulate your readings, as shown in Table 16.1.

Repeat the investigation using first two cells, and then three cells, in place of the original single cell.

Make another heating coil, using the same length of wire as before, but using nichrome wire of diameter 0·46 mm.

TABLE 16.1. THE HEATING EFFECT OF A CURRENT

Time (m)	Temperature (°C)			Temperature Rise (k)		
	One Cell	Two Cells	Three Cells	One Cell	Two Cells	Three Cells

When an electric current flows along a conductor, heat is produced. The rate at which this heat is produced depends on the size of the current. This, in turn, depends on the conductor itself, and on the voltage applied to the circuit.

16.3. Measuring electricity

We have stated that when a stream of electrons flow round a circuit, this constitutes a current of electricity. In order to drive such a stream of electrons round a circuit, an electromotive force must be provided. The strength of an electromotive force is measured in **volts**.

The **coulomb** is the unit for measuring the quantity of electrical charge. A quantity of one coulomb of electricity comprises the charge of almost two-thirds of a million, million, million electrons. If a current of one ampere flows for one second, a charge of one coulomb will have been transferred.

In order that a current of electricity may be kept flowing in a circuit, power is required. Power depends on both the current and the voltage, and is expressed in **watts**.

Watts = Voltage × Current (measured in amperes)

Current is measured by means of an ammeter. When an ammeter is connected in a circuit, it is essential that all the current passes through it. A voltmeter is used to measure voltage. Figure 16.8 shows how a voltmeter and an ammeter may be connected in a circuit.

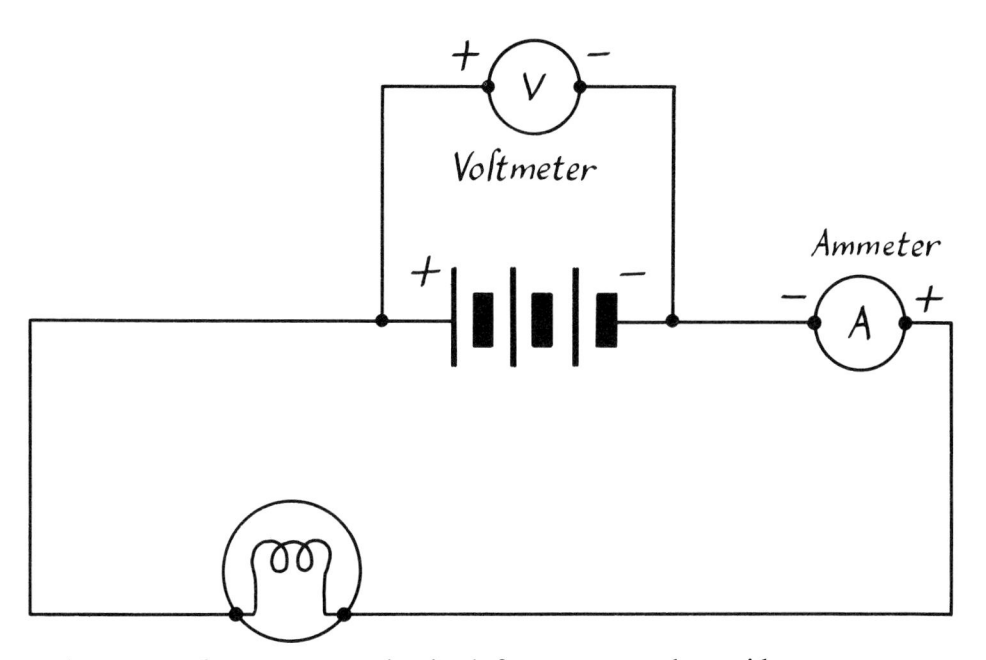

A power of one watt maintained for one second provides an energy of **one joule**.

Figure 16.8 Voltmeter and ammeter connections

16.4. Electrical resistance

In the previous chapter we noted that some materials conduct electricity better than others. To explain this we should also refer to the Rutherford–Bohr model of the atom (see Chapter 14).

A copper atom has a nucleus containing twenty-nine protons and the nucleus is surrounded by twenty-nine electrons. The electrons are regarded as being arranged in orbits, as shown in Figure 16.9. The innermost shell contains two electrons, the second has eight, the next eighteen and the outermost shell only one. This outermost electron, because of its relatively large distance from the nucleus, may be regarded as being loosely bound to the atom, so that only a small

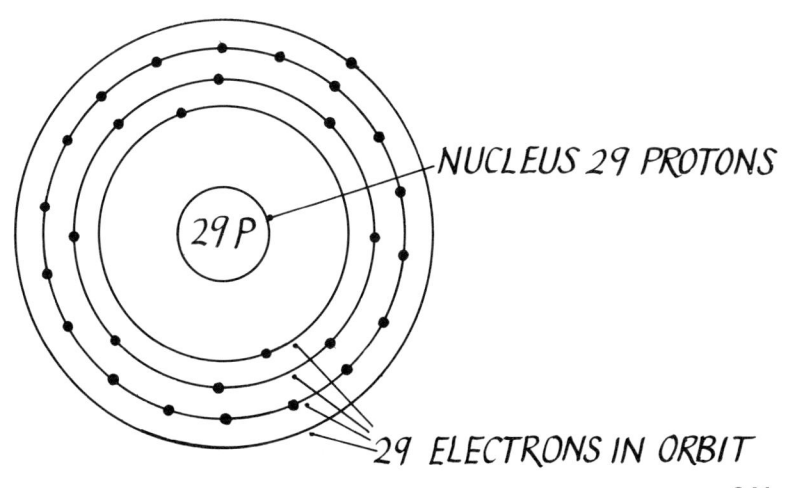

Figure 16.9 The copper atom

241

force is needed to transfer it to the next atom. When no current flows there is a continual random movement of electrons within the metal. The passage of an electric current consists of a large number of such transfers taking place *in one direction* in the atoms making up the conductor.

The good conductors, such as copper, have loosely bound electrons. Bad conductors do not, and so it is much more difficult to pass electric currents through them.

A good conductor of electricity is said to have a low electrical resistance, while a bad conductor has a high electrical resistance.

Investigation 16e. The relationship between the current flowing in a conductor and the applied voltage

Set up the circuit shown in Figure 16.10. PQ is a piece of nichrome wire of diameter 0·56 mm and about 1 m long. The ammeter should

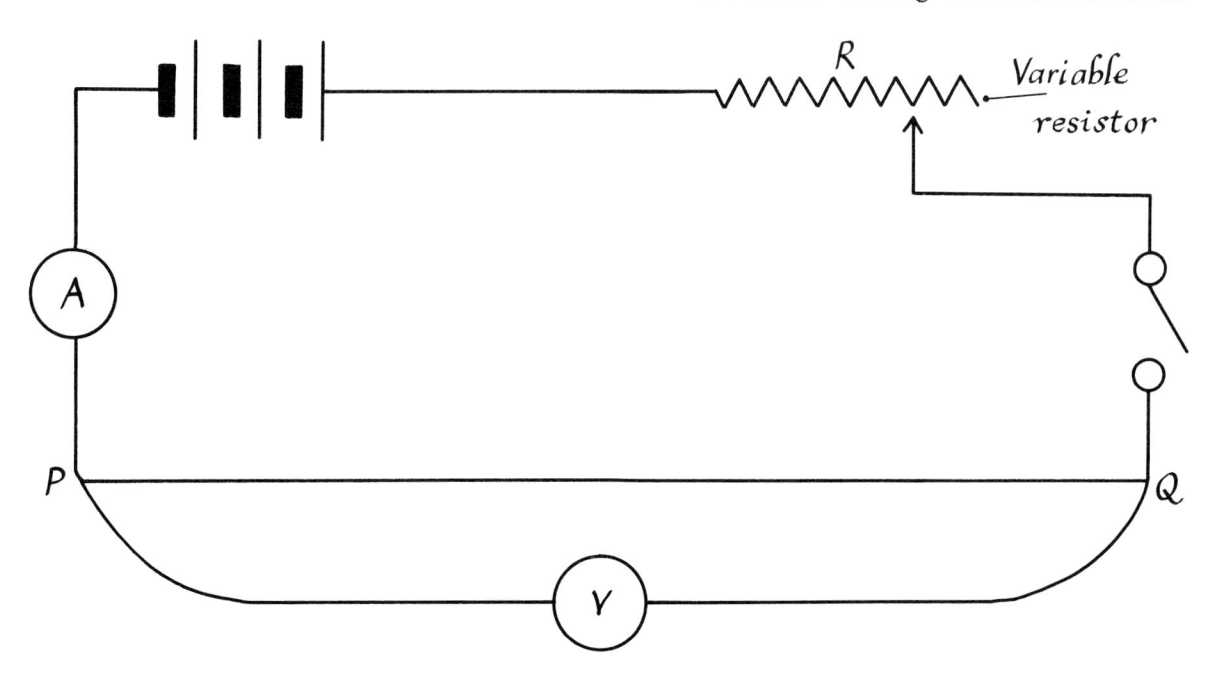

Figure 16.10 Variation of current and voltage

have a range of 0–2 A; the battery should be three lead–acid cells connected in series; the variable resistor should have a maximum value of about 20 ohms; and the voltmeter should read up to 6v.

Close the switch and adjust the variable resistor until the ammeter reading is 0·1 A. Note this value, and also the corresponding voltmeter reading. Switch off.

Switch on and adjust the variable resistor to give a new pair of meter readings. Continue in this way until you have at least six pairs of readings.

Tabulate your results, and for each pair calculate the value obtained by dividing the voltage by the current. Is there a constant relationship between current and voltage?

TABLE 16.2. VARIATION OF VOLTAGE AND CURRENT

Voltage (V)	Current (A)	Voltage/Current

Also plot a graph, using the *x* axis to represent voltage and the *y* axis to represent current (see Figure 16.11).

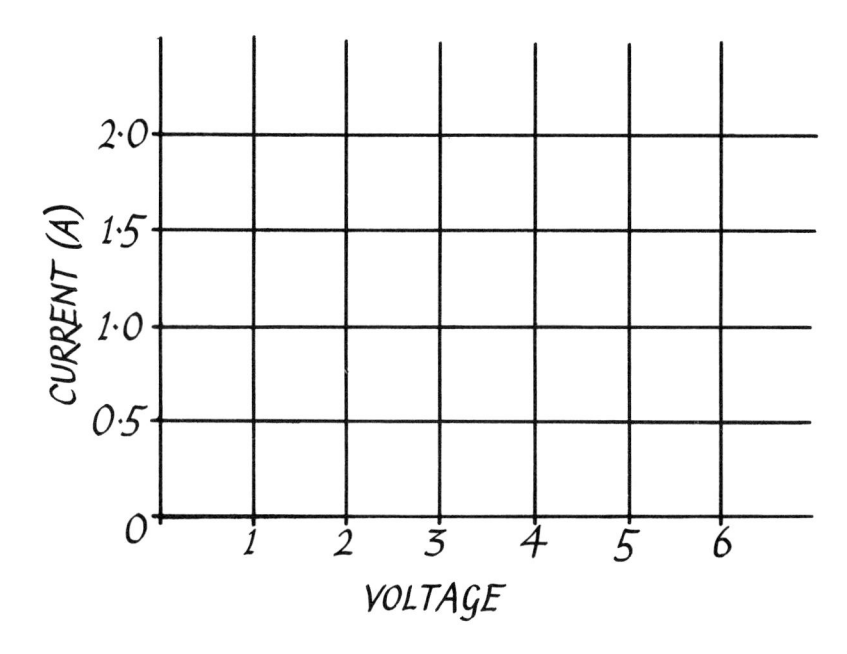

Figure 16.11 Plotting voltage against current

Provided that the conductor does not become hot, the current flowing along a conductor is directly proportional to the voltage applied to the ends of the conductor. This means that, for a particular conductor, the voltage divided by the current will give a constant value. This value is called the **resistance** and is measured in **ohms**. This unit is named after Georg Simon Ohm who first stated the law. The connection between current, voltage and resistance units may be stated thus:

$$\text{Ohms} = \frac{\text{Volts}}{\text{Amperes}}$$

Investigation 16f. The effect of thickness

Repeat Investigation 16e, using nichrome wire of diameter 0·38 mm in place of the wire used previously. Compare the results with those obtained in Investigation 16e.

Investigation 16g. The effect of length

Set up the apparatus shown in Figure 16.12. AB is a piece of nichrome wire of diameter 0·38 mm and 1 m long, C is a crocodile clip, the cell is a lead–acid type and the ammeter has a range of 0–2 A.

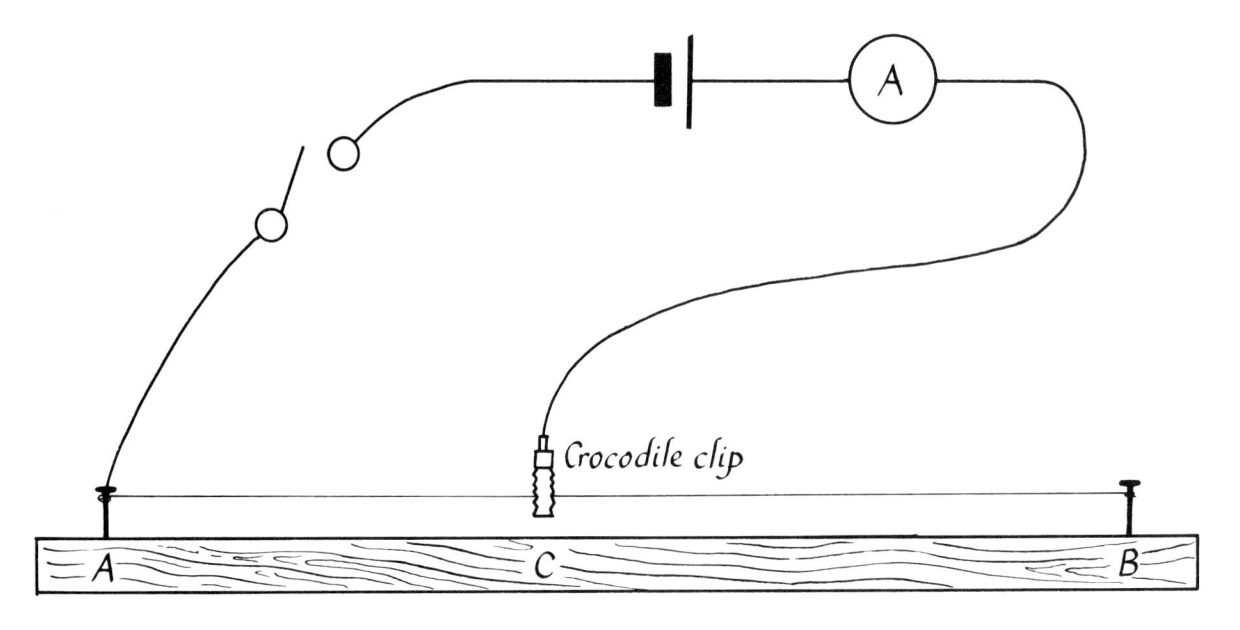

Figure 16.12 The effect of length

Attach the crocodile clip to the wire so that the length AC is 20 cm. Close the switch and note the ammeter reading. Repeat, with the clip attached, to give each of the following lengths in turn: 30 cm, 40 cm, 50 cm, 60 cm, 70 cm, 80 cm, 90 cm and 100 cm.

Tabulate your results, as shown in Table 16.3. What is the connection between length and current?

TABLE 16.3. THE EFFECT OF LENGTH ON RESISTANCE

Length AC (cm)	Current (A)

Draw a graph, plotting length against current.

The resistance of a conductor increases in direct proportion to its length, but inversely as its area of cross-section. Thus for a given material, the thicker the wire, the lower the resistance; and the longer the wire, the greater the resistance.

16.5. The chemical effect

We learned in Chapter 14 that certain compounds, called electrolytes, will conduct electricity when they are molten or are in solution. At the same time, they are decomposed by the passage of the current.

This effect is often used in order to deposit a thin layer of metal on objects made of some other metal. This process is called **electroplating**. (Metals can also be deposited on carbon; see Sections 14.13 and 14.14.)

Investigation 16h. Nickel plating

For this investigation you will need a large pure nickel crucible, a lead–acid cell, an ammeter, a switch, a piece of wood about 1 cm square and long enough to rest across the top of the crucible, a variable resistor (about 10 ohms maximum value), some flexible wire, two crocodile clips and a solution prepared in the following way:

250 g of hydrated nickel sulphate
40 g of hydrated nickel chloride
20 g of hydrated cobalt sulphate
35 g of sodium formate
35 g of boric acid
} Dissolve in distilled water to make 1 litre. (Enough for from six to ten groups.)

Set up the apparatus, as shown in Figure 16.13. Select an object for plating made of brass or copper, and suspend it in the solution, as shown, making sure that it does not touch the sides or bottom of the crucible. Switch on the current and adjust the variable resistor to pass a current of less than 0·5 A. Allow the current to pass for five minutes, and then switch off and examine the object you have plated.

For good results in electroplating, the object to be plated must first be thoroughly cleaned, and it is important not to pass too large a current.

Investigation 16i. Making a secondary cell

For this investigation you will need a beaker, some fairly strong sulphuric acid (density $1·1$ g cm^{-3}), two strips of lead, connecting wire, a switch, a 2·5-volt lamp and holder, a voltmeter (range 0–5 or 6 V), and a 4-volt bench supply, or alternatively a battery of two lead–acid cells.

Figure 16.13 Nickel plating

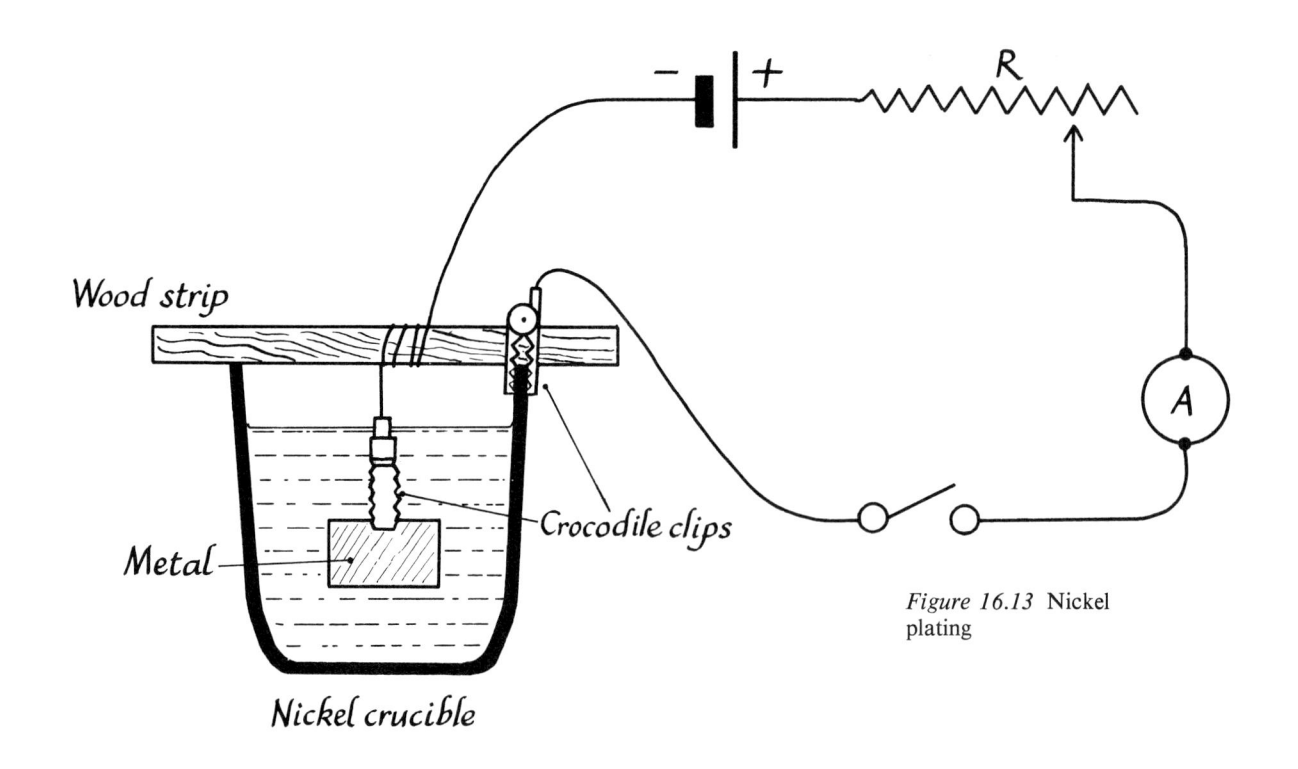

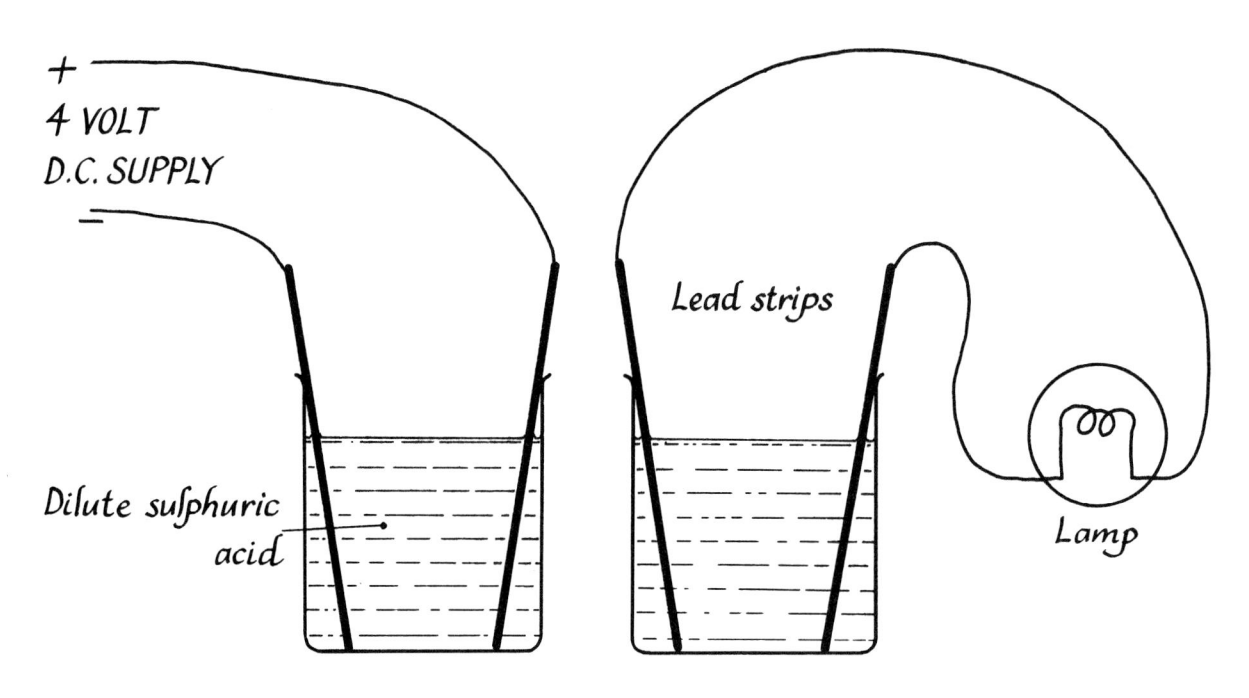

Figure 16.14 Making a secondary cell

Figure 16.15 Connecting the lamp

Set up the circuit, as shown in Figure 16.14, and allow the current to flow for about ten minutes. Observe what happens, and record your observations in your notebook. Then switch off the supply and connect the lamp to the lead plates, as shown in Figure 16.15. Note what happens.

Set up the original circuit again, and after five minutes switch off the supply and connect the voltmeter to the lead strips, noting its initial reading.

In this investigation you have made a lead–acid cell. This kind of cell is called a **secondary cell** because, unlike the cells described in Chapter 15, after it has been in use for some time it can be restored to its original condition by passing an electric current through it.

When you first passed the current through the lead plates, you probably noticed that the lead plate connected to the positive supply became coated with a dark brown substance. This was lead dioxide. You may also have noticed that some gas bubbles were produced. What do you think these were?

Lead–acid cells are used to make up car batteries, a 12-volt battery consisting of six lead–acid cells connected in series. Each cell, when in good condition, has an electromotive force of a little more than two volts.

16.6. Light from electricity

a. The filament lamp

When an electric current flows along a wire the wire becomes hot. If a large enough current is passed, the wire will glow, so that light as well as heat is produced. The colour of the light will depend on the temperature reached by the wire, a temperature of at least 2 000 °C being required in order to give white light.

If a hot wire is surrounded by air, oxidation will take place, i.e. the wire will combine with oxygen from the air to form an oxide (see Chapter 3). This oxidation takes place more rapidly at higher temperatures, so that, in an efficient electric lamp, it is necessary to take steps to prevent it occurring.

In early types of lamp, this was achieved by enclosing the filament in an evacuated glass bulb, but it was found that under these conditions the hot filament tended to vaporize, and be deposited on the bulb.

In the modern type of lamp, the bulb is filled with an inert gas, such as argon, which will not react with the filament. The material used for the filament is tungsten, which has a high melting point (3 400 °C). When heated to about 2 800 °C it gives a very white light. The filament is coiled (and sometimes the coil is coiled again), so that one coil heats the next. This assists in raising the filament temperature by reducing the effects of convection currents within the gas surrounding it (see Chapter 6). It must be noted, however, that even a modern filament lamp is inefficient, in that most of the electrical energy supplied to it is converted into heat, rather than light. Figure 16.16 shows a filament lamp.

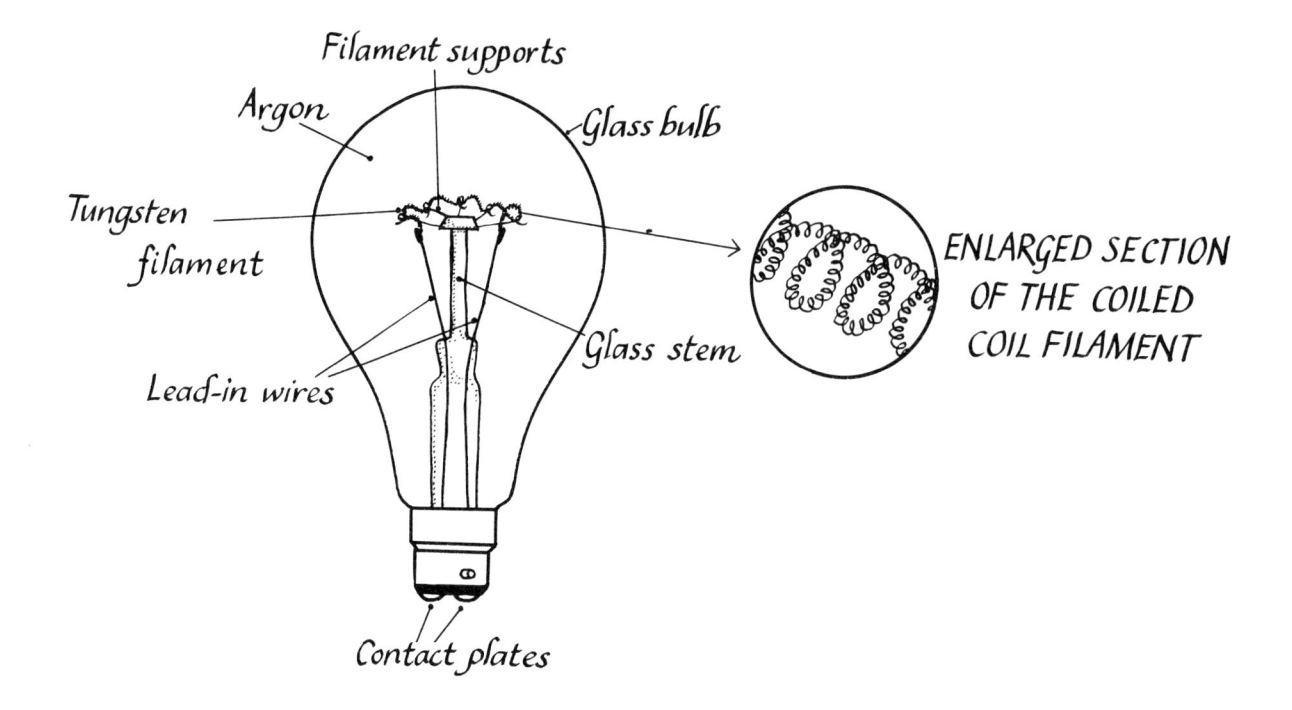

Figure 16.16 The filament lamp

b. The fluorescent lamp

Investigation 16j. The neon lamp

1. Examine a neon lamp of the 'beehive' type. Does it have a filament? Sketch what you can see inside the bulb.
2. Now examine the lamp when it is connected to an alternating current mains supply. Describe what you see.
3. A direct current is passed through the lamp (a potential difference of at least 180 volts will be needed). Describe what you see. The direction of the current is then reversed. Note what happens.

★ WARNING. *This is a demonstration to be performed by the teacher.*

Figure 16.17 shows a neon lamp.

Gases at normal atmospheric pressure are very poor conductors of electricity. At very low pressures, however, it is possible to pass a current easily. Towards the end of the nineteenth century, it was

Figure 16.17 The neon lamp

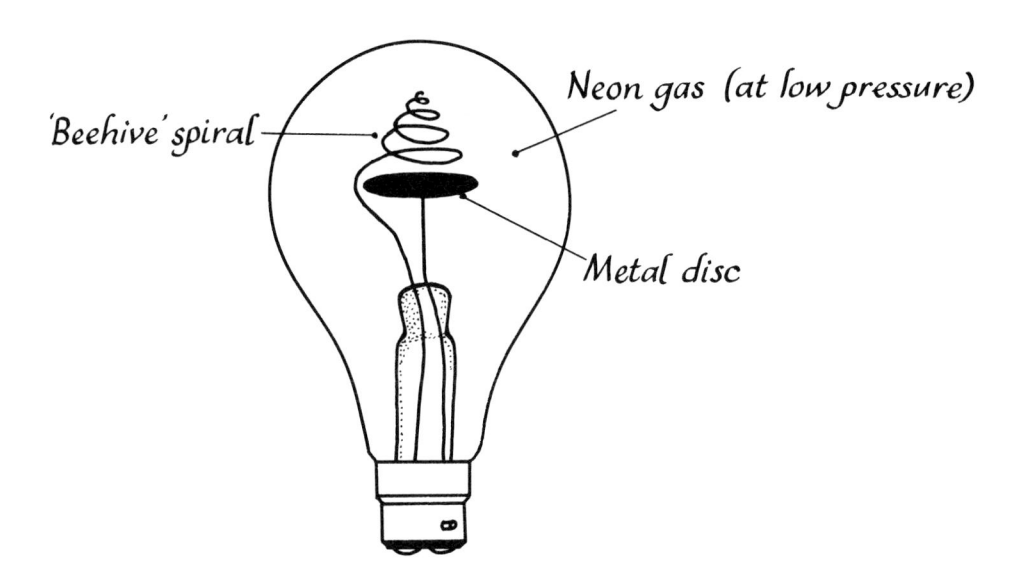

discovered that an electric current would pass between electrodes sealed into the ends of a glass tube, provided that the air in the tube was at a low pressure. Using a tube as shown in Figure 16.18, it was noticed that the air in the tube glowed when a high voltage was applied between the electrodes.

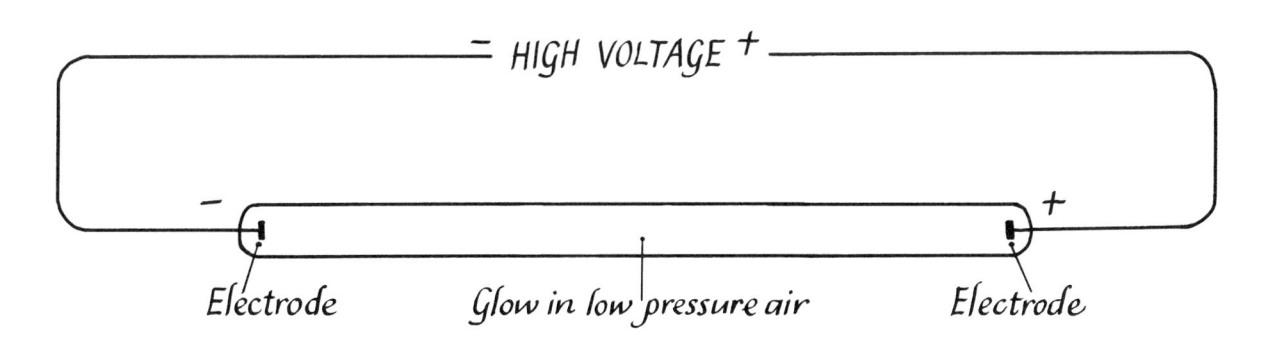

Figure 16.18 Discharge through rarefied air

When the voltage was applied, a stream of electrons was emitted by the cathode (negative electrode) and moved at high speed towards the anode (positive electrode). On the way, many electrons collided with gas molecules, ionizing them and, at the same time, causing them to emit light.

If gases other than air were used in the tube, the colour of the glow was different, each gas producing its own characteristic colour. If mercury vapour was used, a bluish glow was produced but, in addition, ultraviolet radiation occurred. Ultraviolet rays are like light rays except that they are of slightly shorter wavelength and cannot be detected by the eye.

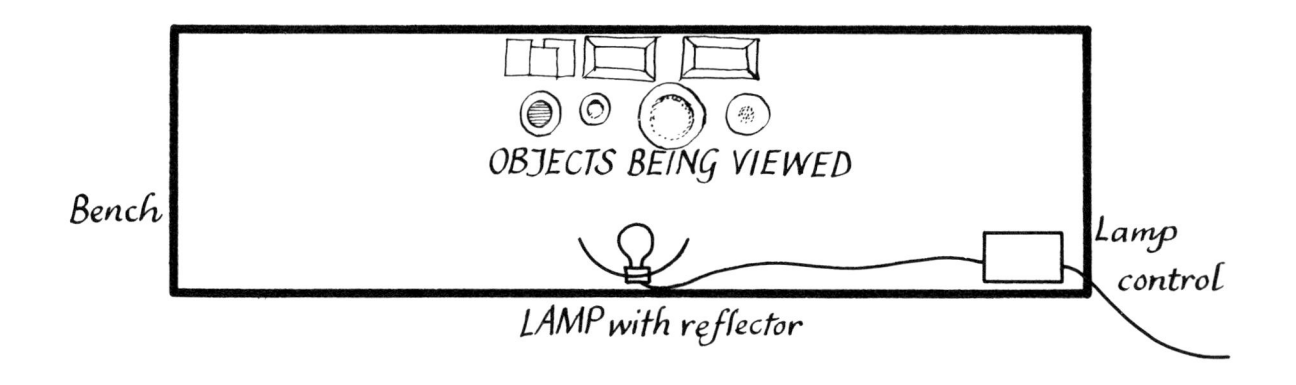

Figure 16.19 Arrangement of the ultraviolet lamp

Investigation 16k. Fluorescence

★ WARNING. *This is a demonstration to be carried out by the teacher. The lamp used in this investigation must be of the safe type similar to the kind used for stage lighting effects. Phillips type MBW/U125W is suitable, but it is essential to use a capacitor and choke in conjunction with the lamp. Types of lamp giving a wide range of ultraviolet frequencies are* **very dangerous and should not be used**.

The ultraviolet lamp is set up so that it shines on a bench. The laboratory should be blacked out and the lamp switched on. The following materials are placed in the beam of the lamp: sheets of paper, pieces of cotton or linen fabric which have previously been washed in soapless detergent, a dish containing a solution of quinine sulphate, some zinc sulphate and, if available, some pictures painted in fluorescent paint (see Figure 16.19).

Certain substances fluoresce when ultraviolet radiation falls upon them. The ultraviolet radiation is absorbed by the material, and visible light (of longer wavelength) is emitted.

The type of fluorescent lighting tube in common use contains mercury vapour so that, when it is switched on, ultraviolet radiation is produced within the tube. The inside wall of the tube is coated with a fluorescent powder and the colour of the light produced depends on this powder. Tubes of this type are more efficient than filament lamps because there is little unwanted heat produced (see Figure 16.20).

The current flows through a starter switch (X) and the filaments for a brief period, and then the starter switch breaks the circuit, causing a high voltage discharge from the choke. This causes electron emission within the tube, giving rise to the production of ultraviolet radiation, which causes the coating to fluoresce.

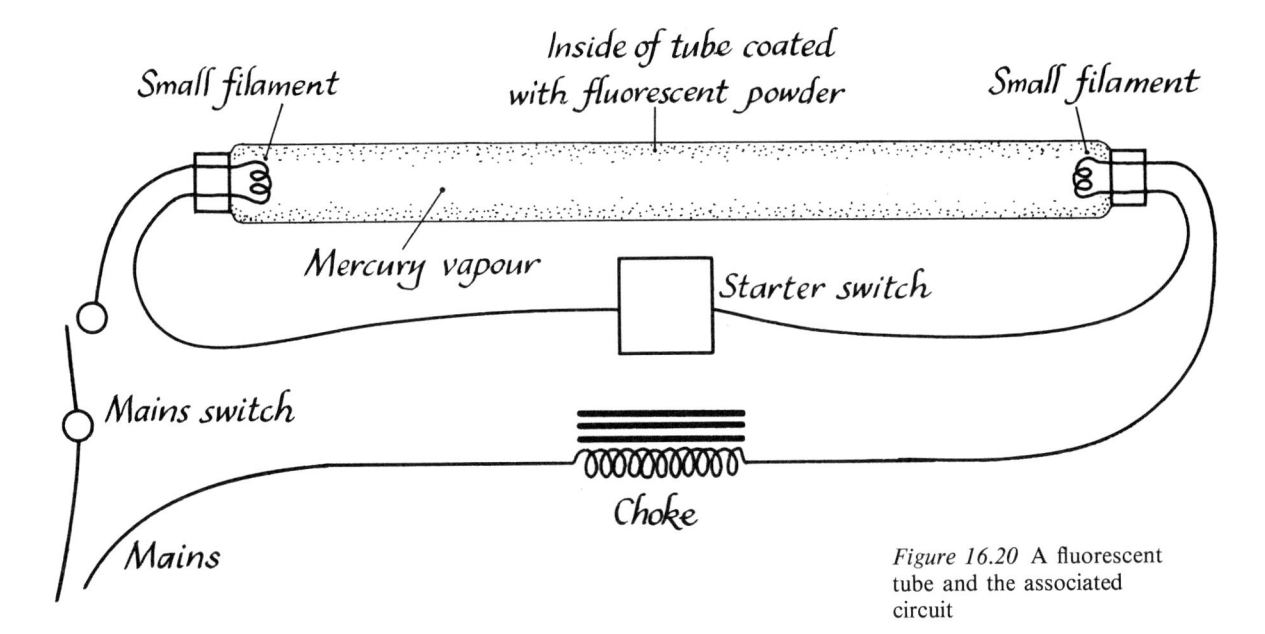

Figure 16.20 A fluorescent tube and the associated circuit

Test your understanding

1. If you were given a compass needle and two similar pieces of metal, one of which was magnetized, how would you find which of the two was a magnet?
2. Make a sketch showing the field you would expect to find around a bar magnet.
3. Sketch a solenoid connected to a battery so that an electric current passes through it. Indicate clearly the direction of the current and the polarity of the solenoid.
4. When in use, what current would you expect to flow through each of the following: (a) a lamp marked 240 V 60 W, (b) a heater marked 3 kW 250 V, (c) a 12-volt car lamp marked 36 W?
5. What is the electrical resistance of each of the appliances named above? (Assume that they are working normally.)
6. Explain the following terms: 'electrolyte', 'anode', 'cathode', 'fluorescence'.
7. Describe how you would make a lead–acid cell.
8. Draw a filament-type electric lamp. Why are lamps gas-filled? What gas is used for this purpose? What is the advantage of a coiled filament?
9. Sketch a fluorescent lamp. Why is this type of lamp more efficient than the filament type?

Chapter 17

Using Electricity

17.1. Direct current and alternating current

When a battery is connected in a circuit such as that shown in Figure 17.1, the current will flow continuously in one direction. Such a flow is called **direct current**.

If a current is made, continually, to change direction, then it is said to be **alternating current**.

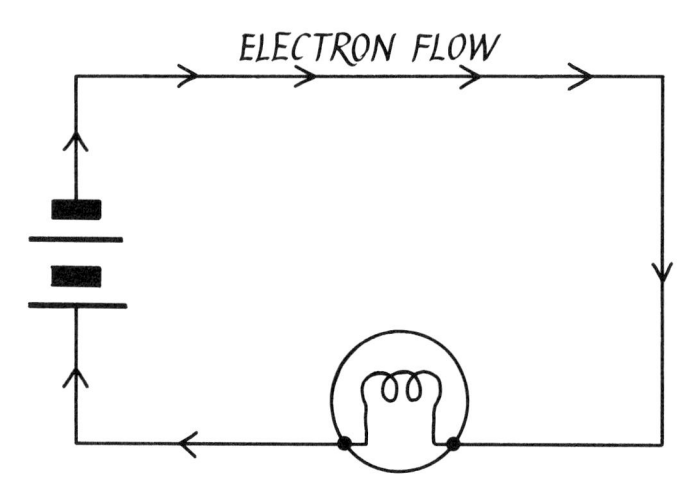

Figure 17.1 Direct current

Investigation 17a. Alternating current

For this investigation you will need a 6-volt battery, a centre-zero voltmeter, a very low frequency alternating current generator and some connecting wire.

Connect the circuit, as shown in Figure 17.2. Turn the handle of the generator slowly, and note how the voltage reading fluctuates.

The current flowing in the voltmeter circuit is alternating, so the voltmeter registers an alternating voltage.

Figure 17.2 Producing alternating current

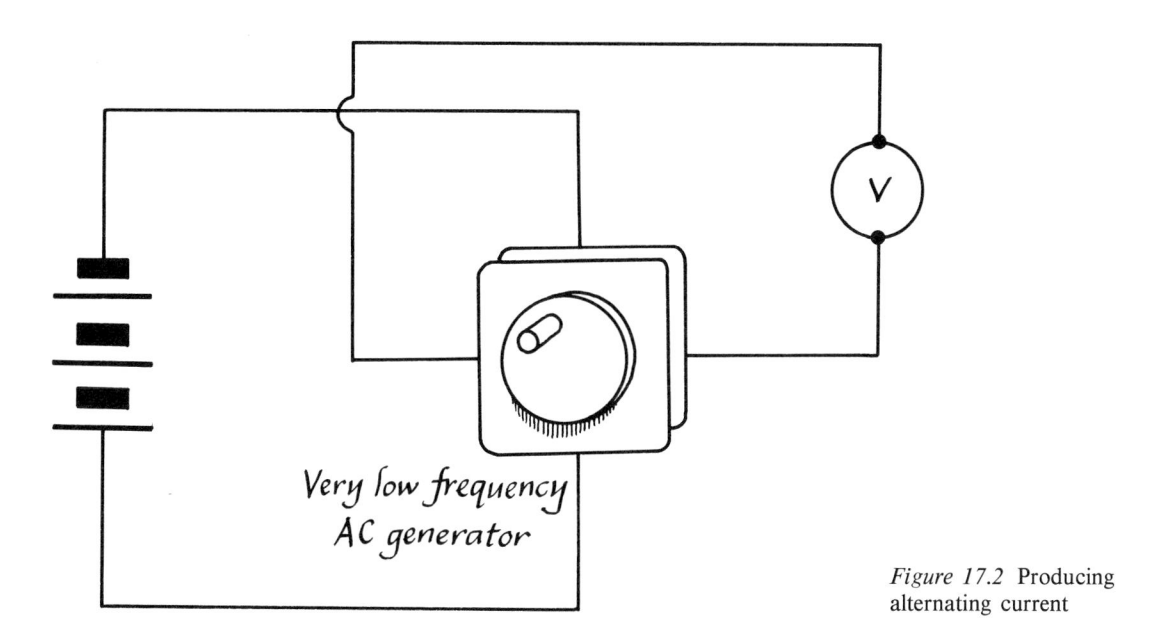

Investigation 17b. Using a cathode-ray oscilloscope

For this investigation a cathode-ray oscilloscope having a direct current input position must be used. The cathode-ray oscilloscope may be regarded as a device for plotting a graph of how a fluctuating voltage varies with time. It is possible to adjust the oscilloscope so that a small spot of light appears on the centre of the screen. By adjusting the appropriate control (the time base control), the spot can be made to move horizontally across the screen. The rate at which the spot moves can be varied by the same control.

With the spot moving across the screen, connect a battery to the terminals marked 'Y plate'. Note how the spot is deflected. Note what happens when the battery connections to these terminals are reversed.

Connect the output from a bell transformer to the Y plates, as shown in Figure 17.3. This is an alternating voltage. Note how the spot fluctuates.

Figure 17.3 Using a cathode-ray oscilloscope to study alternating current

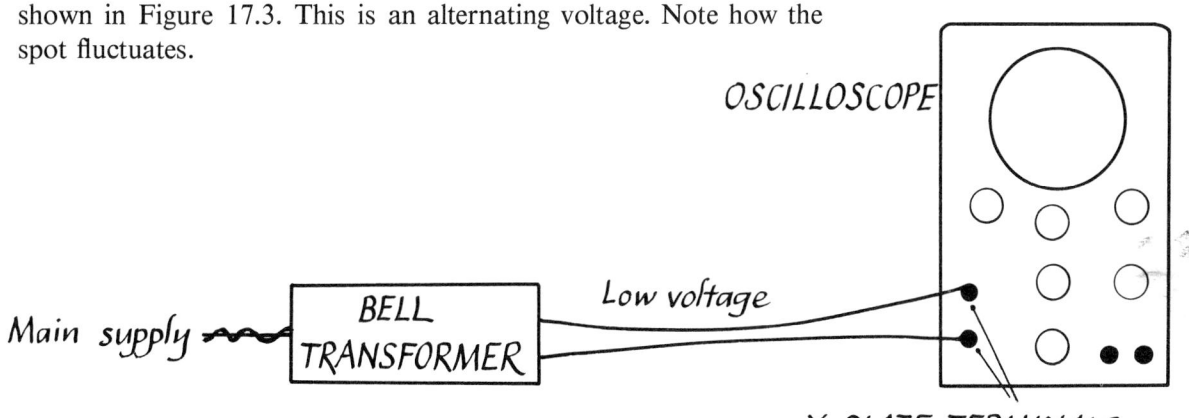

(a) No voltage applied to Y plate terminals

(b) Direct voltage applied

(c) Direct voltage applied but in opposite direction to (b)

(d) Alternating voltage applied to Y plate terminals

(e) Same voltage applied as in (d) but spot moving half as fast along X axis

Figure 17.4 Oscilloscope patterns

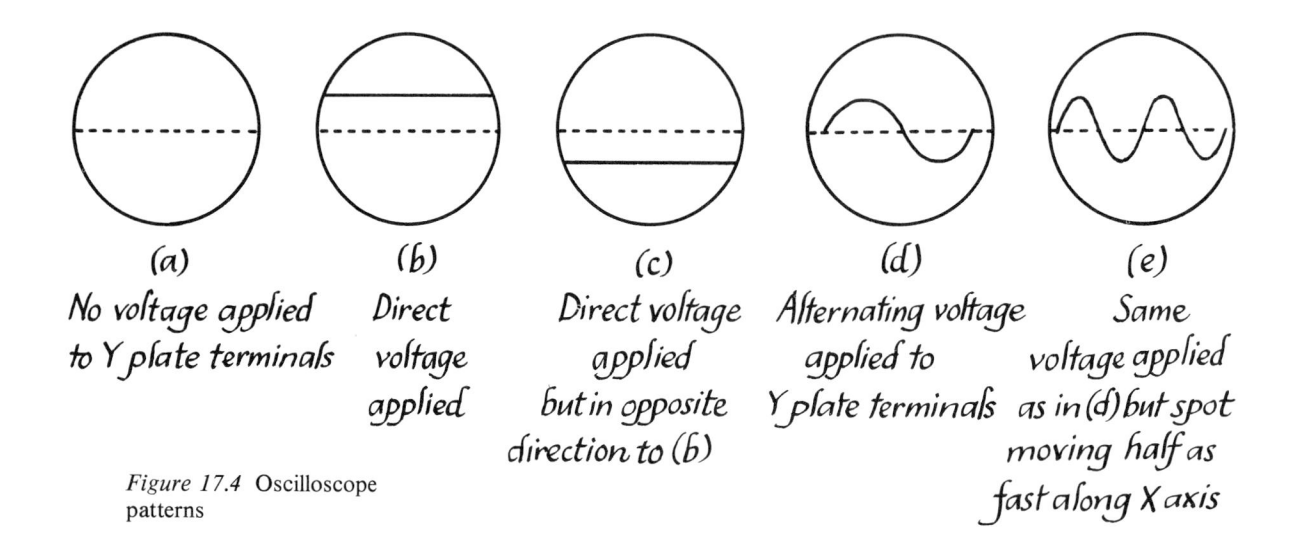

Figure 17.4 shows the patterns produced on the oscilloscope screen under several conditions.

Figure 17.5 shows the kind of pattern produced by the mains supply. This shape is called a **sine wave**. The number of such waves which occurs in one second is called the **frequency** of the supply, and is expressed in **hertz**. The mains supply fluctuates in this way fifty times a second and so is said to have a frequency of fifty hertz (50 Hz).

Figure 17.5 A sine wave

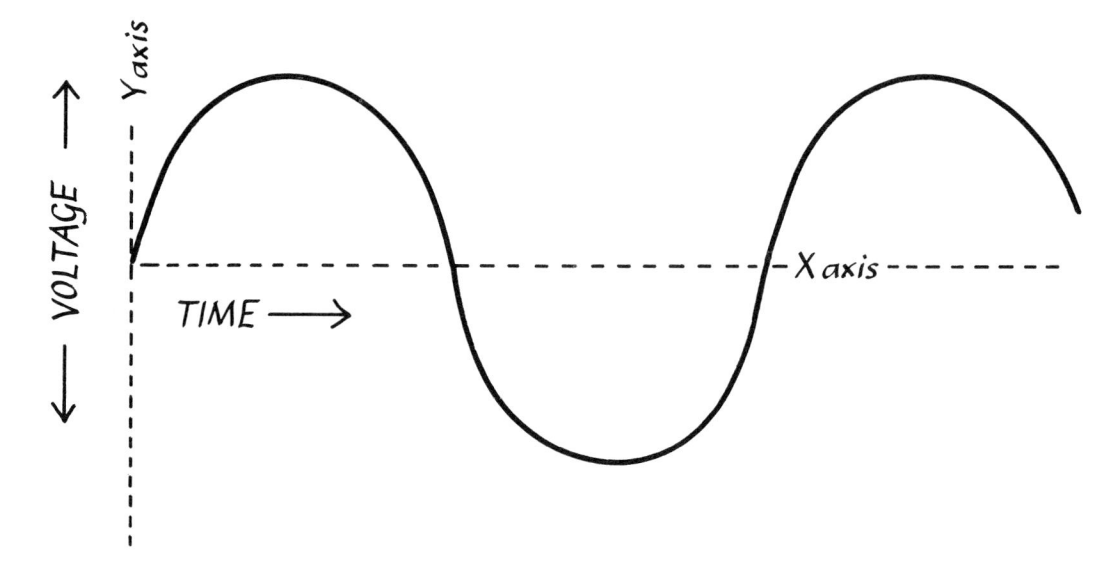

17.2. The transformer

We saw in Chapter 16 that electricity could be generated by moving a magnet inside a solenoid. In Chapter 16 we found that by passing an electric current through a solenoid a magnetic field could be produced. In the next investigation we shall combine these two effects.

Investigation 17c. Electromagnetic induction

For this investigation you will need a solenoid having about 100 turns, a second solenoid having about 250 turns (and of such a diameter that the smaller solenoid will fit inside it), a soft iron rod which will fit inside the smaller solenoid, a centre-zero galvanometer, a voltaic cell, a switch and some connecting wire.

Connect the smaller solenoid to the cell, and the larger solenoid to the galvanometer, as shown in Figure 17.6. Switch on and move the smaller solenoid inside the larger. Note what happens and switch off.

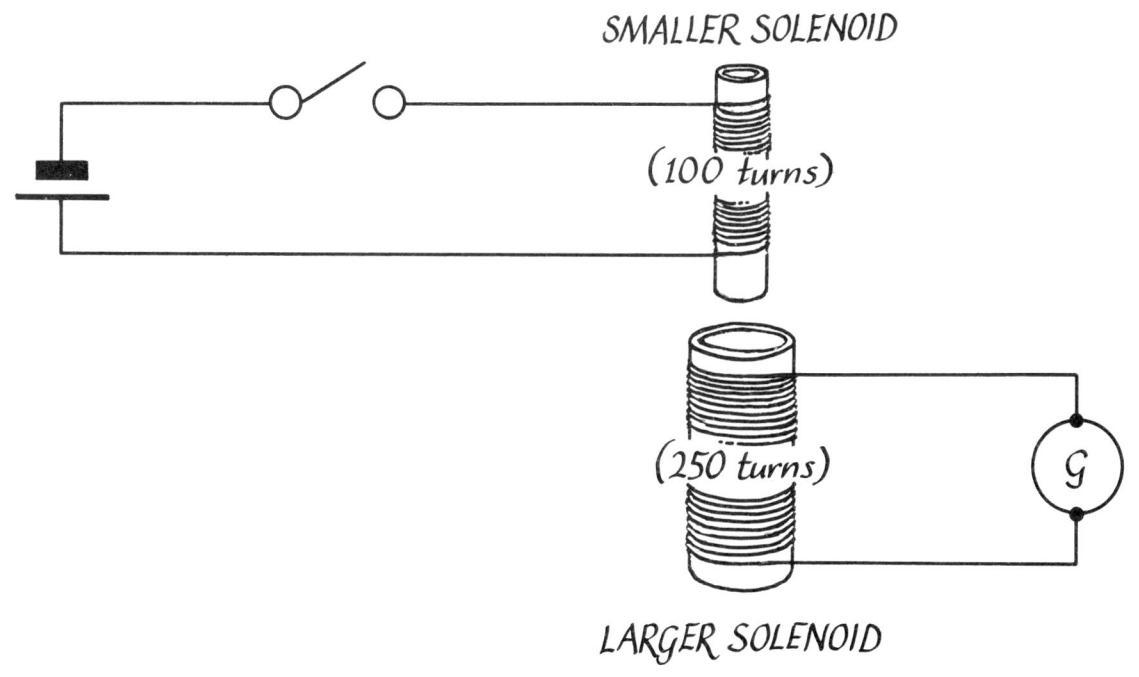

Place the iron rod inside the smaller solenoid and repeat the experiment.

Place the smaller solenoid inside the larger and close the switch, without moving either solenoid. Note what happens. Open the switch. Place the iron rod inside the smaller solenoid and, once more, open and close the switch.

Change the solenoid connections, so that the larger solenoid is connected to the cell and the smaller is connected to the galvanometer. Repeat the whole investigation and record your observations.

Figure 17.6 Apparatus for Investigation 17c

When the current is passed through one of the solenoids, a magnetic field is set up around it. As it builds up, this field cuts across the second solenoid and is said to **induce** a current in it. When the current is switched off, the field collapses and, once more, a current is induced in the other solenoid. Placing an iron rod in the middle tends to concentrate the magnetic field, so that the effect is more pronounced.

The coil through which the electric current is passed is called the **primary** winding and the coil in which the current is induced is called the **secondary** winding. If the primary winding has fewer turns than the secondary, a larger voltage will be induced in the secondary than is applied to the primary. The current flowing in the secondary circuit will, however, be smaller than that flowing in the primary circuit.

The **ignition coil** used in a car engine works on this principle. A small voltage is applied to the primary and this is switched off as a piston reaches the top of its cylinder. As the magnetic field collapses, a high voltage is induced in the secondary circuit. This is used to provide the spark which is required to ignite the mixture of petrol vapour and air in the cylinder (see Chapter 9).

You may have an induction coil in the laboratory. This is fitted with an interrupter, so that the primary current is automatically switched on and off. Figure 17.7 shows this type of induction coil, as well as the car ignition coil.

Figure 17.7 The ignition coil and the induction coil

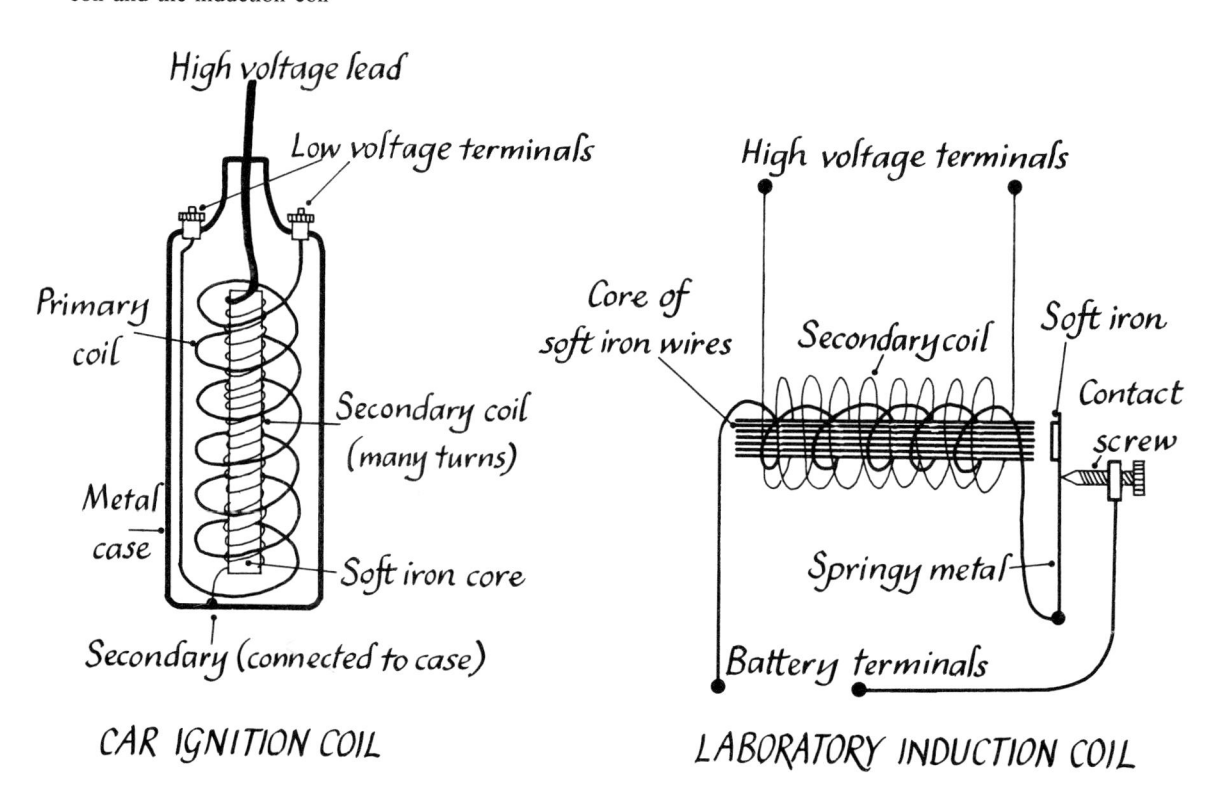

Transformers are used to change the voltage of alternating current supplies. A transformer has primary and secondary windings. An alternating current is supplied to the primary, and since this current is fluctuating, the magnetic field produced by it also fluctuates, so that an alternating current is induced in the secondary circuit. Figure 17.8 shows the general arrangement.

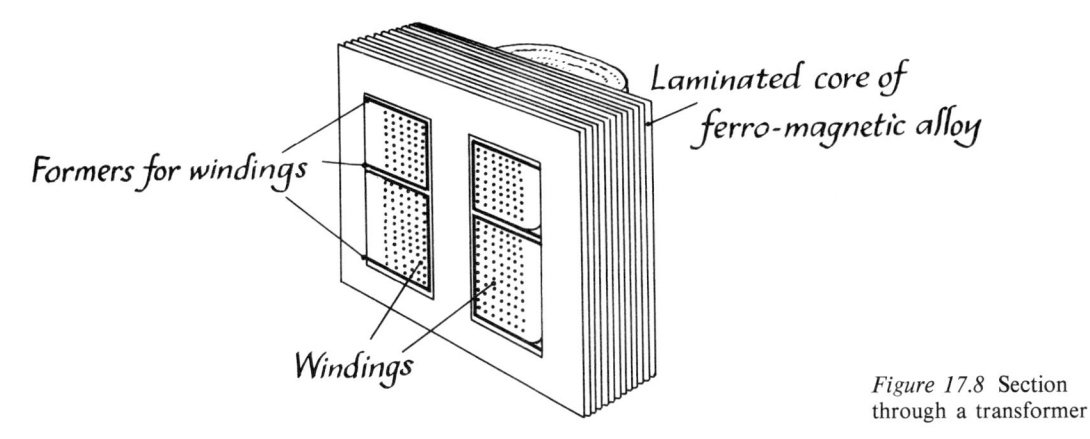

Figure 17.8 Section through a transformer

The windings are wound on formers, so that they may easily be put into place on the core. The core is made up of a large number of thin strips, or laminations, of ferro-magnetic alloy, or is moulded from a ferrite material.

The voltage which may be obtained from the secondary may be estimated by using the following formula:

$$\frac{\text{Secondary Voltage}}{\text{Primary Voltage}} = \frac{\text{Number of Secondary Turns}}{\text{Number of Primary Turns}}$$

It will be seen from this that if the secondary has more turns than the primary the output voltage will be greater than the input voltage. Such an arrangement is called a **step-up** transformer. If the secondary has fewer turns than the primary the output voltage will be lower than the input voltage. This is a **step-down** transformer.

Figure 17.9 shows how a transformer is represented in circuit diagrams.

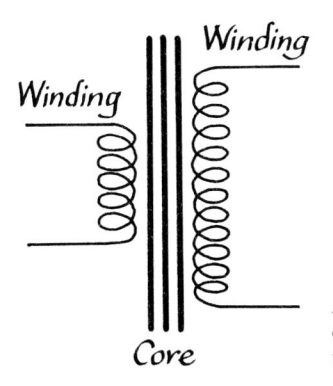

Figure 17.9 Circuit diagram representing a transformer

17.3. Transmission of electricity across country

The electricity which you use in your home may be generated in a distant part of the country. It is carried by wires on the **National Grid System**. Because of the resistance of these wires, heat is produced in them by the passage of the current, causing a loss of electrical energy. It is important to keep this loss to a minimum, and for this reason the current is kept as low as possible, since the amount of heat energy produced is directly proportional to the square of the current. You will remember that electrical power is measured in watts and is the product of the voltage and the current; therefore, if a high power is required and the current is to be kept as low as possible, a very high voltage must be used.

Transformers are used to step up the voltage produced at the power station, so that the electricity may be transmitted across country at a high voltage. Other transformers are used to step down this voltage before it is brought into our homes. Figure 17.10 represents this.

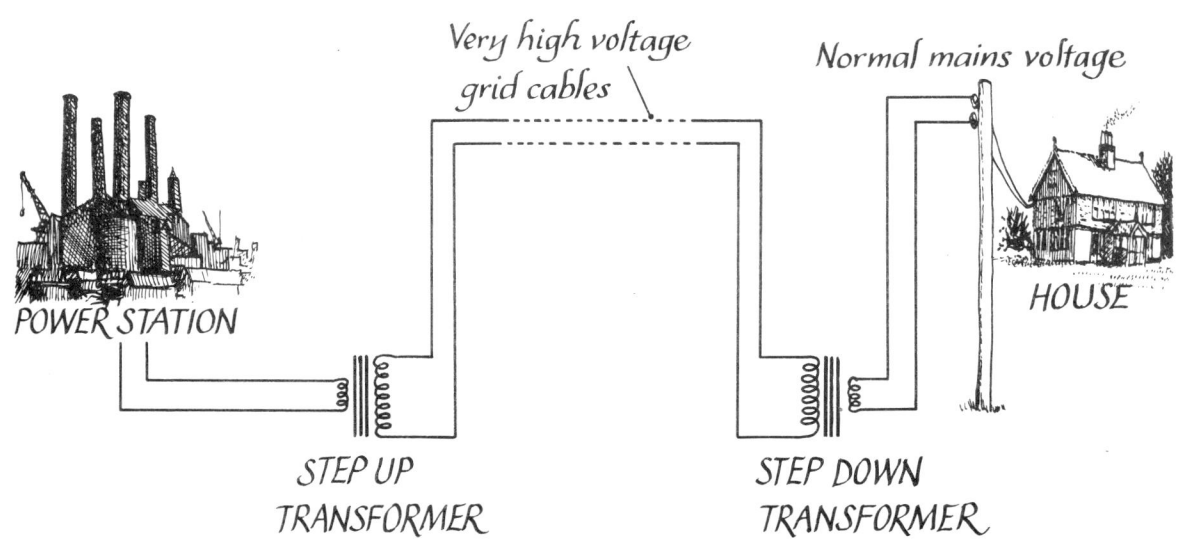

Figure 17.10 Transmission of electricity

17.4. The live and neutral leads

Before it reaches your home, one of the two leads carrying electricity is connected to earth (i.e. it is connected to a metal plate buried in the ground). This is called the **neutral** lead. At your home, a voltmeter connected between the neutral lead and the ground would give little or no reading.

The other lead is called the **live** lead, or the **line**, and a voltmeter connected between this and the ground would register the normal supply voltage (usually 240 V). It would be extremely dangerous to touch the live lead, because you would complete the circuit between it and the ground and the current would run through you to the ground.

17.5. The supply in the home

The leads into the home connect to the Electricity Board fuse box. From here the leads go to the meter, and then through the household main switch and fuse box to the various household circuits. Figure 17.11 shows this arrangement.

A **fuse** may be regarded as a weak link in the electrical circuit. In its simplest form, it is a thin wire which becomes hot and melts if an excessive current passes through it. This causes a break in the circuit, so that no more electricity flows. If fuses were not employed, an excessive current might very easily cause a fire by causing the circuit wiring to become very hot.

Most homes have at least two separate circuits, one for lighting and one for the power points. In addition, an electric cooker would have its own separate circuit. Each of these would be protected by a suitable fuse. For the lighting circuit, the fuse employed will usually melt if a current of more than 5 amperes passes through it, while a power circuit may be protected by a fuse which will pass up to 30 amperes before it melts. Examine some fuse wires of different ratings. Notice that the fuse wires having the higher ratings are the thicker wires.

Also examine specimens of cables and flexible leads. Notice that thicker wires are used for power appliances than are used for lighting purposes.

The main switch is a double-pole switch, i.e. one which breaks both live and neutral leads when it is in the 'off' position, so that the circuit is completely isolated from the supply.

17.6. The lighting circuit

All the lights in the home are connected in parallel, and each has its switch in the live side of the circuit. Why is the switch in the live side, rather than the neutral side? The switches are spring loaded so that when they are operated contact is made, or broken, very quickly. Sometimes a light can be controlled by either one of two switches. This is called a **two-way switching circuit**.

Figure 17.12 shows a typical lighting circuit.

17.7. The power circuit

The power points are also connected in parallel, but are on a separate circuit from the lights. The type of socket in general use is

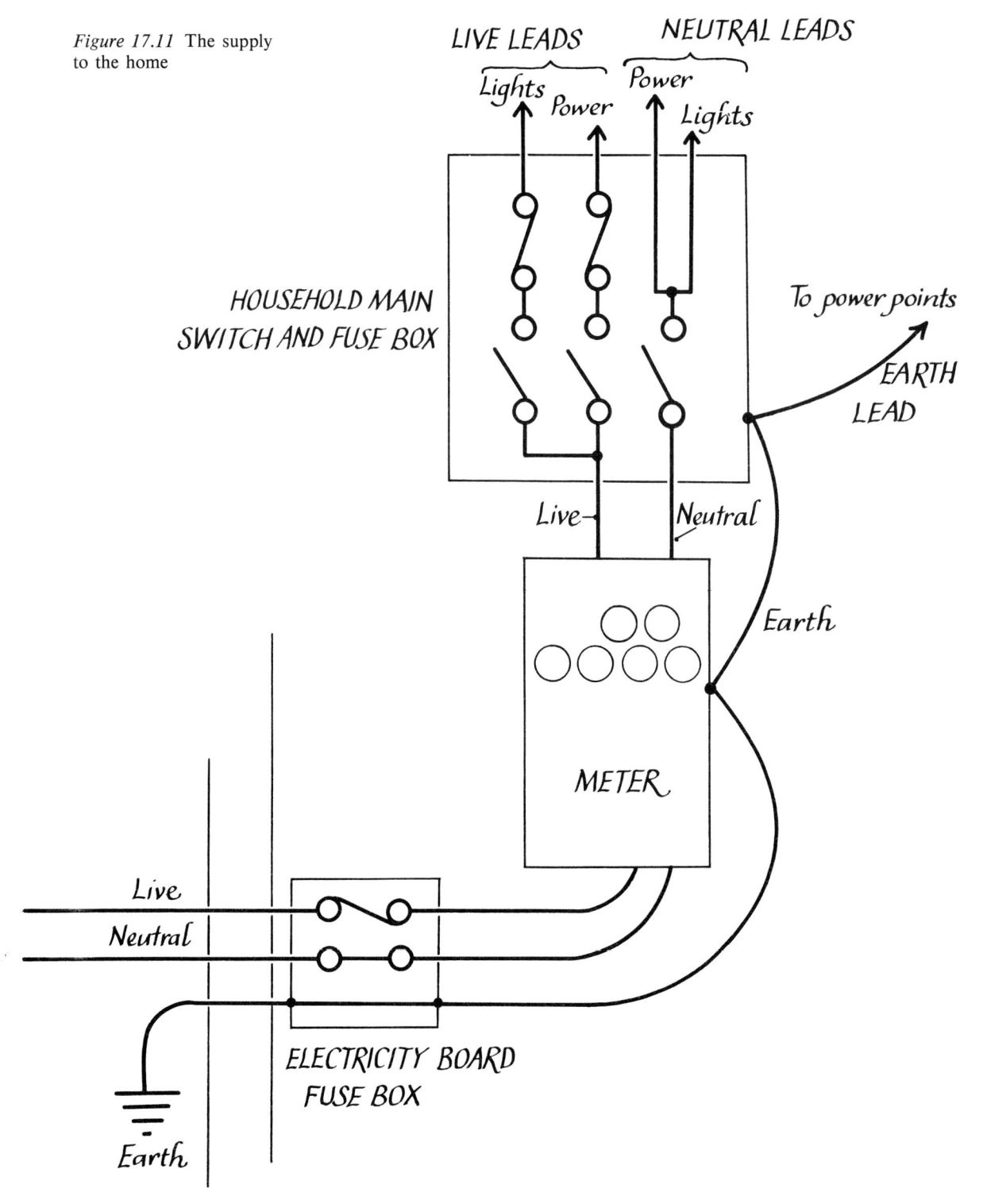

Figure 17.11 The supply to the home

Figure 17.12 A typical lighting circuit

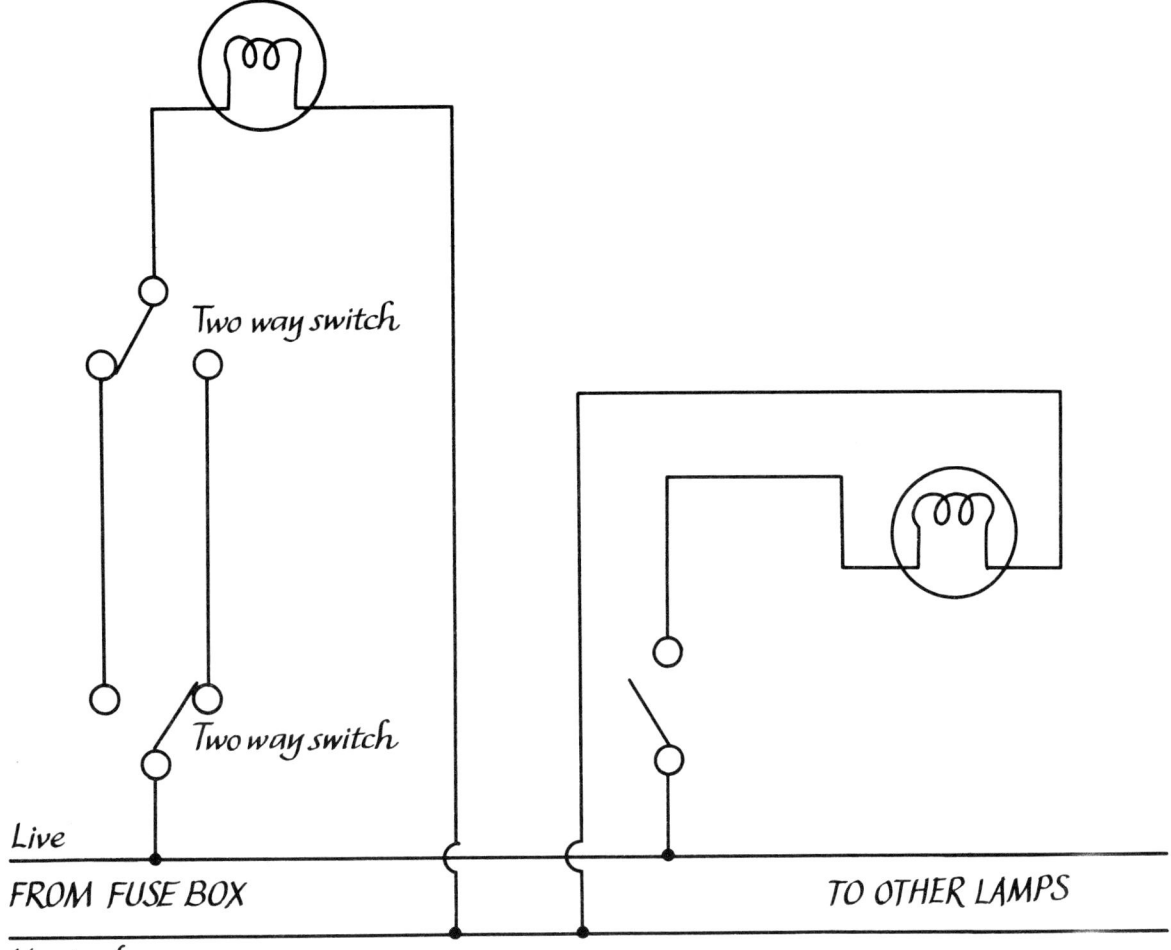

the 13 ampere, square-pin type. Such sockets can supply a maximum current of 13 amperes. A fuse is incorporated in the plug, and it is important that the rating of this fuse should be suited to the appliance to which the plug is fitted.

An electric heater rated at 2 500 watts would pass a current of a little more than 10 amperes on a 240 volt supply; for this a 13 ampere fuse would be suitable. A device rated at 250 watts would pass a little more than 1 ampere, and so a 2 ampere fuse should be used.

In addition to the live and neutral pins, a plug is fitted with a third, larger, pin. This is the **earth** pin and the corresponding hole in the socket is connected to an earthed lead. Appliances having metal cases should have a wire connecting the casing to the earth pin. This is a most important safety precaution. If the live lead to the appliance should become worn, or broken, it could make contact with the metal casing. When the earth lead is correctly fitted, a heavy current will then pass to earth, causing the fuse to melt. If no earth lead were fitted, the case would become live, so that anyone touching it would receive a severe electric shock.

The insulating sleeving on cables is colour coded, the live lead being brown, the neutral lead blue and the earth lead green with a yellow stripe.

Figure 17.13 shows the wiring of a plug.

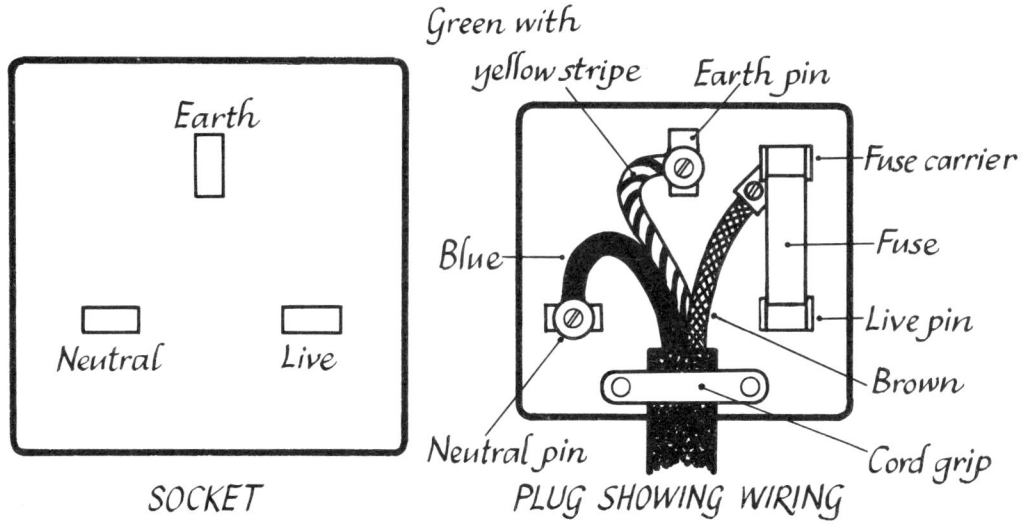

Figure 17.13 Socket connections and plug wiring

When wiring a plug the following points should be borne in mind:
a. Do not strip off too much of the insulating sleeving. When the wire is connected, the sleeving should reach up to the terminal.
b. If the wire has to be be looped around a screw, wind the loop in a clockwise direction, so that the loop tightens as the screw is driven home.
c. Make sure that you have used the colour coding of the lead correctly.
d. If the plug is fitted with a fuse, make sure that this is of a suitable rating for the appliance to which the plug is attached.
e. Remember to secure the cord grip.

17.8. Home electrical appliances

We shall now consider, briefly, a few electrical appliances commonly found in the home.

a. The electric clothes iron

Figure 17.14 shows the electric iron. This has a heating element fixed between a heavy iron block and the sole plate. The heating element is made of nickel-chromium wire wound on a mica sheet. This is sandwiched between two other mica sheets, and a sheet of asbestos is placed on top, to reduce the amount of heat conducted to the upper part of the iron. Heat is, however, readily conducted to the polished sole plate.

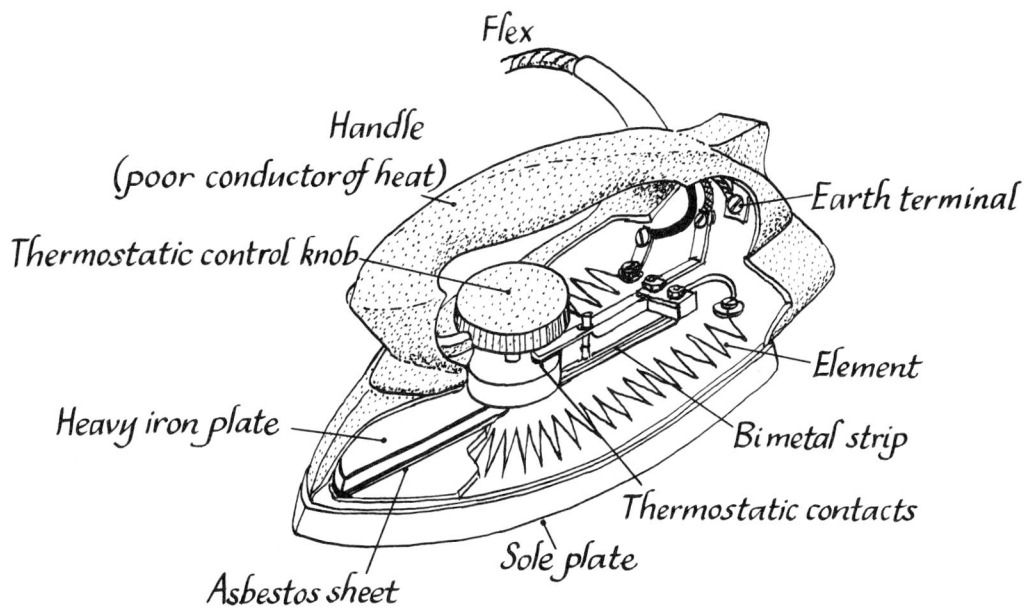

Figure 17.14 The electric iron

The temperature is controlled by means of a thermostat consisting of a bimetal strip, which bends to break the circuit when the required temperature is reached. As the iron cools, this strip straightens, once more making contact and completing the circuit.

b. The electric bell

Figure 17.15 shows the electric bell. When the current is switched on, the electromagnet comes into operation, attracting the iron strip and causing the striker to hit the gong. This causes the circuit to be broken at the contact screw. As the current is no longer flowing the electromagnet ceases to function, and the springy metal strip then brings the striker back to its original position. This again completes the circuit, and the whole sequence begins again.

Domestic electric bells are made to operate at a low voltage and either a battery or a step-down transformer may be used to work them. Figure 17.16 shows the circuit for a bell worked from a transformer.

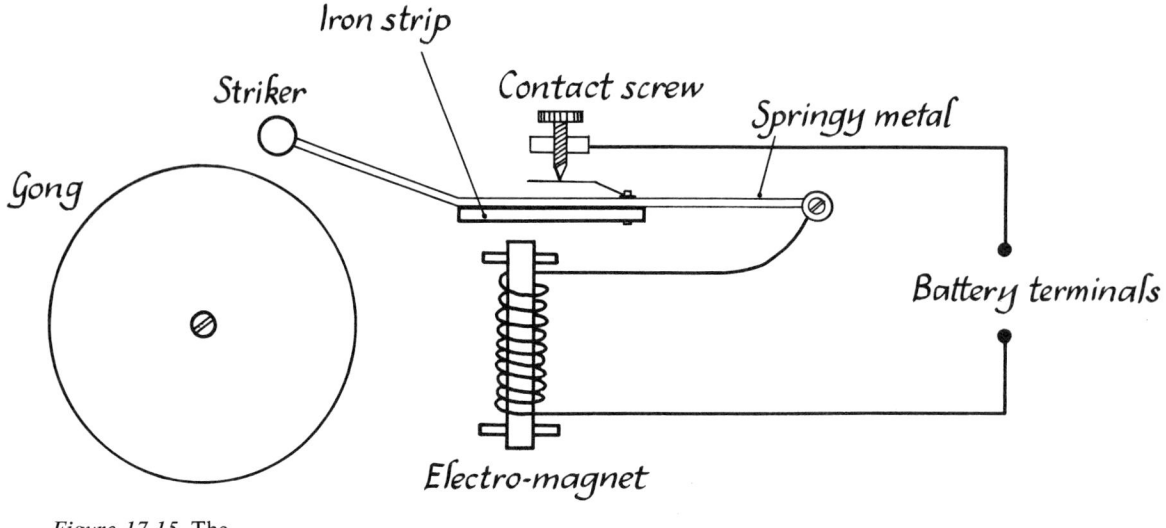

Figure 17.15 The electric bell

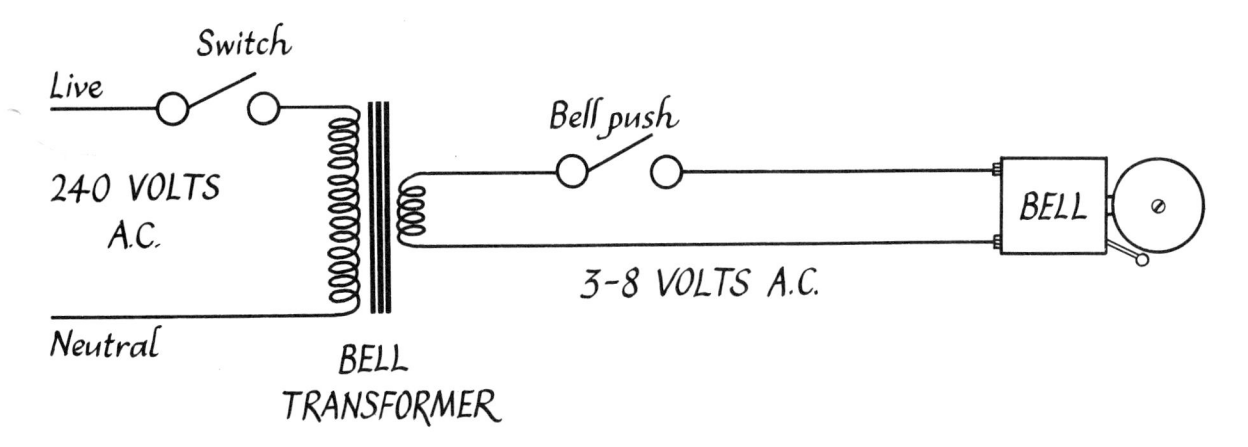

Figure 17.16 The electric bell circuit

c. The series–parallel switch

Electric hot plates are sometimes fitted with switches having four positions: 'low', 'medium', 'high' and 'off'. Such hot plates contain two heating elements. When the switch is in the 'high' position, both of these are connected in parallel; in the 'medium' position, only one of the elements is connected in the circuit; and in the 'low' position, both elements are connected in series (see Figure 17.17).

A similar circuit is used in photographic studios to control the lighting. The lights are switched to the series position while they are being arranged, and are switched to the parallel position while the photograph is being taken. This arrangement ensures that the lamps last longer.

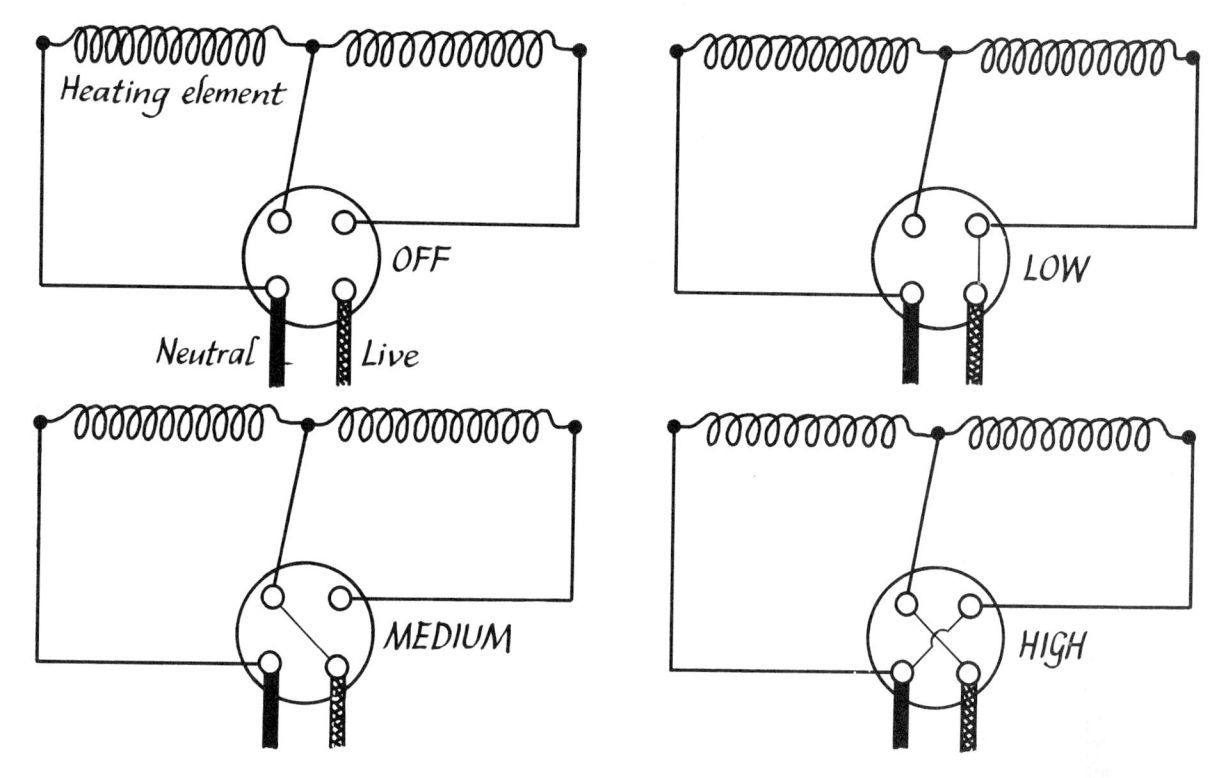

Figure 17.17 The series–parallel switch in a hot-plate circuit

d. The vacuum cleaner

This is essentially a fan, driven by an electric motor, which causes air to be blown in one direction. The fan is mounted in a cylinder and the incoming air is drawn through a flexible tube attached to a suitable nozzle. When the nozzle is placed on a carpet, air is drawn through the carpet, and any dust which is on the carpet is drawn through with the air. A bag is fitted in the vacuum cleaner to separate the dust from the air. The material from which the bag is made is woven too finely for the dust to travel through it, although the air can do so. (See Figure 17.18.)

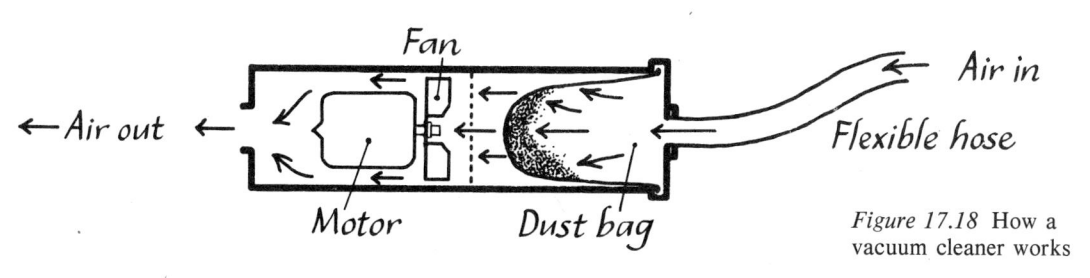

Figure 17.18 How a vacuum cleaner works

17.9. Care of home appliances

It is important to maintain electrical apparatus in good condition. A few simple routine checks will assist in this.

a. Flexible leads

Frayed or worn leads should be renewed.

b. Cable grips

Plugs are fitted with cable grips. These should be examined periodically to ensure that they have not become loosened. The cable grip prevents the wires becoming disconnected from the terminals if the cable is accidentally pulled.

c. Loading of power points

Power points should *never* have appliances connected to them taking a total current greater than the rating of the power point. Particular care should be given to this when adaptors are used to allow several appliances to be connected to the same socket.

Appliances taking a large current or which are fitted with an earth lead should *never* be connected to a lighting circuit.

If a plug becomes hot in use it should be checked to see if the wires are firmly connected to the terminals and that the plug fits the socket properly. A further check should be made to ensure that the socket is not overloaded.

d. Leads to pendant (hanging) lights

These require occasional examination and need to be renewed periodically. Heat from the lamp causes the insulation to become brittle and to crack.

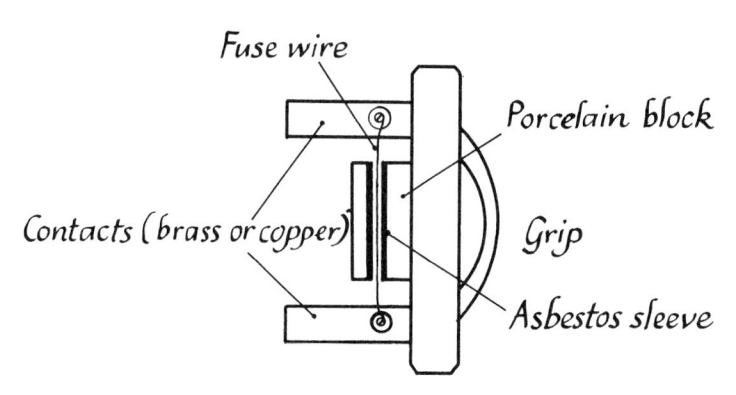

Figure 17.19 A fuse carrier

e. Fuses

When a fuse blows, it is important to find out why it has done so, before attempting to replace it. When the fault has been rectified, the fuse should be replaced by a wire or cartridge of the correct rating. Some homes now have an electromagnetic cut-out instead of fuses in the main box. If this cuts out, the cause must be established and the fault corrected, before the circuit is re-connected. Figure 17.19 shows a typical wire fuse in its carrier.

17.10. Electric motors

The electric motor is a device for converting electrical energy into mechanical energy. In order to do so, it makes use of the magnetic effect of an electric current.

The simple electric motor (see Figure 17.20) consists of a rotating electromagnet, called the **armature**, placed in a magnetic field. A

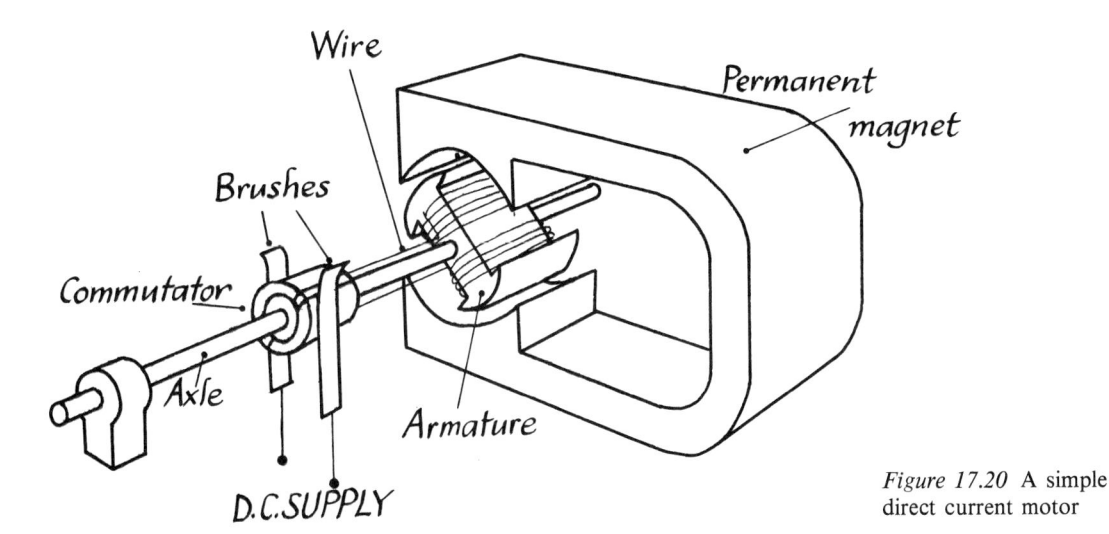

Figure 17.20 A simple direct current motor

series of attractions and repulsions causes the armature to rotate. The direction in which the current passes through the armature must be reversed periodically so that the polarity of the field is reversed, thus enabling the armature to rotate continuously. In order to bring about the change in current direction, a **commutator** is used. The current is passed to the commutator by means of contacts called **brushes**. These are usually made of carbon.

If such a motor is required to work from alternating current as well as direct current, the permanent magnet is replaced by an electromagnet, as shown in Figure 17.21.

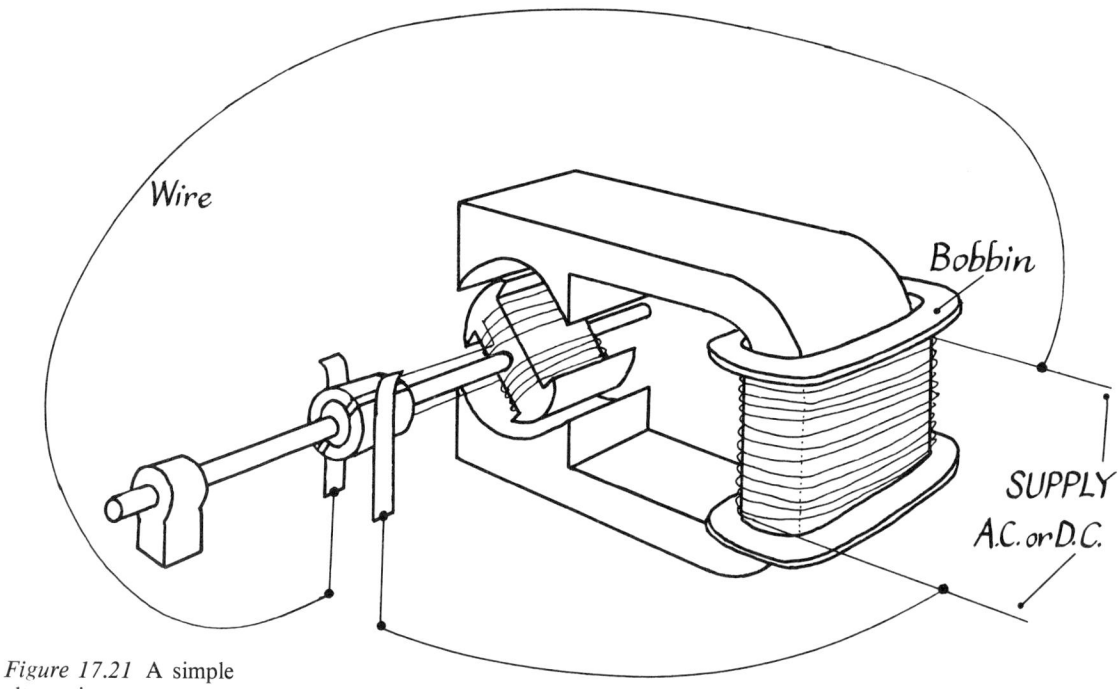

Figure 17.21 A simple alternating current motor

Figures 17.20 and 17.21 show motors having the simplest kind of armature. This really consists of a single winding and is connected to a commutator having only two segments. It is common practice, even in small motors, to employ rather more complex armatures having a larger number of windings and commutators having a correspondingly larger number of segments. Examine the armature of an old electric motor removed from an electric drill or a vacuum cleaner.

There are types of electric motor other than those described here. **Synchronous motors** work only on alternating current and rotate at a speed which depends on the frequency of the supply. Such motors are generally used in electric clocks. For heavy duty work, a motor working on an entirely different principle is used. This is the **induction**, or **squirrel cage**, **motor.**

17.11. Electrical communications systems

Before the invention of electrical systems, one of the most efficient systems of passing messages over a distance was by means of **semaphore** towers. These were situated on high ground and had two wooden arms which could be rotated. Each letter of the alphabet was represented by a pair of arm positions. During the Napoleonic wars, a chain of semaphore towers existed between the Admiralty, in London, and the Royal Naval Dockyard, in Portsmouth. These places are about 100 km apart. By means of semaphore, a message could be passed from one to the other in a matter of minutes. The disadvantage with this system was that in conditions of poor visibility it could not be used.

The first electrical system was **telegraphy**. In a telegraph system, a series of electrical impulses are transmitted along cables to a suitable receiver. Samuel Morse introduced the code of short and long impulses (dots and dashes) given below.

TABLE 17.1. THE MORSE CODE

Letters				Numerals, etc.	
A	· —	N	— ·	1	· — — — —
B	— · · ·	O	— — —	2	· · — — —
C	— · — ·	P	· — — ·	3	· · · — —
D	— · ·	Q	— — · —	4	· · · · —
E	·	R	· — ·	5	· · · · ·
F	· · — ·	S	· · ·	6	— · · · ·
G	— — ·	T	—	7	— — · · ·
H	· · · ·	U	· · —	8	— — — · ·
I	· ·	V	· · · —	9	— — — — ·
J	· — — —	W	· — —	0	— — — — —
K	— · —	X	— · · —	Full stop	· — · — · —
L	· — · ·	Y	— · — —	Apostrophe	· — — — — ·
M	— —	Z	— — · ·	Inverted commas	· — · · — ·
				Question mark	· · — — · ·

Figure 17.22 shows a simple telegraph system which you can set up for yourself.

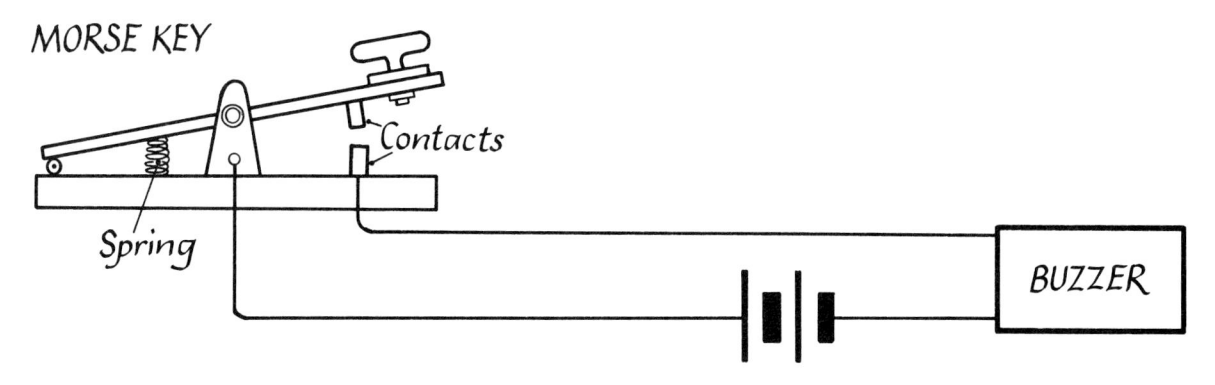

Figure 17.22 A simple telegraph system

In 1874, Alexander Graham Bell, working with his assistant Thomas Watson, discovered that it was possible to transmit sounds by means of an electric current. This system was called **telephony**, and Bell displayed his telephone at an exhibition in Philadelphia in 1876.

Figure 17.23 shows a telephone transmitter (microphone) connected to a telephone receiver. The microphone consists of a box containing carbon granules between two plates. One of these plates

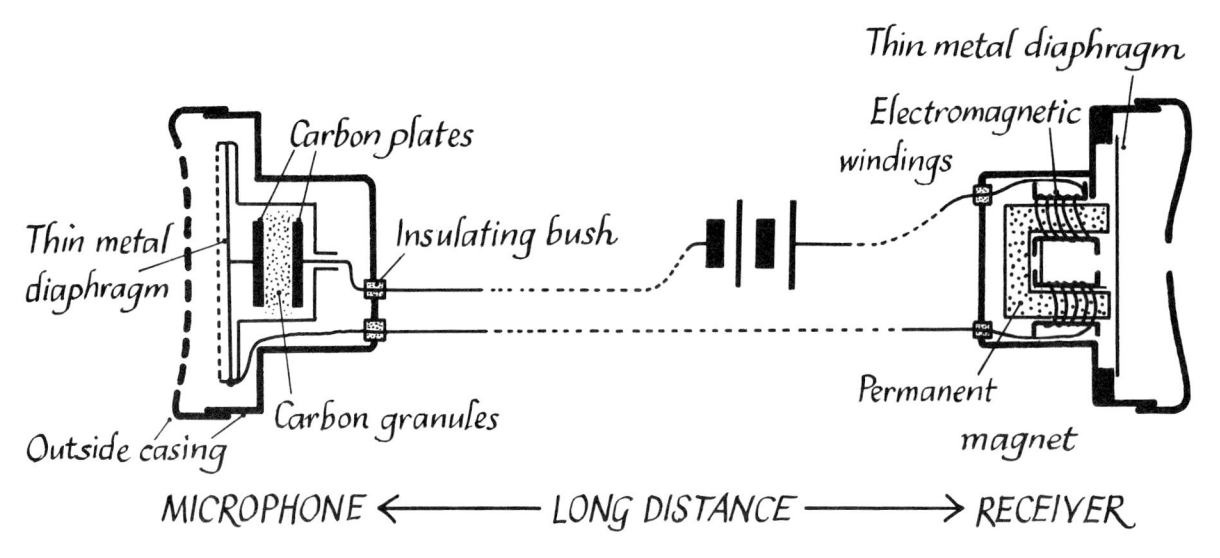

Figure 17.23 Telephone transmitter and receiver

is fixed and the other is free to vibrate. This plate is attached to a thin metal diaphragm. The receiver consists of windings on a permanent magnet which attracts a thin diaphragm of magnetic metal supported close to it (but not touching it).

From the diagram, it can be seen that the receiver is connected in series with the microphone and a battery, so that the current has to flow through the carbon granules in the microphone. When we speak into the microphone, its diaphragm vibrates, and as it does so, the pressure on the granules varies. When they are compressed, the electrical resistance between the plates decreases, so that a larger current flows. When the pressure decreases, the resistance increases and a smaller current flows.

These changes in current cause the strength of the magnetic field produced by the electromagnet windings in the receiver to fluctuate. This causes the attraction on the receiver diaphragm to fluctuate so that it vibrates in a similar way to the microphone diaphragm, producing sound waves.

If some telephone handsets are available, you should examine them. Set up a simple telephone system between two rooms.

At the beginning of the century, an Italian engineer, Guglielmo Marconi, devised a method of transmitting a message over a long distance, without the need for the receiver and transmitter to be connected by wires. The transmitter produced electromagnetic waves which were picked up by an aerial connected to the receiver. The first messages transmitted in this way were in Morse code, and the system is called **wireless telegraphy**. Later, it was found possible to transmit vocal messages in a similar way. This is called **wireless telephony**.

It was then found that it was possible to use a similar system to transmit pictures. Images in the television camera in a studio are used to cause fluctuations in a transmission of electromagnetic waves. These fluctuations are made to produce an image in the cathode-ray tube of a receiver. John Logie Baird, a Scottish scientist, pioneered the work in television.

Test your understanding

1. What is the difference between alternating current and direct current? What is meant by the 'frequency' of an alternating current? What is the frequency of the supply to your home?
2. Explain why the mains supply is alternating current, rather than direct current.
3. What is the purpose of the ignition coil fitted to an internal combustion engine?
4. An electric radiator is connected to a 13 A plug fitted with a fuse. Draw a diagram showing the wiring of the plug, and explain the importance of the fuse and the earth wire.
5. Draw a diagram showing how a lamp may be controlled by a two-way switching system.
6. Draw diagrams to show how a 'high–medium–low' switch may be used to control a hot plate.
7. Draw a diagram showing the construction of an electric bell. Show how it is connected to work normally. How could you modify the wiring so that the striker struck the gong once when the bell push was depressed?
8. Draw a diagram of a simple direct current motor and explain briefly how it works.
9. Make a diagram showing a microphone connected to a receiver. Explain how the system is used to transmit sound.
10. With what inventions do you associate the following names: Baird, Marconi, Bell, Morse?

Appendix 1

Units of Measurement

Quantity	Unit	Symbol	
Mass	kilogramme gramme	kg g	1 kg = 1 000 g
Length	metre millimetre kilometre	m mm km	1 m = 1 000 mm 1 km = 1 000 m (*Note.* The centimetre (cm) is sometimes used. 1 m = 100 cm)
Time	second hour	s h	1 h = 3 600 s
Area	square millimetre square centimetre square metre square kilometre	mm^2 cm^2 m^2 km^2	
Volume	cubic millimetre cubic centimetre cubic metre litre millilitre	mm^3 cm^3 m^3 l ml	 Used for fluids 1 ml = 1 cm^3
Density	kilogramme per cubic metre gramme per cubic centimetre	kg m^{-3} g cm^{-3}	
Velocity	metre per second kilometre per second kilometre per hour	m s^{-1} km s^{-1} km h^{-1}	
Acceleration	metre per second squared	m s^{-2}	
Force	newton	N	A force of 1 N will produce an acceleration of 1 m s^{-2} in a mass of 1 kg 1 MN = 1 000 000 N
Work and Energy	joule kilojoule	J kJ	Energy is the capacity for doing work 1 MJ = 1 000 000 J
Power	watt kilowatt	W kW	Power is the rate of doing work 1 W = 1 joule per second 1 kW = 1 000 W 1 MW = 1 000 000 W
Electromotive force	volt kilovolt	V kV	1 kV = 1 000 V
Electric current	ampere	A	1 milliampere = $\frac{1}{1000}$ ampere
Electric charge	coulomb	C	
Electrical resistance	ohm	Ω	1 K Ω = 1 000 Ω 1 M Ω = 1 000 000 Ω

Appendix 2

The Elements

Element	Symbol	Mass No.	Atomic No.	Element	Symbol	Mass No.	Atomic No.
Actinium	Ac	277	89	Neodymium	Nd	144	60
Aluminium	Al	27	13	Neon	Ne	20	10
Antimony	Sb	122	51	Nickel	Ni	59	28
Argon	A	40	18	Niobium	Nb	93	41
Arsenic	As	75	33	Nitrogen	N	14	7
Astatine	At	210	85	Osmium	Os	190	76
Barium	Ba	137	56	Oxygen	O	16	8
Beryllium	Be	9	4	Palladium	Pd	107	46
Bismuth	Bi	209	83	Phosphorus	P	31	15
Boron	B	11	5	Platinum	Pt	195	78
Bromine	Br	80	35	Plutonium	Pu	239	94
Cadmium	Cd	112	48	Polonium	Po	209	84
Caesium	Cs	133	55	Potassium	K	39	19
Calcium	Ca	40	20	Praseodymium	Pr	141	59
Carbon	C	12	6	Promethium	Pm	145	61
Cerium	Ce	140	58	Protoactinium	Pa	231	91
Chlorine	Cl	35	17	Radium	Ra	226	88
Chromium	Cr	52	24	Radon	Rn	222	86
Cobalt	Co	59	27	Rhenium	Re	186	75
Copper	Cu	64	29	Rhodium	Rh	103	45
Dysprosium	Dy	162	66	Rubidium	Rb	85	37
Erbium	Er	167	68	Ruthenium	Ru	101	44
Europium	Eu	152	63	Samarium	Sm	150	62
Fluorine	F	19	9	Scandium	Sc	45	21
Francium	Fr	223	87	Selenium	Se	79	34
Gadolinium	Gd	157	64	Silicon	Si	28	14
Gallium	Ga	70	31	Silver	Ag	108	47
Germanium	Ge	73	32	Sodium	Na	23	11
Gold	Au	197	79	Strontium	Sr	88	38
Hafnium	Hf	179	72	Sulphur	S	32	16
Helium	He	4	2	Tantalum	Ta	181	73
Holmium	Ho	165	67	Technetium	Tc	99	43
Hydrogen	H	1	1	Tellurium	Te	128	52
Indium	In	115	49	Terbium	Tb	159	65
Iodine	I	127	53	Thallium	Tl	204	81
Iridium	Ir	192	77	Thorium	Th	232	90
Iron	Fe	56	26	Thulium	Tm	169	69
Krypton	Kr	84	36	Tin	Sn	119	50
Lanthanum	La	139	57	Titanium	Ti	48	22
Lead	Pb	207	82	Tungsten	W	184	74
Lithium	Li	7	3	Uranium	U	238	92
Lutetium	Lu	175	71	Vanadium	V	51	23
Magnesium	Mg	24	12	Xenon	Xe	131	54
Manganese	Mn	55	25	Ytterbium	Yb	173	70
Mercury	Hg	201	80	Yttrium	Y	89	39
Molybdenum	Mo	96	42	Zinc	Zn	65	30
				Zirconium	Zr	91	40

The mass numbers quoted are those of the most common isotope.

Appendix 3

Solutions

The table below gives the quantities of solutes required to make one litre of solution.

Solution	Strength (molarity)	Quantity of Solute
Sodium hydroxide	1·0 M	40 g solid sodium hydroxide
Potassium hydroxide	1·0 M	56 g solid potassium hydroxide
Ammonia solution	1·0 M	52 cm^3 ammonia fortis (0·88)
Sulphuric acid	1·0 M	55 cm^3 conc. acid. (Add acid to water. *Care!*)
Nitric acid	1·0 M	64 cm^3 conc. acid
Hydrochloric acid	1·0 M	95 cm^3 conc. acid
Silver nitrate		15 g (Store in dark glass bottle.)
Barium chloride		100 g
Methyl orange		1 g
Phenolphthalein		2 g (Dissolve in 500 cm^3 industrial spirit, then add water.)
Limewater		Saturated solution (Mix calcium hydroxide with water, then filter.)

Note. Where the strength of a solution is given as 1·0 M (molar), the gramme formula mass of solute is dissolved in water to make 1 litre of solution.

If a solution of different strength is required, a proportionately different mass of solute is used. Thus, 4 g of solid sodium hydroxide would make 1 litre of solution of strength 0·1 M, and 80 g of solid sodium hydroxide would make 1 litre of solution of strength 2·0 M.

Appendix 4

Common Chemical Substances

Common Name	Chemical Name	Formula
Baking soda	Sodium hydrogen carbonate	$NaHCO_3$
Borax	Sodium tetraborate	$Na_2B_4O_7 \cdot 10H_2O$
Caustic potash	Potassium hydroxide	KOH
Caustic soda	Sodium hydroxide	$NaOH$
Chalk (limestone and marble)	Calcium carbonate	$CaCO_3$
Chrome alum	Chromium potassium sulphate	$K_2SO_4Cr_2(SO_4)_3 \cdot 24H_2O$
Common salt (table salt)	Sodium chloride	$NaCl$
Condy's crystals	Potassium permanganate	$KMnO_4$
Epsom salts	Magnesium sulphate	$MgSO_4$
Glauber's salts	Sodium sulphate	Na_2SO_4
Hypo	Sodium thiosulphate	$Na_2S_2O_3$
Nitre	Potassium nitrate	KNO_3
Oil of vitriol	Sulphuric acid	H_2SO_4
Plaster of Paris	Calcium sulphate	$CaSO_4$
Potash alum	Potassium aluminium sulphate	$K_2SO_4Al_2(SO_4)_3 \cdot 24H_2O$
Quicklime	Calcium oxide	CaO
Sal ammoniac	Ammonium chloride	NH_4Cl
Sal volatile	Ammonium carbonate (soln.)	$(NH_4)_2CO_3$
Slaked lime	Calcium hydroxide	$Ca(OH)_2$
Water glass	Sodium silicate	Na_2SiO_3

Analytical Contents List

1 Fundamental Ideas

MEASUREMENT

1.1 Standards of measurement 1.2 Methods of measurement

ENERGY

1.3 What is force? 1.4 Forms of energy
1.5 Energy chains 1.6 Conservation of energy

2 Matter

2.1 What is matter? 2.2 All matter possesses heat
2.3 The behaviour of materials when they are heated
2.4 Matter in motion 2.5 Heat and temperature
2.6 Heat radiation 2.7 Pressure in gases
2.8 Expansion of gases

3 Chemistry

3.1 Different kinds of matter 3.2 Formulae and equations
3.3 Oxides 3.4 Acidic and basic oxides
3.5 Acids, bases and salts 3.6 Preparation of salts
3.7 Solutions 3.8 Crystals
3.9 Water of crystallization

4 The Measurement of Temperature

4.1 'Hotness' 4.2 The laboratory thermometer
4.3 Mercury and alcohol as thermometric liquids
4.4 The clinical thermometer
4.5 The Six's maximum–minimum thermometer
4.6 The gas thermometer 4.7 The Bourdon principle
4.8 The bimetal strip thermometer 4.9 The thermocouple
4.10 The resistance pyrometer 4.11 Seger cones

5 Heat Measurement

5.1 The difference between heat and temperature
5.2 Specific heat capacity
5.3 Heat losses and their prevention
5.4 Latent heat 5.5 Sensible heat and latent heat

6 Producing Heat

6.1 The coke fire 6.2 The coal fire
6.3 Heat from gas 6.4 The electric fire
6.5 The domestic hot water system 6.6 Central heating

7 Maintaining Heat

7.1 Heat insulation 7.2 Heat and colour
7.3 Cooling by evaporation 7.4 The refrigerator
7.5 Heat balance

8 Mechanics

8.1 Speed and velocity 8.2 Mass and inertia
8.3 Weight 8.4 Forces and acceleration
8.5 Measuring forces 8.6 Work 8.7 Energy
8.8 Power 8.9 The moment of a force 8.10 Machines
8.11 Some other simple machines 8.12 Friction 8.13 Momentum
8.14 Newton's laws of motion 8.15 Gravity 8.16 Projectiles

9 Engines

9.1 Hero's engine 9.2 Windmills and water-wheels
9.3 The steam engine 9.4 The steam turbine
9.5 The internal combustion engine 9.6 The carburettor
9.7 The four-stroke cycle of a petrol engine
9.8 Cooling the engine 9.9 Producing the spark
9.10 Operating the valves 9.11 Starting the engine
9.12 Other types of petrol engine 9.13 The diesel engine
9.14 Gas turbine and jet propulsion engines
9.15 Rocket propulsion

10 The Basic Foodstuffs

10.1 The fuel foods 10.2 The body-building foods
10.3 The non-nutrient foodstuffs 10.4 Food-testing
10.5 A balanced diet 10.6 The importance of protein
10.7 Burning food 10.8 Food energy values
10.9 How much food do we need?

11 Food Sources

11.1 Water in plants
11.2 Oxygen and carbon dioxide in plants
11.3 The leaf as a factory 11.4 Photosynthesis
11.5 The oxygen cycle 11.6 The carbon cycle
11.7 The nitrogen cycle 11.8 The water cycle
11.9 Water supply

12 Digestion

12.1 Digestion in the earthworm
12.2 The alimentary canal in man
12.3 Digestion in man

13 Cooking

13.1 Why do we cook food?
13.2 Low-temperature cooking
13.3 High-temperature cooking
13.4 Making bread rise 13.5 Making cakes rise
13.6 Cooking utensils

14 Atomic Structure

14.1 Early ideas 14.2 Dalton's atomic theory
14.3 Chemical combination 14.4 Twentieth-century ideas
14.5 Electrons and protons 14.6 Neutrons
14.7 Atomic structure: the Rutherford–Bohr model
14.8 Atomic number 14.9 Mass number 14.10 Isotopes
14.11 Ions 14.12 Formation of compounds
14.13 Electrolytes 14.14 Electrolysis
14.15 Matter and energy 14.16 Radioactivity
14.17 Radioactive decay
14.18 The harmful effects of radiation
14.19 Uses of radiation
14.20 Nuclear fission 14.21 Nuclear fusion

15 Electricity

15.1 Electric charges
15.2 Positive and negative charges
15.3 Conductors and insulators
15.4 Current electricity 15.5 Generating electricity

16 Effects of Electricity

16.1 The magnetic effect 16.2 The heating effect
16.3 Measuring electricity
16.4 Electrical resistance
16.5 The chemical effect
16.6 Light from electricity

17 Using Electricity

17.1 Direct current and alternating current
17.2 The transformer
17.3 Transmission of electricity across country
17.4 The live and neutral leads
17.5 The supply in the home
17.6 The lighting circuit 17.7 The power circuit
17.8 Home electrical appliances
17.9 Care of home appliances 17.10 Electric motors
17.11 Electrical communications systems

Appendices

1 Units of Measurement
2 The Elements 3 Solutions
4 Common Chemical Substances

Analytical Contents List for Book II

For the convenience of the users of this book, we are setting out below the contents of the companion volume, *Man and His Environment*, CSE General Science Book II.

1 Classification
Living and non-living things   Plant and animal life   Plants without seeds   Plants with seeds   Invertebrates   Vertebrates   Webs of dependence   The soil   Soil formation   Types of rock

2 Water
The impurities in water   Hardness of water   Stalactites and stalagmites   Softening water   Surface tension   Detergents

3 Density and Water Pressure
Floating and sinking   Hydrometers   Water pressure   The effect of water pressure   Making water move   Harnessing water power

4 Air
Air pressure   Some uses of air pressure   Flight without wings   Flight with wings   The constituents of air   Pollution of air   Air and the weather   Air and life

5 Carbon and its Compounds
The element carbon   The oxides of carbon   Carbon dioxide   Uses of carbon dioxide   Carbon monoxide   Chalk   Sodium carbonate and sodium hydrogen carbonate   Hydrocarbons   Alkanes   Ethene   Benzene   Carbohydrates   Sugars   Starch   Cellulose   Alcohols   Distillation   Fats and oils

6 Metals
What are metals?   Sources of metals   Extraction of metals from their ores   Steel   Alloys   Some non-ferrous alloys   Grouping metals   Corrosion of metals   Anodizing

7 Light Energy
The nature of light   Light and shadow   Eclipses   Reflection   Mirror images   The laws of reflection   Images   Curved mirrors   Parabolic reflectors

8 Refraction
Refraction Internal reflection Waves Wavelength, frequency and velocity Light waves and rays Coloured light The spectral colours Mixing coloured lights Colour filters Pigments

9 Some Optical Instruments
The camera The eye Long sight and short sight The projector Telescopes Prismatic binoculars The microscope

10 Sound
Sources of sound How sound travels Sound waves The ear Musical instruments The velocity of sound

11 How the Body Works
The skeletal system The muscular system The digestive system The respiratory system The blood Blood groups, the A–B–O system The heart and circulation The nervous system The endocrine system The liver The excretory system

12 Reproduction
Asexual reproduction Sexual reproduction Flowers Pollination Fertilization in flowers Fruits Fruit and seed dispersal Seeds Germination of seeds Sexual reproduction in animals Reproduction in human beings Early growth of the baby Growing up Menstruation

13 Genetics
What is genetics? The work of Mendel Some terms used in genetics The cell Mitosis Meiosis Incomplete dominance The sex chromosomes Sex linkage The functions of the genes Heredity and environment Eye colour and heredity Height and heredity Disease and heredity Mutation

14 Evolution
Earth and its crust The three main types of rock Dating the rocks Fossils Coal and oil Geological time The meaning of evolution Comparative embryology The evidence of vestiges Comparative anatomy Lamarck's theory Darwin's theory of evolution Extinction Artificial selection The evolution of man and the future

15 The Earth in Space
Stars and planets The solar system The galaxies The age of the earth How old is the universe? Is there life anywhere else?

Appendices
Units of measurement The elements Solutions Common chemical substances Some famous scientists

Index

Where the subject is illustrated the page number is shown in bold type

Absolute zero   24
Acceleration   113
Acid, definition   30
Acidic oxide   29
Alimentary canal   187, **189**
Alternating current   252
Ampere   240
Amylase   192
Anhydrous   40
Appendix   194
Atom   209
Atomic number   212
Atomic structure
    Dalton model   209
    Rutherford–Bohr model   210, **211**

Baking   199
Baking powder   205
Baking soda   204
Basic oxide   29
Bile   192
Bimetal strip   **16**
Boiling   198
Bourdon principle   50, **51**
Braising   199
Bread   202
Bunsen burner   70, **71,** 72

Caecum   194
Carbohydrates   155
Carboxyhaemoglobin   68
Carburettor   143, **144**
Catalyst   28
Cathode-ray oscilloscope   253
Cavity wall   **90**
Cell, secondary   245
Celsius, Anders   43
Centimetre   2
Central heating   80, **82, 86**
Chain reaction   221, **222**
Charges, law of   226
Chemical effect of
    electric current   245
Chemical symbols   27
Chlorine, preparation of   35, **36**

Chlorophyll   174
Chloroplasts   174
Chyle   193
Chyme   192
Clothes iron   262, **263**
Colon   **189**
Colour coding of flex   262
Compound   26
Condensation   182
Conduction of heat   18
Conservation of energy   12
Conservation of mass   209
Constant composition   209
Convection of heat   18
Copper sulphate,
    preparation of   34
Coulomb   240
Covalency   215
Critical mass   221
Current, electric   227
    chemical effect   245
    heating effect   239
    magnetic effect   235
Cycle, carbon   178, **179**
    four-stroke   144, **145, 151**
    nitrogen   179, **181**
    oxygen   **178**
    rain   **182**
    two-stroke   148, **149**
    water   180, **183**

Dalton, John   209
Dew, formation of   182
Dielectric cooking   201
Diesel engine   149, **151**
Diet   160
Diffusion   **169**
Direct current   252
Distillation, destructive   **69,** 70
Distributor   147
Double decomposition   35
Double glazing   **91**
Dry cell   **231**
Duodenum   192
Dynamo   232, **233**

Eccentric **142**
Efficiency 124
Einstein, Albert 217
Electric bell 263, **264**
  lamp 247, **248**
  motor **267, 268**
Electrolysis 216
Electrolyte 215
Electromagnetic
  induction 255
Electromotive force 227
Electron 210
Electrophorus **226**
Electro-plating 254
Electroscope **226**
Electrovalency 213
Element 26
Energy chains 12
Energy, conservation of 12
Energy value of foods 163
Engine, Diesel 149, **151**
  four-stroke 144, **145, 151**
  two-stroke 148, **149**
  Wankel 148, **150**
Enzymes 190
Erepsin 193
Evaporation, cooling by 98, **102**
Expansion 15

Fats 155
Fehling's solution 159
Fission, nuclear 221
Fixed points 43
Fluorescent tube 248, **251**
Food energy values 163
Food tests 158
Force 9
Four-stroke engine 144, **145, 151**
Friction 130
Frost, formation of 163
Frying 200
Fuse **266,** 267
Fusion, nuclear 224

Gas turbine 149, **152**
Gears **126,** 127
Gliadin 202
Glucose 193
Gluten 202
Glutenin 202
Gramme 2
Gravity 134
Grilling 201

Haemoglobin 68
Half-life 219
Heating effect
  of electric current 239
Hero's engine 137
Hot water system 77, **78, 79, 81**

Hydraulic jack 129, **130**
Hydrogen 30, **31**

Ice box **103**
Ignition coil 147, **256**
Inclined plane **127**
Indicators 30
Inertia 112
Insulation, heat 88
  electrical 227
Internal combustion engine 142
Intestine 191
Iodine solution 158
Ion 212
Isotope 212

Jet propulsion
  engine **152**
Joule 2, 118, 241

Kelvin scale 24
Kilogramme 2, 112
Kilojoule 2, 164
Kilometre 2
Kinetic energy **119, 120**
Kinetic theory
  of heat 21, 24

Lactase 193
Lacteals 194
Latent heat 61
Lead-acid cell 247
Lead sulphate,
  preparation of 35
Leclanché cell 230, **231**
Lever **122,** 123
Lipase 192
Lymph vessels 194

Machines 123, 125
Magnesium sulphate,
  preparation of 35
Magnetic effect
  of electric current 235
Magnetic fields **236,** 237
  poles 235
Maltase 193
Mass 2, 112
Mass number 212
Matter 15
Mechanical advantage 124
Meniscus 4
Metre 2
Millilitre 2
Millimetre 2
Mineral salts 157
Moment of a force 121
Momentum 131
Morse code 269

National grid system **258**
Neon lamp 248, **249**
Neutron 210
Newton 2, 113
Newton's laws of motion 132
Nickel plating 245, **246**
Nitrogen cycle 179, **181**
Nuclear energy 221

Ohm 243
Ohm's law 243
Osmosis **168, 169**
Oxidation of coke 65, **67**
 of food 162
Oxide 28
Oxygen cycle **178**
Oxygen, preparation of **28**
Oxyhaemoglobin 68

Pancreatic juice 192
Pendulum 6, 7
Pepsin 192
Peristalsis **191**
Photosynthesis 176, **177**
Photo-voltaic cell 231, **232**
Plug wiring **262**
Polarization of cells 230
Potential energy **119, 120**
Power 119
Power points, loading of 266
Preserving 197
Pressure 129
Projectile 134
Protein 156
Proton 210
Proving bread 203
Ptyalin 190
Pulley system 123, **124**
Pyrometer 52

Quantum 212

Radiation of heat 23, 95
Radioactive decay 219
 radiation 218
Rain cycle **182**
Refrigeration 102, **104, 105**
Rennin 192
Resistance, electrical 241
Roasting 199
Root hairs 167, **168**
Roughage 195

Saliva 190
Salt, definition of 33
 preparation of 34
Scaling bread 203
Seger cones 54, **55**
Sensible heat 62, **63**

Series-parallel switch 264, **265**
Simmering 197
Sodium chloride,
 preparation of 34
Solar heating **85, 86**
Solute 36
Solution 36
Solvent 36
Specific heat
 capacity 23, 57
Starch 156
Static charge 227
Steam engine 140
Steaming 198
Steam turbine **142**
Stomach 192
Stomata 172
Sucrase 193
Sugar 156

Telegraphy 269
Telephone **270**
Thermocouple **52**, 232
Thermometer, bimetal 51, **52**
 clinical **46, 47**
 gas 48, **49**
 laboratory 43
 maximum-minimum 47, **48**
Thickness gauging 220
Tracer technique 220
Trajectory **135**
Transformer **257**
Transpiration 170
Trypsin 192
Turgidity 170
Two-stroke engine 148, **149**
Two-way switching 259, **261**

Vacuum cleaner **265**
Vacuum flask 58, **59**
Velocity 111
Velocity ratio 124
Vitamins 158
Volt 227
Voltaic cell **229**

Wankel engine 148, **150**
Water, as a foodstuff 157
 in plants 166
 of crystallization 40
Watt 119, 240
Weight 113, 117
Wheel and axle **125**
Work 118

X-rays 221

Yeast 202